METHODS IN MOLECULAR BIOLOGY

For further volumes:
http://www.springer.com/series/7651

For over 35 years, biological scientists have come to rely on the research protocols and methodologies in the critically acclaimed *Methods in Molecular Biology* series. The series was the first to introduce the step-by-step protocols approach that has become the standard in all biomedical protocol publishing. Each protocol is provided in readily-reproducible step-by-step fashion, opening with an introductory overview, a list of the materials and reagents needed to complete the experiment, and followed by a detailed procedure that is supported with a helpful notes section offering tips and tricks of the trade as well as troubleshooting advice. These hallmark features were introduced by series editor Dr. John Walker and constitute the key ingredient in each and every volume of the *Methods in Molecular Biology* series. Tested and trusted, comprehensive and reliable, all protocols from the series are indexed in PubMed.

Virus-Induced Gene Silencing in Plants

Methods and Protocols

Edited by

Vincent Courdavault and Sébastien Besseau

EA2106 Biomolécules et Biotechnologies Végétales, Université de Tours, Tours, France

Humana Press

Editors
Vincent Courdavault
EA2106 Biomolécules et
Biotechnologies Végétales
Université de Tours
Tours, France

Sébastien Besseau
EA2106 Biomolécules et
Biotechnologies Végétales
Université de Tours
Tours, France

ISSN 1064-3745 ISSN 1940-6029 (electronic)
Methods in Molecular Biology
ISBN 978-1-0716-0750-3 ISBN 978-1-0716-0751-0 (eBook)
https://doi.org/10.1007/978-1-0716-0751-0

This Humana imprint is published by the registered company Springer Science+Business Media, LLC part of Springer Nature.
The registered company address is: 1 New York Plaza, New York, NY 10004, U.S.A.

Preface

Elucidating the roles and functions of genes has always been, and still remains, a fascinating endeavor within the plant research field and beyond. This is particularly true in the post-genomic era that has resulted in an abundance of gene-coding and non-coding sequences. While many of the first gene function discoveries have relied on classical approaches developed during the initial rise of molecular biology, the ever-growing set of transcriptomics/genomics data and the resulting need of gene characterization have triggered the evolution of new methodologies such as reverse genetic procedures. These techniques involve the invalidation of expression of targeted genes and the analysis of their consequential developmental, metabolic, and/or reproductive effects. For plant model species, e.g., *Arabidopsis thaliana*, libraries of insertion mutants have been generated to systematically knockdown genes that are not essential to the plants survival. However, this long-term task cannot be applied to all plant species, particularly not to non-model plants including those of agronomic, pharmaceutical, or cosmetic importance. The emergence of transient invalidation procedures, such as virus-induced gene silencing (VIGS), has provided a solution for elucidating gene functions in many of these plant species. By exploiting the plant's endogenous defense system against viruses, VIGS allows for the downregulation of a specific plant gene following the degradation of its corresponding transcripts. Such a targeted transcript elimination relies on the integration of a short fragment of the gene of interest into the genome of the virus used to inoculate plants. By degrading viral transcripts after inoculation, plants produce siRNA against the inserted plant gene fragment. These siRNA will target and degrade the transcripts of the corresponding endogenous plant gene. In most of the cases, silencing reaches around 80–90% of the "normal" gene expression level allowing studying the consequences of this strong gene downregulation within 2 or 3 weeks after virus inoculation. Since first descriptions in the early 2000s, VIGS has become a popular technique deployed in many laboratories for many plants, using different viruses and distinct inoculation modalities. Getting access to this technique thus constitutes a cornerstone in the understanding of the functions of the genes you will study in the plant you are working on.

With this issue of Methods in Molecular Biology, we wanted to gather classical and newly developed protocols of VIGS to allow readers to initiate and/or optimize their own silencing experiments according to the methods described herein.

We would like to thank all the authors who shared their expertise in VIGS by contributing a chapter to this volume. The protocols they described were created and used in their labs and thus include niceties of correct procedures that make the difference between a poorly working protocol and an efficient one. We thus wish success to researchers who try silencing genes based on the step-by-step and readily reproducible protocols described in this issue.

Tours, France

Sébastien Besseau
Vincent Courdavault

Contents

Preface .... *v*
Contributors .... *ix*

1 Virus-Induced Gene Silencing in Lilies Using Cucumber Mosaic Virus Vectors 1
Keisuke Tasaki, Masumi Yamagishi, and Chikara Masuta

2 Virus-Induced Gene Silencing in Poaceae Using a *Foxtail Mosaic Virus* Vector 15
Ying-Wen Huang, Chao-Yuan Chang, and Yau-Heiu Hsu

3 Virus-Induced Gene Silencing (VIGS) in Chili Pepper (*Capsicum* spp.) 27
Magda Lisette Arce-Rodríguez and Neftalí Ochoa-Alejo

4 Virus-Induced Gene Silencing in Diploid and Tetraploid Potato Species 39
Jinping Zhao, Haolang Jiang, Guanyu Wang, Zonghua Wang, Jingao Dong, and Junqi Song

5 Virus-Induced Gene Silencing (VIGS) in Cassava Using Geminivirus Agroclones 51
Syed Shan-e-Ali Zaidi, Kumar Vasudevan, Ezequiel Matias Lentz, and Hervé Vanderschuren

6 Virus-Induced Gene Silencing of Cell Wall Genes in Flax (*Linum usitatissimum*) 65
Maxime Chantreau and Godfrey Neutelings

7 Virus-Induced Gene Silencing to Investigate Alkaloid Biosynthesis in Opium Poppy 75
Rongji Chen, Xue Chen, Jillian M. Hagel, and Peter J. Facchini

8 A Biolistic-Mediated Virus-Induced Gene Silencing in Apocynaceae to Map Biosynthetic Pathways of Alkaloids 93
Pamela Lemos Cruz, María Isabel Restrepo, Thomas Dugé de Bernonville, Audrey Oudin, Thibaut Munsch, Arnaud Lanoue, Sébastien Besseau, Lucia Atehortùa, Nathalie Giglioli-Guivarc'h, Nicolas Papon, Marc Clastre, Inês Carqueijeiro, and Vincent Courdavault

9 Virus-Induced Gene Silencing in *Nepeta* 111
Lira Palmer and Sarah E. O'Connor

10 Virus-Induced Gene Silencing in Sweet Basil (*Ocimum basilicum*) 123
Rajesh Chandra Misra, Shubha Sharma, Anchal Garg, and Sumit Ghosh

11 Virus-Induced Gene Silencing for Functional Genomics in *Withania somnifera*, an Important Indian Medicinal Plant 139
Dikki Pedenla Bomzan, H. B. Shilpashree, P. Anjali, Sarma Rajeev Kumar, and Dinesh A. Nagegowda

12 A *Prunus necrotic ringspot virus* (PNRSV)-Based Viral Vector for Characterization of Gene Functions in *Prunus* Fruit Trees 155
Hongguang Cui, Yinzi Li, and Aiming Wang

13 Virus-Induced Gene Silencing in Olive Tree (Oleaceae) 165
Konstantinos Koudounas, Margarita Thomopoulou, Elisavet Angeli, Dikran Tsitsekian, Stamatis Rigas, and Polydefkis Hatzopoulos

14 ALSV-Based Virus-Induced Gene Silencing in Apple Tree (*Malus × domestica* L.) 183
Carolina Werner Ribeiro, Thomas Dugé de Bernonville, Gaëlle Glévarec, Arnaud Lanoue, Audrey Oudin, Olivier Pichon, Benoit St-Pierre, Vincent Courdavault, and Sébastien Besseau

15 Virus-Induced Gene Silencing for Functional Analysis of Flower Traits in Petunia 199
Shaun R. Broderick, Laura J. Chapin, and Michelle L. Jones

16 Virus-Induced Gene Silencing in Rose Flowers 223
Huijun Yan, Zhao Zhang, Jean-Louis Magnard, Benoît Boachon, Sylvie Baudino, and Kaixue Tang

17 Virus-Induced Gene Silencing (VIGS) in Flax (*Linum usitatissimum* L.) Seed Coat: Description of an Effective Procedure Using the *transparent testa 2* Gene as a Selectable Marker. 233
Christophe Hano, Samantha Drouet, and Eric Lainé

18 Virus-Based microRNA Silencing in Plants. 243
Jinping Zhao, Guanyu Wang, Haolang Jiang, Tingli Liu, Jingao Dong, Zonghua Wang, Baolong Zhang, and Junqi Song

Index *259*

Contributors

ELISAVET ANGELI • *Laboratory of Molecular Biology, Department of Biotechnology, Agricultural University of Athens, Athens, Greece*

P. ANJALI • *Molecular Plant Biology and Biotechnology Lab, CSIR-Central Institute of Medicinal and Aromatic Plants Research Centre, Bengaluru, India*

MAGDA LISETTE ARCE-RODRÍGUEZ • *Departamento de Ingeniería Genética, Centro de Investigación y de Estudios Avanzados del Instituto Politécnico Nacional, Unidad Irapuato, Gto., Mexico*

LUCIA ATEHORTÙA • *Laboratorio de Biotecnología, Universidad de Antioquia, Medellin, Colombia*

SYLVIE BAUDINO • *Université de Lyon, UJM-Saint-Etienne, CNRS, BVpam FRE 3727, Saint-Etienne, France*

SÉBASTIEN BESSEAU • *EA2106 Biomolécules et Biotechnologies Végétales, Université de Tours, Tours, France*

BENOÎT BOACHON • *Université de Lyon, UJM-Saint-Etienne, CNRS, BVpam FRE 3727, Saint-Etienne, France*

DIKKI PEDENLA BOMZAN • *Molecular Plant Biology and Biotechnology Lab, CSIR-Central Institute of Medicinal and Aromatic Plants Research Centre, Bengaluru, India*

SHAUN R. BRODERICK • *Department of Plant and Soil Sciences, Mississippi State University, Truck Crops Experiment Station, Crystal Springs, MS, USA*

INÊS CARQUEIJEIRO • *EA2106 Biomolécules et Biotechnologies Végétales, Université de Tours, Tours, France*

CHAO-YUAN CHANG • *Graduate Institute of Biotechnology, National Chung Hsing University, Taichung, Taiwan*

MAXIME CHANTREAU • *Umeå Plant Science Centre, Department of Plant Physiology, Umeå University, Umeå, Sweden*

LAURA J. CHAPIN • *Department of Horticulture and Crop Science, The Ohio State University, The Ohio Agricultural Research and Development Center, Wooster, OH, USA*

RONGJI CHEN • *Epimeron Inc., Calgary, AB, Canada; Department of Biological Sciences, University of Calgary, Calgary, AB, Canada*

XUE CHEN • *Epimeron Inc., Calgary, AB, Canada; Department of Biological Sciences, University of Calgary, Calgary, AB, Canada*

MARC CLASTRE • *EA2106 Biomolécules et Biotechnologies Végétales, Université de Tours, Tours, France*

VINCENT COURDAVAULT • *EA2106 Biomolécules et Biotechnologies Végétales, Université de Tours, Tours, France*

PAMELA LEMOS CRUZ • *EA2106 Biomolécules et Biotechnologies Végétales, Université de Tours, Tours, France*

HONGGUANG CUI • *Key Laboratory of Green Prevention and Control of Tropical Plant Diseases and Pests, Ministry of Education and College of Plant Protection, Hainan University, Haikou, Hainan, China*

THOMAS DUGÉ DE BERNONVILLE • *EA2106 Biomolécules et Biotechnologies Végétales, Université de Tours, Tours, France*

JINGAO DONG • *Key Laboratory of Hebei Province for Plant Physiology and Molecular Pathology, Mycotoxin and Molecular Plant Pathology Laboratory, College of Life Sciences, Hebei Agricultural University, Baoding, China*

SAMANTHA DROUET • *Laboratoire de Biologie des Ligneux et des Grandes Cultures (LBLGC INRA USC1328), Université d'Orléans—Pôle Universitaire d'Eure et Loir, Chartres, France*

PETER J. FACCHINI • *Epimeron Inc., Calgary, AB, Canada; Department of Biological Sciences, University of Calgary, Calgary, AB, Canada*

ANCHAL GARG • *Biotechnology Division, Council of Scientific and Industrial Research-Central Institute of Medicinal and Aromatic Plants, Lucknow, India*

SUMIT GHOSH • *Biotechnology Division, Council of Scientific and Industrial Research-Central Institute of Medicinal and Aromatic Plants, Lucknow, India*

NATHALIE GIGLIOLI-GUIVARC'H • *EA2106 Biomolécules et Biotechnologies Végétales, Université de Tours, Tours, France*

GAËLLE GLÉVAREC • *EA2106 Biomolécules et Biotechnologies Végétales, Université de Tours, Tours, France*

JILLIAN M. HAGEL • *Epimeron Inc., Calgary, AB, Canada; Department of Biological Sciences, University of Calgary, Calgary, AB, Canada*

CHRISTOPHE HANO • *Laboratoire de Biologie des Ligneux et des Grandes Cultures (LBLGC INRA USC1328), Université d'Orléans—Pôle Universitaire d'Eure et Loir, Chartres, France*

POLYDEFKIS HATZOPOULOS • *Laboratory of Molecular Biology, Department of Biotechnology, Agricultural University of Athens, Athens, Greece*

YAU-HEIU HSU • *Graduate Institute of Biotechnology, National Chung Hsing University, Taichung, Taiwan; Advanced Plant Biotechnology Center, National Chung Hsing University, Taichung, Taiwan*

YING-WEN HUANG • *Graduate Institute of Biotechnology, National Chung Hsing University, Taichung, Taiwan; Advanced Plant Biotechnology Center, National Chung Hsing University, Taichung, Taiwan*

HAOLANG JIANG • *Texas A&M AgriLife Research Center at Dallas, Texas A&M University System, Dallas, TX, USA; State Key Laboratory of Ecological Pest Control for Fujian and Taiwan Crops, College of Plant Protection, Fujian Agriculture and Forestry University, Fuzhou, China*

MICHELLE L. JONES • *Department of Horticulture and Crop Science, The Ohio State University, The Ohio Agricultural Research and Development Center, Wooster, OH, USA*

KONSTANTINOS KOUDOUNAS • *Laboratory of Molecular Biology, Department of Biotechnology, Agricultural University of Athens, Athens, Greece; EA2106 Biomolécules et Biotechnologies Végétales, Université de Tours, Tours, France*

SARMA RAJEEV KUMAR • *Molecular Plant Biology and Biotechnology Lab, CSIR-Central Institute of Medicinal and Aromatic Plants Research Centre, Bengaluru, India*

ERIC LAINÉ • *Laboratoire de Biologie des Ligneux et des Grandes Cultures (LBLGC INRA USC1328), Université d'Orléans—Pôle Universitaire d'Eure et Loir, Chartres, France*

ARNAUD LANOUE • *EA2106 Biomolécules et Biotechnologies Végétales, Université de Tours, Tours, France*

EZEQUIEL MATIAS LENTZ • *Laboratorio de Biotecnología, Instituto de Educación Superior Alvear IdESA-UGACOOP, General Alvear, Mendoza, Argentina*

TINGLI LIU • *Provincial Key Laboratory of Agrobiology, Jiangsu Academy of Agricultural Sciences, Nanjing, China*

YINZI LI • *London Research and Development Centre, Agriculture and Agri-Food Canada, London, ON, Canada*
JEAN-LOUIS MAGNARD • *Université de Lyon, UJM-Saint-Etienne, CNRS, BVpam FRE 3727, Saint-Etienne, France*
CHIKARA MASUTA • *Research Faculty of Agriculture, Hokkaido University, Sapporo, Hokkaido, Japan*
RAJESH CHANDRA MISRA • *Biotechnology Division, Council of Scientific and Industrial Research-Central Institute of Medicinal and Aromatic Plants, Lucknow, India; Metabolic Biology Department, John Innes Centre, Norwich, UK*
THIBAUT MUNSCH • *EA2106 Biomolécules et Biotechnologies Végétales, Université de Tours, Tours, France*
DINESH A. NAGEGOWDA • *Molecular Plant Biology and Biotechnology Lab, CSIR-Central Institute of Medicinal and Aromatic Plants Research Centre, Bengaluru, India*
GODFREY NEUTELINGS • *UMR CNRS 8576—UGSF, Université de Lille, Villeneuve d'Ascq, France*
SARAH E. O'CONNOR • *Department of Natural Product Biosynthesis, Max Planck Institute for Chemical Ecology, Jena, Germany*
NEFTALÍ OCHOA-ALEJO • *Departamento de Ingeniería Genética, Centro de Investigación y de Estudios Avanzados del Instituto Politécnico Nacional, Unidad Irapuato, Gto., Mexico*
AUDREY OUDIN • *EA2106 Biomolécules et Biotechnologies Végétales, Université de Tours, Tours, France*
LIRA PALMER • *Department of Natural Product Biosynthesis, Max Planck Institute for Chemical Ecology, Jena, Germany*
NICOLAS PAPON • *EA3142 Groupe d'Etude des Interactions Hôte-Pathogène, Université d'Angers, Angers, France*
OLIVIER PICHON • *EA2106 Biomolécules et Biotechnologies Végétales, Université de Tours, Tours, France*
MARÍA ISABEL RESTREPO • *Laboratorio de Biotecnología, Universidad de Antioquia, Medellin, Colombia*
CAROLINA WERNER RIBEIRO • *EA2106 Biomolécules et Biotechnologies Végétales, Université de Tours, Tours, France*
STAMATIS RIGAS • *Laboratory of Molecular Biology, Department of Biotechnology, Agricultural University of Athens, Athens, Greece*
SHUBHA SHARMA • *Biotechnology Division, Council of Scientific and Industrial Research-Central Institute of Medicinal and Aromatic Plants, Lucknow, India*
H. B. SHILPASHREE • *Molecular Plant Biology and Biotechnology Lab, CSIR-Central Institute of Medicinal and Aromatic Plants Research Centre, Bengaluru, India*
JUNQI SONG • *Texas A&M AgriLife Research Center at Dallas, Texas A&M University System, Dallas, TX, USA; Department of Plant Pathology and Microbiology, Texas A&M University, College Station, TX, USA*
BENOIT ST-PIERRE • *EA2106 Biomolécules et Biotechnologies Végétales, Université de Tours, Tours, France*
KAIXUE TANG • *Flower Research Institute of Yunnan Academy of Agricultural Sciences, Kunming, Yunnan, China*
KEISUKE TASAKI • *Research Faculty of Agriculture, Hokkaido University, Sapporo, Hokkaido, Japan; Tokyo University of Agriculture, Atsugi, Kanagawa, Japan*
MARGARITA THOMOPOULOU • *Laboratory of Molecular Biology, Department of Biotechnology, Agricultural University of Athens, Athens, Greece*

Dikran Tsitsekian • *Laboratory of Molecular Biology, Department of Biotechnology, Agricultural University of Athens, Athens, Greece*

Hervé Vanderschuren • *Plant Genetics Laboratory, TERRA Research and Teaching Centre, Gembloux Agro BioTech, University of Liège, Gembloux, Belgium; Tropical Crop Improvement Laboratory, Division of Crop Biotechnics, Biosystems Department, KU Leuven, Leuven, Belgium*

Kumar Vasudevan • *Plant Genetics Laboratory, TERRA Research and Teaching Centre, Gembloux Agro BioTech, University of Liège, Gembloux, Belgium*

Aiming Wang • *London Research and Development Centre, Agriculture and Agri-Food Canada, London, ON, Canada*

Guanyu Wang • *Key Laboratory of Hebei Province for Plant Physiology and Molecular Pathology, Mycotoxin and Molecular Plant Pathology Laboratory, College of Life Sciences, Hebei Agricultural University, Baoding, China*

Zonghua Wang • *State Key Laboratory of Ecological Pest Control for Fujian and Taiwan Crops, College of Plant Protection, Fujian Agriculture and Forestry University, Fuzhou, China; Institute of Oceanography, Minjiang University, Fuzhou, China*

Masumi Yamagishi • *Research Faculty of Agriculture, Hokkaido University, Sapporo, Hokkaido, Japan*

Huijun Yan • *Flower Research Institute of Yunnan Academy of Agricultural Sciences, Kunming, Yunnan, China*

Syed Shan-e-Ali Zaidi • *Plant Genetics Laboratory, TERRA Research and Teaching Centre, Gembloux Agro BioTech, University of Liège, Gembloux, Belgium*

Baolong Zhang • *Provincial Key Laboratory of Agrobiology, Jiangsu Academy of Agricultural Sciences, Nanjing, China*

Zhao Zhang • *Beijing Key Laboratory of Development and Quality Control of Ornamental Crops, Department of Ornamental Horticulture, China Agricultural University, Beijing, China*

Jinping Zhao • *Texas A&M AgriLife Research Center at Dallas, Texas A&M University System, Dallas, TX, USA*

Chapter 1

Virus-Induced Gene Silencing in Lilies Using Cucumber Mosaic Virus Vectors

Keisuke Tasaki, Masumi Yamagishi, and Chikara Masuta

Abstract

Virus-induced gene silencing (VIGS) systems are effective for rapid analysis of gene functions in plants that require a long period of growth such as *Lilium*. We successfully developed a VIGS system using the cucumber mosaic virus (HL strain, CMV-HL) vector to induce RNA silencing of the *L. leichtlinii* phytoene desaturase gene (*LlPDS*), where at 30 days postinoculation (dpi), photo-bleaching was observed in the upper leaves of *L. leichtlinii*, and at 57 dpi, white regions appeared on flower tepals that accumulate orange carotenoids. This vector spreads in bulbs, and it could induce silencing on emerged shoots in the following year. The CMV-HL vector can be easily constructed by insertion of a 30–60 nt fragment into the cloning site of the RNA3 genome. In this chapter, we describe how to use the CMV-HL vector system in the context of *Lilium* plants.

Key words *Lilium leichtlinii*, CMV-HL, Gene function analysis, Monocots, Phytoene desaturase (*PDS*), Stem-loop PCR, VIGS

1 Introduction

Lilies are among the most important cut flowers and flower bulbs worldwide [1, 2]. Recently, the transcriptomes of several lily organs have been sequenced using next-generation sequencing, and a variety of potentially important genes have been identified [3–7]. To determine the function of the isolated genes, it will be necessary to develop reverse genetics tools that are capable of facilitating the functional screening of expressed sequences. In many plants, stable genetic transformation is an essential reverse genetics tool that is used to validate the functions of endogenous genes; however, in *Lilium*, *Agrobacterium*-mediated transformation remains challenging, although the method has been successfully developed in a small number of laboratories [8, 9]. Additionally, stable lily transformation is quite time-consuming, as lily plants typically require three or more years to progress from acclimation to flowering. Therefore,

Vincent Courdavault and Sébastien Besseau (eds.), *Virus-Induced Gene Silencing in Plants: Methods and Protocols*, Methods in Molecular Biology, vol. 2172, https://doi.org/10.1007/978-1-0716-0751-0_1,

stable transformation is not an ideal solution for the functional screening of lily genes. Alternatively, transient gene expression systems, such as agroinfiltration [10], particle bombardment [11], and virus-induced gene silencing (VIGS), have been developed to overexpress or knock down lily genes within a short timeframe.

Virus-induced gene-silencing systems provide for rapid analysis of gene functions in plants. To use VIGS technology, the choice of a viral vector is important, as VIGS is only possible when the viral vector can spread systemically in infected plants [12]. Cucumber mosaic virus (CMV) is one of the major viruses that cause viral disease in lilies. A typical feature of CMV is its wide host range, and the strain CMV-HL, which is isolated from *Lilium leichtlinii*, can spread systemically in lily plants [13, 14]. Recently, we successfully induced endogenous gene silencing in lily plants using a CMV-HL-derived vector [15]. This VIGS system is consisted of three steps: (1) introduction of a lily gene fragment into the cloning site of the CMV-HL genome, (2) amplification of the CMV-HL vector in *Nicotiana benthamiana* plants, and (3) inoculation onto lily plants using the leaf sap of infected *N. benthamiana*. This vector could induce gene silencing in the leaves and flowers of lilies within 2 months after inoculation. Additionally, this vector transmits via storage organs (bulbs) to shoots that develop in the year following inoculation, and the silencing phenotype is observed in the second year. Here, we describe the methods of VIGS by the CMV-HL vector.

2 Materials

2.1 Plant Materials and Growth Conditions

1. *Nicotiana benthamiana* seeds.
2. Peat pellets.
3. Growth incubator at 23 °C under a 16:8 h light/dark cycle photoperiod (35 μmol/m^2/s white fluorescent light).
4. Bulbs from *L. leichtlinii* 'Hakugin' that were derived from shoot tip culture and propagated in vitro (*see* **Note 1**).
5. Greenhouse at 20–25 °C under natural sunlight.

2.2 Viral Vector of CMV-HL

Cucumber mosaic virus possesses tripartite positive-sense RNA genomes (RNA1, 2, and 3). Full-length cDNA sequences of RNA1, RNA2, and RNA3 of CMV-HL are cloned into pPCR-Script Amp SK(+) vectors (pCHL1), pPCR-Script Amp SK(+) vectors (pCHL2), and pGEM-T easy vectors (pCHL3), respectively [14, 16]. In the RNA3 sequence, restriction sites for StuI/BglII or EcoRV/BglII are introduced to allow for the cloning of target gene fragments of lilies, and these sites are located in the intercistronic region immediately upstream of the coat protein (CP) gene (Fig. 1).

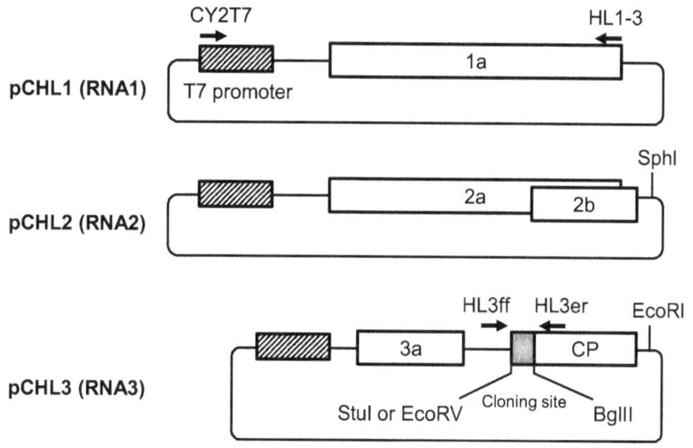

Fig. 1 Structure of the plasmids pCHL1, pCHL2, and pCHL3 containing CMV-HL RNA1, RNA2, and RNA3, respectively. The open reading frame (open boxes) encodes 1a, 2a, 2b, and 3a proteins or the coat protein (CP). Boxes with hatched lines indicate the T7 promoter. A dotted box indicates the site for lily gene cloning. Arrows indicate primer annealing sites

2.3 Cloning into pCHL3

1. Primers to amplify target gene fragments from lilies: at the 5′ termini, the forward primers contain the recognition sequences for StuI or EcoRV, and the reverse primers possess recognition sequences for BglII (*see* **Notes 2** and **3**).
2. Restriction enzymes: StuI, EcoRV, and BglII.
3. 2× Ligation mixture that includes DNA ligase and buffer.
4. Gel Extraction Kit.
5. *E. coli* DH5α competent cells.
6. Plasmid DNA Purification Kit.
7. DNA Polymerase with buffer and dNTP mixture.
8. Primers HL3ff 5′-CTCCGCGAGATTGCGTTAT-3′ and HL3er. 5′-ACGACCAGCTGCTAACGTCT-3′. These are designed to amplify fragments including the cloning sites to allow for confirmation of the presence of insert sequences (Fig. 1).

2.4 In Vitro Transcription

1. Restriction enzymes SphI and EcoRI.
2. Primers CY2T7 5′-CTCCGCGAGATTGCGTTAT-3′ and HL-1-3 5′-ACGACCAGCTGCTAACGTCT-3′ (Fig. 1).
3. DNA polymerase possessing 3′ → 5′ exonuclease (proofreading) activity.
4. 25:24:1 (v/v/v) Phenol.chloroform/isoamyl alcohol.
5. Chloroform.
6. 100% and 70% ethanol (RNase-free).
7. 3 M sodium acetate, pH 5.2 (RNase-free).

8. Precipitation carrier (e.g., Dr. GenTLE™ Precipitation Carrier, Takara Bio).
9. T7 RNA polymerase with buffer.
10. 20× NTP mixture with capping analog containing 10 mM ATP, 10 mM CTP, 10 mM UTP, 1 mM GTP, and 2.5 mM m7G(5′)ppp(5′)G cap analog. Mix 10 μL of 100 mM ATP, 10 μLof 100 mM CTP, 10 μL of 100 mM UTP, 1 μL of 100 mM GTP, and 25 μL of 10 mM m7G(5′)ppp(5′)G cap analog, and then create a final volume of 100 μL by adding nuclease-free water. Store at −20 °C.
11. 100 mM dithiothreitol (DTT).
12. Ribonuclease (RNase) inhibitor.
13. 0.1 M Sodium phosphate buffer, pH 7.

2.5 Inoculation onto Tobacco

1. Rubber gloves.
2. 400 Mesh carborundum.
3. Reverse Transcription Kit including random primers.

2.6 Inoculation onto Lily

1. 0.1 M Sodium phosphate buffer, pH 7.
2. Celite 545.

2.7 Detection of CMV-HL

1. RNA Extraction Kit.
2. Reverse Transcription Kit including random primers.

2.8 Detection of Expression Levels of Target Gene

1. RNA Extraction Kit.
2. Reverse Transcription Kit including random primers.
3. Primers Lh18SrRNAaf 5′-TGCAACAAACCCCGACTTTC-3′ and Lh18SrRNAbr 5′-CCGTCACCCGTCAATACCAT-3′ for amplification of lily *18S rRNA*, as a reference.

2.9 Detection of siRNA

1. Low-molecular-weight (LMW) RNA extraction kit.
2. Estimate siRNA sequences (usually 21 nt) by following the algorithm of Ui-Tei et al. [17] and Reynolds et al. [18]. Specifically, incorporate any A or U at the 5′ terminal end of the guide strand, G or C at the 5′ terminal end of the passenger strand, and base preferences at positions 3 (A), 10 (U), and 13 (A, U, or C) of the passenger strand.
3. Design and synthesize RT primers; 5′-GTTGGCTCTGGTG CAGGGTCCGAGGTATTCGCACCAGAGCCAAC+[six-nucleotide antisense sequences of 3′ terminal of the siRNA]-3′ (*see* **Note 4**) [19, 20].
4. Design and synthesize RT forward primers; 5-′-TCGCG + [15-nucleotide sense sequences from 5′ end of the siRNA]-3′ (*see* **Note 4**).

5. Universal reverse primer 5′-GTGCAGGGTCCGAGGT-3′.
6. RNase inhibitor.
7. RT enzyme with buffer and dNTP mixture.
8. Primers U6af 5′-CGGGGACATCCGATAAAATTGGAACG-3′ and U6br 5′-CGATTTGTGCGTGTCATCCTTGC-3′ to amplify U6 small nuclear RNA.

2.10 Other Reagents and Equipment

1. Micro-centrifuge.
2. Thermal cycler.
3. Agarose gel.
4. Electrophoresis equipment.
5. −80 °C Deep freezer.
6. Mortar and pestle.

3 Methods

3.1 Insertion of Lily Gene Fragments into pCHL3

1. Amplify the target gene fragments of lilies by polymerase chain reaction (PCR) using primers containing restriction sites (StuI and BglII, or EcoRV and BglII) and lily cDNA. Small fragments (30–60 bp) are recommended to avoid the deletion or rearrangement of the fragments during viral spreading in plants.
2. Digest PCR products using StuI and BglII or EcoRV and BglII.
3. Linearize the plasmid containing RNA3 cDNA (pCHL3) using the same restriction enzymes.
4. Electrophorese the restriction-digested products on an agarose gel, recover gel blocks containing the gene fragments or the linearized plasmid, and then purify them from gels using an appropriate kit (*see* **Note 5**).
5. Mix 50 ng of plasmid, the PCR product (its amount is three times that of the plasmid in molecular ratio), and water up to 5 μL. Heat at 65 °C for 5 min, and cool on ice. Add 5 μL of a 2× ligation mixture and incubate at 16 °C for at least 30 min.
6. Transform *E. coli* DH5α strain using the ligated plasmid vector.
7. Isolate plasmids from transformed *E. coli* using a Plasmid Purification Kit.
8. Confirm the presence of lily gene fragments in the cloning site by PCR using the primers HL3ff and HL3er and subsequent electrophoresis. The reaction conditions are as follows: preheat at 94 °C for 2 min, 30 cycles at 94 °C for 10 s, 50 °C for 30 s, and 72 °C for 1 min 30 s and extension at 68 °C for 2 min.
9. Check the insert fragment by Sanger sequencing.

3.2 In Vitro Transcription

1. Prior to synthesis of CMV RNAs by in vitro transcription, linearize 1.5 μg of plasmids containing RNA2 and RNA3 cDNA using the restriction enzymes SphI and EcoRI, respectively. After the digestion, bring the total volume to 100 μL by adding water.
2. PCR-amplify the fragment that includes the T7 promoter and RNA1 sequences from the plasmid containing RNA1 cDNA using appropriate DNA polymerase possessing $3' \rightarrow 5'$ exonuclease (proof-reading) activity with the primers CY2T7 and HL-1-3. The reaction conditions are as follows: preheat at 94 °C for 2 min, 32 cycles at 98 °C for 10 s, 55 °C for 30 s, and 68 °C for 2 min and extension at 68 °C for 2 min. Transfer PCR products to a new 1.5 mL microcentrifuge tube, and bring the total volume to 100 μL by adding water.
3. Add 100 μL volume of phenol/chloroform/isoamyl alcohol (25:24:1) to the linearized plasmids and PCR products, and then vortex thoroughly. Centrifuge at room temperature for 5 min at 15,000 × *g*. Carefully transfer the upper aqueous phase (approximately 100 μL) to a new 1.5 mL microcentrifuge tube.
4. Add 100 μL of chloroform to the collected aqueous solution and vortex. Centrifuge at room temperature for 5 min at 15,000 × *g*. Carefully transfer the upper aqueous phase (approximately 100 μL) to a new 1.5 mL microcentrifuge tube.
5. Add 1/10 volume of 3 M sodium acetate pH 5.2 and a precipitation carrier (e.g., 4 μL of precipitation carrier) to the reaction mixture, and then vortex. Add 2.5 volumes of 100% ethanol, and then vortex. Centrifuge at room temperature for 5 min at 15,000 × *g*. After removing the supernatant, add 400 μL of 70% ethanol. Centrifuge for 2 min at 15,000 × *g*. Carefully remove the supernatant. Dry the DNA pellet at room temperature for 5 min. Resuspend the DNA pellet in 4.5–10 μL of RNase-free water by pipetting.
6. Perform the in vitro synthesis of CMV RNA1, RNA2, and RNA3 using 1.0–1.5 μg of linearized vector DNA and PCR product (*see* **Note 6**). Mix 4 μL of T7 polymerase buffer, 2 μL of T7 RNA polymerase (10 U), 0.5 μL of 20 U of RNase inhibitor, 1 μL of 100 mM DTT (5 mM final conc.), 8 μL of 20× NTP mixture with capping analog, and 4.5 μL volume with DNA template. Incubate at 37 °C for 60 min. Transcribed RNA should immediately be placed on ice and then always kept on ice to promote stability.
7. Confirm the synthesis of CMV-HL RNAs by electrophoresis. Run 5 μL of the solution on a 1% agarose gel (Fig. 2).
8. Mix the three transcription products at the ratio of 1:1:1, and then add 0.1 M sodium phosphate buffer pH 7 at equal volume

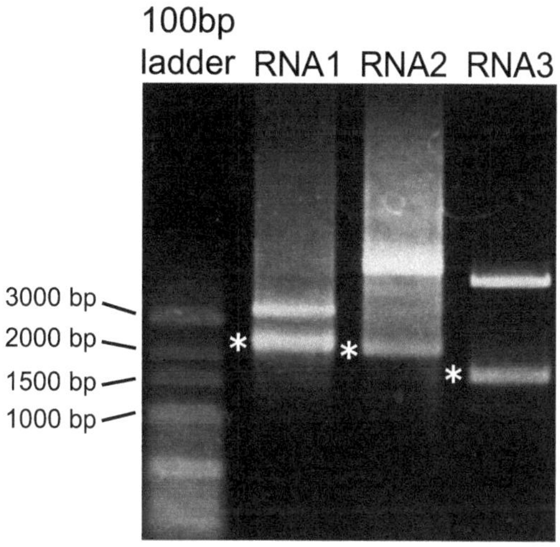

Fig. 2 In vitro transcribed RNA genome (RNA1, RNA2, or RNA3) of CMV-HL. Asterisks indicate transcribed products, and larger bands are template DNA

of the mixed transcriptional products. The RNA mixture is kept on ice. Approximately 15 μL RNA mixture is required for each *N. benthamiana* plant.

3.3 Inoculation onto N. benthamiana

1. Add *N. benthamiana* seeds onto peat pellets, and grow them in a growth incubator at 23 °C.
2. Dust carborundum onto the leaves of 4-week-old *N. benthamiana* plants, and add the RNA mixture dropwise onto the leaves by pipetting. Then, softly rub the leaf surfaces with a rubber-gloved finger.
3. After 30 s, use water to wash away the RNA mixture and carborundum on the inoculated leaves by gently squirting with a wash bottle. Then, softly dry the water by paper towels.
4. Cultivate the *N. benthamiana* plants in a growth chamber. Mild viral symptoms will appear at 6–10 days postinoculation (dpi) (Fig. 3).
5. Harvest the systemic leaves of *N. benthamiana* plants harboring CMV with mild mosaic symptoms (*see* **Note 7**). The harvested leaves can be kept at −80 °C.
6. Isolate total RNA from the leaves using a commercial kit.
7. Reverse-transcribe the total RNA by utilizing random primers.
8. PCR-amplify RNA3 fragments, including the cloning sites, using the primers HL3ff and HL3er. PCR conditions are as follows: preheat at 94 °C for 2 min, 30 cycles at 94 °C for 10 s, 50 °C for 30 s, and 72 °C for 1 min 30 s and extension at 68 °C for 2 min. Then, confirm the presence of inserts by electrophoresis.

Fig. 3 The leaves exhibiting mild symptoms (indicated by white arrows) found in *N. benthamiana* plants inoculated with CMV-HL empty vector at 8 days postinoculation (dpi)

Fig. 4 *Lilium leichtlinii* plants inoculated with CMV-HL harboring a 33-nt *LlPDS* fragment. (**a**) Lily plants at the first inoculation (0 dpi). (**b**) A *PDS*-silencing phenotype found in the tip of an outer tepal (white arrow) at 57 dpi

3.4 Inoculation onto Lily

1. After cold treatment to break dormancy, plant the lily bulbs in pots filled with soil, and cultivate in a closed greenhouse.
2. Grind the *N. benthamiana* leaves harboring CMV-HL vectors using a mortar and pestle in the presence of 0.1 M phosphate buffer (use 100 mg fresh weight leaves/mL) to produce leaf sap.
3. A few weeks after sprouting (Fig. 4a), dust Celite 545 onto the all leaves of lily plants, and add the leaf sap onto the leaves of young seedlings by pipetting.
4. Thoroughly rub the surface of leaves using rubber-gloved fingers. Be careful not to tear a leaf.

5. After 30 s, wash away the leaf sap and Celite 545 on the surface of inoculated leaves by squirting with water from a wash bottle, and softly dry the water by paper towels.
6. Cultivate the lily plants in a closed greenhouse. Apply fertilizer (e.g., 1000-fold diluted Hyponex solution) once a week if necessary.
7. Perform the second inoculation 7 days after the first inoculation according to **steps 3–5** of this section. In this second inoculation, inoculate using all leaves.

3.5 Detection of CMV

1. To confirm the systemic infection by CMV in lily plants, harvest newly emerged leaves or flowers (Fig. 4b).
2. Isolate total RNA from the harvested organs, and synthesize cDNA using random primers.
3. Perform PCR using the primers HL3ff and HL3er to confirm infection by the CMV-HL vector and the presence of target gene fragments in the cloning site. PCR conditions are shown in Subheading 3.3.
4. Run the products on 1% agarose gels.

3.6 Detection of Silencing Phenotype and Expression of Target Gene

1. Gene-silencing phenotype will appear on organs 1 or 2 months after inoculation (*see* **Note 8**).
2. Isolate total RNA from the organs showing silencing phenotype, and synthesize cDNA using random primers.
3. Perform qRT-PCR using a primer set designed for the target gene and 18S rRNA as a reference gene to detect the relative expression levels of the target gene.

3.7 Detection of siRNA

1. The production of siRNA is a hallmark of the occurrence of VIGS. Isolate low-molecular-weight (LMW) RNA from the leaves or flowers.
2. Synthesize cDNA including short interfering RNAs (siRNAs) sequence using the pulsed RT protocol. Create a mixture of 20–50 ng LMW RNA, 1 μL of the RT primer (1 μM), 0.5 μL of 10 mM dNTP mixture, and water (14.4 μL total). Heat the mixture at 65 °C for 5 min. Add 4 μL 5× buffer, 0.5 μL RNase inhibitor, and 1 μL RT enzyme. Then, perform pulsed RT. Specifically, maintain the reaction at 16 °C for 30 min, and follow this by pulsed RT of 60 cycles at 30 °C for 30 s, 42 °C for 30 s, and 50 °C for 1 s. Inactivate the RT enzyme at 85 °C for 5 min [19].
3. Conduct endpoint PCR by creating a mixture of PCR enzyme, buffer, dNTP, 1 μL of the pulsed RT product (template), a single RT forward primer, and the universal reverse primer. Incubate at 94 °C for 5 min, followed by 40 cycles of 94 °C for 15 s and 60 °C for 1 min.

4. Run the PCR products (60 bp) on a 4% agarose gel.
5. Confirm the quality of LMW RNA by amplifying U6 small nuclear RNA fragments (77 bp) by reverse transcribing LMW RNA with the U6br primer and then PCR amplification with the primers U6af and U6br.

3.8 Silencing Phenotype in the Second Year

1. Harvest bulbs from CMV-inoculated plants. Store them in a refrigerator (at 4 °C) for longer than 2 (Asiatic hybrid lilies) or 3 (Oriental hybrid lilies) months to break dormancy.
2. Plant the bulbs in pots and allow them to grow. Phenotypes caused by gene silencing should appear on emerged shoots in the year following inoculation (*see* **Note 9**).

4 Notes

1. Use virus-free bulbs to avoid coinfection with another virus.
2. If an extra three nucleotides are added to the 5′ end, PCR products are cut directly by restriction enzymes.
3. The gene fragments are inserted in opposite orientations if the restriction sites of primers are exchanged.
4. When we silenced the *LlPDS* gene in *L. leichtlinii* plants, we detected siRNA using the following primers (Fig. 5).

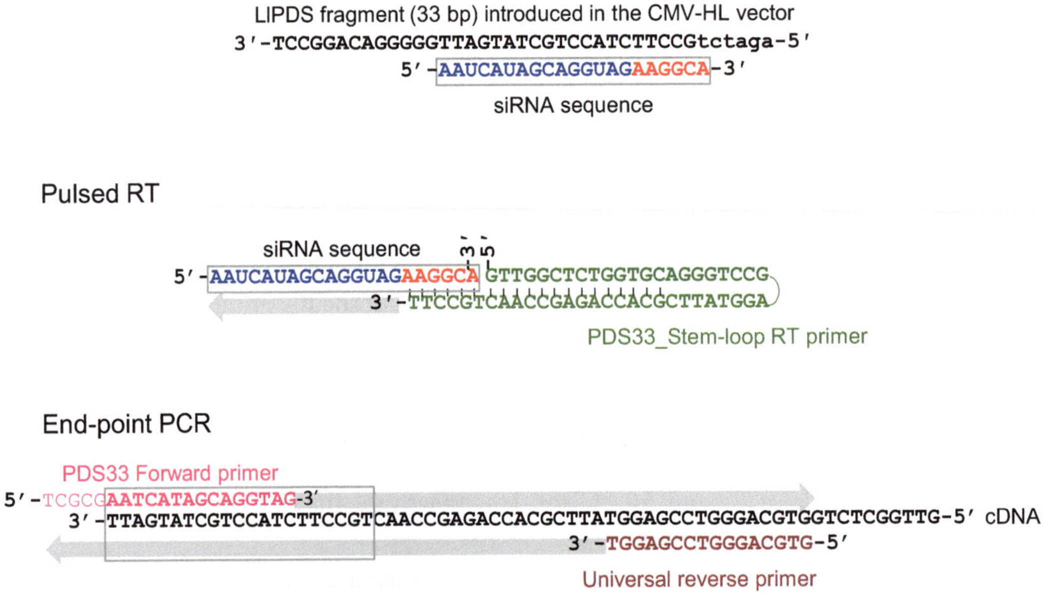

Fig. 5 Schematic representation of the stem loop RT-PCR. Pulsed RT followed by endpoint PCR amplifies the 60-nt fragment including the target siRNA sequence (boxed). The stem loop RT primers (green letters) bind to the 3′ portion of siRNA molecules (red letters), initiating reverse transcription of the siRNA. Then, the RT product is amplified using a siRNA-specific forward primer (pink letters) and the universal reverse primer (brown letters)

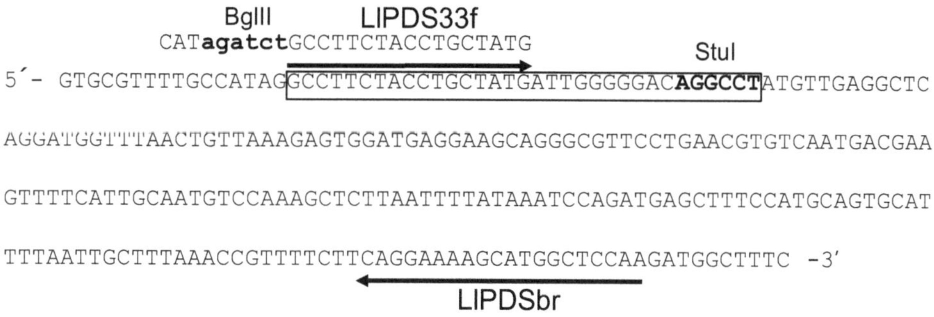

Fig. 6 Partial nucleotide sequence of *LlPDS*. The open box shows the *LlPDS* 33-bp fragment. Allows indicate primer annealing sites. Bold letters indicate restriction enzyme recognition sites (StuI and BglII)

LlPDS fragment (33 bp) introduced using the CMV-HL vector: GCCTTCTACCTGCTATGATTGGGGGACAGGCCT.

Expected siRNA sequence: *AAUCAUAGCAGGUAG* AAGGCA.

PDS33_Stem-loop RT primer; 5′-GTTGGCTCTGGTGCAGGGTCCGAGGTATTCGCACCAGAGCCAACTGCCTT-3′.

PDS33_Forward primer; 5′-TCGCG *AATCATAGCAGGTAG*-3′.

5. If a purification of small PCR fragments (less than 40 bp) is difficult, prepare the fragment using an alternate method. We created the *LlPDS* 33 bp insert using a StuI site found in lily *PDS* sequences (Fig. 6). First, we amplified the 229-bp fragment using primers LlPDS33f (5′-CATagatctGCCTTCTACCTGCTATG-3′, lowercase letters indicate BglII recognition site) and LlPDSbr (5′-TTGGAGCCATGCTTTTCCTG-3′). Then, the PCR products were digested by BglII and purified. The pCHL3 plasmid was digested by BglII and StuI and ligated with the purified PCR products at the BglII site. Next, the ligated plasmid was purified and digested by StuI. Finally, the plasmid was self-ligated to make a circular pCHL3 plasmid containing the *LlPDS* 33-bp fragment.
6. Transcription of RNA 2 is often inefficient for unknown reasons. In such a case, use greater than 1.5 μg of linearized pCHL2 plasmid for transcription.
7. We recommend collecting the leaves exhibiting symptoms as early as possible to avoid the risk of deletion of the insert sequence.

8. Under our experiment conditions targeting *LlPDS*, photo-bleaching phenotype was observed at 30 and 57 dpi in leaves and flowers, respectively.
9. Under our experiment conditions targeting *LlPDS*, one *L. leichtlinii* plant with CMV infection exhibited photo-bleaching in both the first and succeeding years, but, compared with the symptoms observed in the first year, the photo-bleaching phenotype was mild in a succeeding year, while in another *L. leichtlinii* plant infected with CMV, photo-bleaching phenotypes were not observed in the first year but observed in the successive season [15].

Acknowledgments

We thank the Hokuren Federation of Agricultural Cooperatives, Sapporo, Japan for offering the virus-free *Lilium leichtlinii* bulbs.

References

1. Vilsack T, Clark CZF (2009) US census of agriculture, 2007, vol 1. United States Department of Agriculture, Washington, DC, p 51
2. Ministry of Agriculture, Forestry and Fisheries of Japan (2016) The 90th statistical yearbook of Ministry of Agriculture, Forestry and Fisheries. http://www.maff.go.jp/e/data/stat/90th/index.html#20. Accessed 6 Apr 2019
3. Huang J, Liu X, Wang J, Lü Y (2014) Transcriptomic analysis of Asiatic lily in the process of vernalization via RNA-seq. Mol Biol Rep 41:3839–3852
4. Suzuki K, Suzuki T, Nakatsuka T, Dohra H, Yamagishi M, Matsuyama K, Matsuura H (2016) RNA-seq-based evaluation of bicolor tepal pigmentation in Asiatic hybrid lilies (*Lilium* spp.). BMC Genomics 17:611
5. Villacorta-Martin C, Núñez de Cáceres González FF, de Haan J, Huijben K, Passarinho P, Hamo MLB et al (2015) Whole transcriptome profiling of the vernalization process in *Lilium longiflorum* (cultivar White Heaven) bulbs. BMC Genomics 16:550
6. Wang J, Yang Y, Liu X, Huang J, Wang Q, Gu J et al (2014) Transcriptome profiling of the cold response and signaling pathways in *Lilium lancifolium*. BMC Genomics 15:203
7. Yamagishi M, Uchiyama H, Handa T (2018) Floral pigmentation pattern in Oriental hybrid lily (*Lilium* spp.) cultivar 'Dizzy' is caused by transcriptional regulation of anthocyanin biosynthesis genes. J Plant Physiol 228:85–91
8. Azadi P, Chin DP, Kuroda K, Khan RS, Mii M (2010) Macro elements in inoculation and co-cultivation medium strongly affect the efficiency of *Agrobacterium*-mediated transformation in *Lilium*. Plant Cell Tissue Organ Cult 101:201–209
9. Ogaki M, Furuichi Y, Kuroda K, Chin DP, Ogawa Y, Mii M (2008) Importance of co-cultivation medium pH for successful *Agrobacterium*-mediated transformation of *Lilium*×*formolongi*. Plant Cell Rep 27:699–705
10. Fatihah HNN, Moñino López D, Arkel G, Schaart JG, Visser RGF, Krens FA (2019) The *ROSEA1* and *DELILA* transcription factors control anthocyanin biosynthesis in *Nicotiana benthamiana* and *Lilium* flowers. Sci Hortic 243:327–337
11. Yamagishi M, Shimoyamada Y, Nakatsuka T, Masuda K (2010) Two *R2R3-MYB* genes, homologs of petunia *AN2*, regulate anthocyanin biosyntheses in flower tepals, tepal spots and leaves of Asiatic hybrid lily. Plant Cell Physiol 51:463–474
12. Pflieger S, Richard MMS, Blanchet S, Meziadi C, Geffroy V (2013) VIGS technology: an attractive tool for functional genomics studies in legumes. Funct Plant Biol 40:1234–1248
13. Hagita T, Kodama F, Akai J (1989) The virus diseases of lily in Hokkaido. Jpn J Phytopathol 55:1–8. (in Japanese)

14. Masuta C, Seshimo Y, Mukohara M, Jung HJ, Ueda S, Ryu KH et al (2002) Evolutionary characterization of two lily isolates of *Cucumber mosaic virus* isolated in Japan and Korea. J Gen Plant Pathol 68:163–168
15. Tasaki K, Terada H, Masuta C, Yamagishi M (2016) Virus-induced gene silencing (VIGS) in *Lilium leichtlinii* using the *Cucumber mosaic virus* vector. Plant Biotechnol 33:373–381
16. Yamaguchi N, Seshimo Y, Yoshimoto E, Ahn HI, Ryu KH, Choi JK, Masuta C (2005) Genetic mapping of the compatibility between a lily isolate of *Cucumber mosaic virus* and a satellite RNA. J Gen Virol 86:2359–2369
17. Ui-Tei K, Naito Y, Takahashi F, Haraguchi T, Ohki-Hamazaki H, Juni A et al (2004) Guidelines for the selection of highly effective siRNA sequences for mammalian and chick RNA interference. Nucleic Acids Res 32:936–948
18. Reynolds A, Leake D, Boese Q, Scaringe S, Marshall WS, Khvorova A (2004) Rational siRNA design for RNA interference. Nat Biotechnol 22:326–330
19. Varkonyi-Gasic E, Wu R, Wood M, Walton EF, Hellens RP (2007) Protocol: a highly sensitive RT-PCR method for detection and quantification of microRNAs. Plant Methods 3:12
20. Czimmerer Z, Hulvely J, Simandi Z, Varallyay E, Havelda Z, Szabo E, Varga A, Dezso B, Balogh M, Horvath A, Domokos B, Torok Z, Nagy L, Balint BL (2013) A versatile method to design stem-loop primer-based quantitative PCR assays for detecting small regulatory RNA molecules. PLoS One 8:e55168

Chapter 2

Virus-Induced Gene Silencing in Poaceae Using a *Foxtail Mosaic Virus* Vector

Ying-Wen Huang, Chao-Yuan Chang, and Yau-Heiu Hsu

Abstract

Virus-induced gene silencing (VIGS) is a powerful tool for rapidly knocking down the expression of plant genes to elucidate functional genomics. We have established a VIGS vector for monocot plants derived from *Foxtail mosaic virus* (FoMV), a positive-sense single-stranded RNA virus. For silencing a targeted gene, plant gene fragment was inserted into the vector between open reading frame 4 (ORF4) and ORF5 under the control of a duplicated coat protein promoter. Plants of different monocot species were infected by mechanical inoculation with sap from FoMV derivative-infected *Chenopodium quinoa* leaves. Gene silencing was typically observed within 2–3 weeks after inoculation. In this chapter, we describe the detailed protocol for silencing a target gene in various Poaceae plants by using FoMV-based vectors.

Key words Posttranscriptional gene silencing (PTGS), Virus-induced gene silencing (VIGS), *Foxtail mosaic virus* (FoMV), FoMV-based VIGS vector, Monocot, Poaceae

1 Introduction

With the advancement of whole-genome sequencing, a vast amount of sequence information has become available. Virus-induced gene silencing (VIGS) provides a means to unravel the functions of candidate genes identified in large genomic or transcriptomic screens [1]. Upon virus infection, plants defend themselves by activating antiviral RNA silencing pathways. These pathways involve the ribonucleases dicer-like protein (DCL) and Argonaute protein. DCL cleaves double-stranded RNAs (dsRNAs) such as self-folded viral genomes or dsRNAs of replication intermediates into small interfering RNA (siRNA) fragments of about 21–24 nt [2]. Subsequently, one strand of the siRNA is loaded into Argonaute to form an RNA-induced silencing complex that targets complementary viral RNAs for inactivation by slicing or translational inhibition [3, 4]. Additionally, the plant's RNA-dependent RNA polymerase 6 uses siRNAs as primers to synthesize new

Vincent Courdavault and Sébastien Besseau (eds.), *Virus-Induced Gene Silencing in Plants: Methods and Protocols*, Methods in Molecular Biology, vol. 2172, https://doi.org/10.1007/978-1-0716-0751-0_2,

dsRNAs for amplifying gene-silencing signals, which can spread from the site of infection throughout the entire plant to trigger systemic gene silencing [5, 6].

In VIGS, the plant is infected with an engineered virus that carries a fragment of a plant gene. The recombinant virus replicates and moves throughout the plant and triggers gene silencing to cleave the corresponding mRNA transcripts. VIGS has become a widely used technique for analyzing plant gene functions because it is easy to operate and rapid, with high specificity and high efficiency [7].

Eight plant viruses have been used for developing viral vectors for VIGS in Poaceae plants: *Barley stripe mosaic virus* in barley (*Hordeum vulgare*), *Brachypodium distachyon*, and wheat (*Triticum aestivum*) [8–11]; *Brome mosaic virus* in rice (*Oryza sativa*), barley, and maize (*Zea mays*) [12]; *Rice tungro bacilliform virus* in rice [13, 14]; *Bamboo mosaic virus* and its associated satellite RNA in *B. distachyon* [15]; *Cucumber mosaic virus* in maize [16]; *Foxtail mosaic virus* (FoMV) in maize, barley, wheat, and foxtail millet (*Setaria italica*) [17–19]; *Tobacco rattle virus* in wheat and maize [20]; and, most recently, *Chinese wheat mosaic virus* in wheat [21]. Among them, FoMV has the following advantages as an applicable VIGS vector for studying gene function in Poaceae: (1) it has a broad host range in Poaceae, including 56 species [22, 23], (2) it induces only mild symptoms in native or adaptive hosts [17, 18], and (3) preparing the inoculum is very easy.

FoMV, a member of the genus *Potexvirus* in the family Alphaflexiviridae, has a single-stranded, positive-sense RNA genome of approximately 6200 nt [excluding the 3′ poly(A) tail]. The five open reading frames (ORF1 to 5) of the FoMV genome encode ORF1 protein (152 kDa), triple-gene-block protein 1 (TGBp1, 26 kDa), TGBp2 (11 kDa), TGBp3 (6 kDa), and capsid protein (CP, 24 kDa), respectively. Unlike other potexviruses, FoMV has an additional ORF5A, which initiates 144 nt upstream of ORF5 and causes an N-terminal CP extension with dispensable function in FoMV infection (Fig. 1a) [24]. ORF1 encodes a viral replicase that is expressed from genomic RNA (gRNA), whereas TGBp1-3 and CP are expressed from subgenomic RNA1 (sgRNA1) and sgRNA2, respectively. The synthesis of sgRNAs is individually driven by subgenomic promoter 1 (SGP1) and cpSGP [25].

Two FoMV vectors for VIGS had been applied in monocot plants. In the first one, FoMV infectious cDNA was cloned under the control of the *Cauliflower mosaic virus* (CaMV) 35S promoter and the nopaline synthase terminator. The vector was designed by engineering the cloning sites after the stop codon of ORF5. This vector successfully silenced endogenous genes in maize by delivering the plasmid of recombinant FoMV by bombardment [18]. The second cloning strategy involved duplicating cpSGP between ORF4 and ORF5 of FoMV vector for expressing homologous

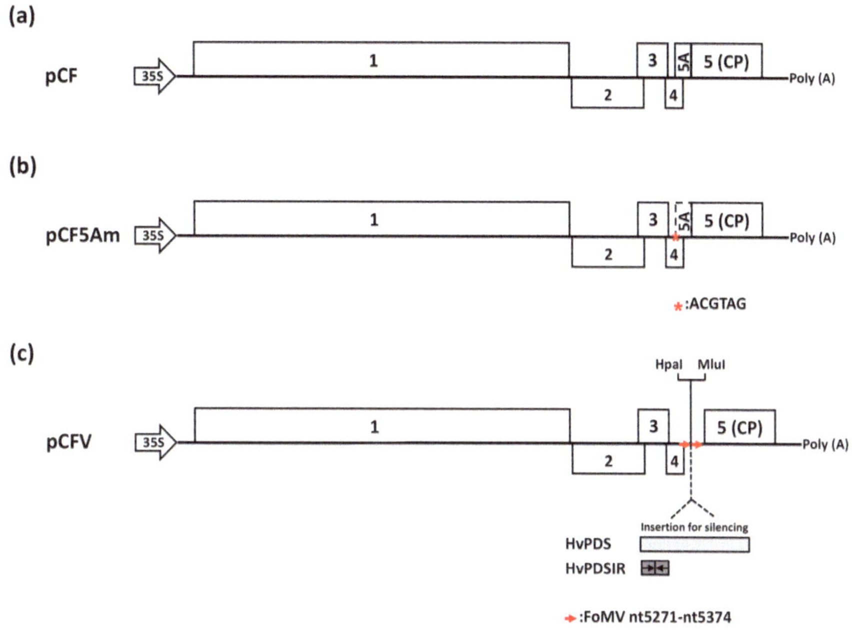

Fig. 1 Schematic representation of construction of the FoMV cDNA infectious clone and derivative mutants. (**a**, **b**) Genome organization of FoMV (pCF) and FoMV-derived mutant (pCF5Am). Rectangles represent open reading frames (ORFs), ORF1-ORF5, encoded by FoMV genomic RNA. The FoMV genome is cloned under control of the CaMV 35S promoter (35S). Red star (∗) in frame B: the start codon and second codon of ORF5A were altered to ACGTAG for disrupting translation of 5A protein. (**c**) Schematic diagram of the constructed FoMV silencing vector, pCFV. Plasmid of pCFV was derived from pCF5Am and duplicated the promoter region of CP (red arrowhead). Gene fragments for silencing are inserted between the cut sites of HpaI and MluI. The location of the CP promoter in FoMV genome is indicated below the diagram

fragments of host genes. Endogenous genes in barley, wheat, and foxtail millet were efficiently silenced after sap inoculation of recombinant FoMV in agroinfiltrated *Nicotiana benthamiana* leaves [17].

Here, we present the use of a new FoMV vector pCFV designed for VIGS. The cDNA of the FoMV genome (AY121833.1) was cloned under the control of the CaMV 35S promoter to generate pCF (Fig. 1a). To develop an FoMV VIGS vector, we employed the strategy of cpSGP duplication between ORF4 and ORF5. Because duplicated cpSGP and insertion of foreign fragments would interfere with the translation codon of protein 5A, we stopped the expression of protein 5A by substitution of the initiation and second codons of ORF5A with ACGTAG and named as pCF5Am (Fig. 1b). The final vector pCFV is a pCF5Am derivative that contains a putative FoMV cpSGP (nt5271-nt5374 of AY121833.1) and cloning sites (HapI and MluI) inserted between ORF4 and ORF5 (Fig. 1c) [26]. In this chapter, we present a detailed protocol for constructing the target sequence into pCFV and silencing a target gene in various Poaceae plants by using an

FoMV-based vector. It may be an efficient strategy for high-throughput functional genomics in important cash crops.

2 Materials

2.1 Plant Materials and Plant Growth

1. *Brachypodium distachyon* seeds: lines Bd3-1, Bd21-1, and Bd21-3.
2. *Chenopodium quinoa* seeds.
3. *Hordeum vulgare* seeds.
4. *Oryza sativa* L. seeds: Taikeng No. 9, Taitung Glutinous No. 31, Taoyuan No. 3, and Tatsuen No. 2.
5. *Setaria viridis* seeds.
6. *Sorghum bicolor* seeds: Taichung No. 3.
7. *Sorghum sudanense* seeds.
8. *Triticum aestivum* seeds.
9. Soil mix and pots.
10. Greenhouse at 25 °C under a 16-h/8-h light/dark photoperiod.

2.2 Creating Silencing Constructs

1. *Escherichia coli* strain DH5α competent cells.
2. pCFV-based expression vector [26]: an infectious cDNA clone of FoMV (AY121833.1) driven by the 35S promoter of *Cauliflower mosaic virus* (CaMV) and containing a duplicated cpSGP and cloning sites (HapI and MluI) inserted between ORF4 and ORF5 (Fig. 1c).
3. Restriction enzymes and buffers: *HpaI* and *MluI*.
4. 1% Agarose gel containing 0.5 mg/mL of ethidium bromide, 0.5× TBE buffer (44.5 mM Tris–HCl pH 8.0, 44.5 mM boric acid, 1 mM EDTA), and electrophorese equipment.
5. Gel Extraction Kit and PCR Purification Kit.
6. cDNAs of Poaceae plants.
7. Oligonucleotide primers for amplifying target sequences.
8. *Taq* DNA polymerase and buffer.
9. T4 DNA ligase and buffer.
10. Luria Broth (LB) liquid media and agar plates.
11. 50 mg/mL Ampicillin stock solution.
12. High-volume plasmid purification kit (e.g., NucleoBond Xtra Maxiprep kit, Macherey-Nagel).
13. Thermal cycler.
14. 37 and 16 °C water bath.

15. Fo5137F primer 5′-GCATGCACGAATCACAC-3′ and Fo5655R primer 5′-GCTAGCGCCAGACTTTTC-3′.

2.3 Plant Inoculation with FoMV Plasmids

1. pCFV Empty vector plasmid (Fig. 1c).
2. pCFHvPDS or pCFHvPDSIR (*H. vulgare phytoene desaturase* target sequence) as a positive control for VIGS [26] (Fig. 1c, *see* **Note 1**).
3. Recombinant pCFV plasmids with target sequence insertion.
4. Carborundum.
5. Cotton swabs.
6. 0.01 M sodium phosphate, pH 7.0.
7. Pestle and mortar.
8. Celite 545.

2.4 Validation of Silencing

1. Total RNA extraction reagent (e.g., TriPure Isolation Reagent, Roche Life Science).
2. Reverse transcription kit including oligo dT(18) primer.
3. qPDSF Primer 5′-AATCCTCCTGAAAGGCTATG-3′ and qPDSR primer 5′-TCTCCAGTTATTTGAGTCCC-3′.
4. qActinF Primer 5′-GAAGATCCTCACCGAGAGAGGTTA-3′ and qActinR primer 5′-GTCACACTTCATGATGGAGTTG TA-3′.
5. SYBR Green Master Mix.
6. qPCR Thermal cycler.

3 Methods

3.1 Construction of pCFV Containing the Target Gene for Silencing

3.1.1 Preparation of Target Gene Fragment

1. Select a 300–500 base pair (bp) region of the target gene for silencing (*see* **Note 2**).
2. Design oligonucleotide primers to amplify the target region for silencing. The primers should be designed to include the site *HpaI* and/or *MluI* at the 5′ end of the forward and reverse primer, respectively, for cloning into pCFV.
3. Amplify the target region by PCR using the oligonucleotide primers, Taq DNA polymerase, and cDNA of *Poaceae* plant as template in 25 μL volume.
4. Electrophorese 3 μL of the PCR product on 1% agarose gel to confirm amplification of the correct-sized product. Use the rest of the sample to purify the amplified PCR fragment by a PCR cleanup kit.
5. Digest 1 μg of the purified PCR fragment with *HpaI* and/or *MluI* according to the design.

6. Electrophorese the digested PCR product on 1% agarose gel, and purify the band by using a gel extraction kit.

3.1.2 Preparation of pCFV Vector for Cloning

1. Digest 2 μg of pCFV plasmid with *HpaI* and/or *MluI* according to the design.
2. Electrophorese the digested product on 1% agarose gel, and purify the band by using a gel extraction kit.

3.1.3 Cloning of pCFV Containing the Target Gene

1. Ligate the digested PCR product (insert fragment) into digested pCFV (the vector) using a 1:3 molar ratio of vector/insert DNA in 10 μL volume. Incubate reaction at room temperature for 1 h.
2. Transform the ligation product into competent *E. coli* DH5α by heat shock.
3. Spread the cells onto LB agar plates supplemented with 50 μg/mL ampicillin, and incubate overnight at 37 °C.
4. Confirm the presence of pCFV carrying the target gene sequence by colony PCR with the primers Fo5137F and Fo5655R on individual transformants.
5. Positive clones are grown in 2 mL LB liquid containing 50 μg/mL ampicillin overnight at 37 °C.
6. Extract plasmid DNAs of the confirmed silencing constructs by using a miniprep procedure and digest with *HpaI* and/or *MluI* to determine that the target fragment for silencing is present.
7. Sequence the positive clones with Fo5137F sequencing primer (*see* **Note 3**).
8. Prepare the silencing construct in large-scale cultures by using a high-volume plasmid purification kit to minimum concentration of 2 μg plasmid per mL culture.

3.2 Plant Inoculation with FoMV Plasmids

3.2.1 Mechanical Inoculation of FoMV Plasmid into C. quinoa

1. Sow *C. quinoa* seeds in pots with soil mix. Place the pots in a greenhouse at 25 °C under a 16-h/8-h light/dark photoperiod. Plants are ready for inoculation at five-leaf stage (approximately 30 days old under our condition).
2. Dust every *C. quinoa* leaf with carborundum.
3. Rub 10 μL of each 200 ng/μL pCFV (negative control), pCFHvPDS or pCFHvPDSIR (positive controls), and pCFV carrying the target gene onto *C. quinoa* leaves by using a cotton swab (*see* **Note 4**). After 5 min the inoculated leaf was washed with sterile water to remove carborundum.
4. Symptoms of infection (local lesions) are typically observed between 7 and 14 days postinoculation on inoculated *C. quinoa* leaves (Fig. 2).

Fig. 2 Local lesions of *Chenopodium quinoa* inoculated with water (Mock) or FoMV infectious clone, pCFV. The photographs were taken at 7 days postinoculation (dpi) and 14 dpi

5. Harvest the *C. quinoa* leaves showing FoMV infection as inoculum.

3.2.2 Mechanical Inoculation of C. quinoa *Sap into Poaceae Plants*

1. Germinate seeds of Poaceae plants in 3-in. diameter pots containing the soil (*see* **Note 5**). Place the pots in a greenhouse at 25 °C under a 16-h/8-h light/dark photoperiod. Seedlings at the two-leaf stage are ready for inoculation.
2. Prepare VIGS inoculum by grinding infected leaf tissue of *C. quinoa* in five volumes of 0.01 M sodium phosphate buffer (pH 7.0) by using a pestle and mortar. Add 10 mg/mL Celite to the mixtures.
3. Inoculate two leaves of seedlings by rubbing with the inoculum mixture between the thumb and index finger. Care is taken to not use too much pressure that would damage the leaf.
4. Place the rub-inoculated plants in a greenhouse at 25 °C under a 16-h/8-h light/dark photoperiod.

3.2.3 Validation of Silencing

1. Isolate total RNA from the leaf tissues showing the desired phenotype by using RNA extraction reagent, as per the manufacturer's instructions. If there is no obvious phenotype, harvest the fourth leaf of infected plants after 14 or 21 days of VIGS for RNA isolation. For *PDS*-silenced plants (positive controls), the leaves should show obvious photo-bleaching in various Poaceae plants after 3 weeks (Fig. 3a).
2. Synthesize first-strand cDNA by using the oligo dT(18) primer and a reverse transcriptase according to the manufacturer's protocol.
3. Use real-time quantitative PCR (qPCR) to test for the targeted gene silencing with designed primers (*see* **Note 6**). Prepare qPCR reaction by mixing 6.2 μL PCR-grade water, 0.4 μL forward primer (10 mM stock), 0.4 μL reverse primer

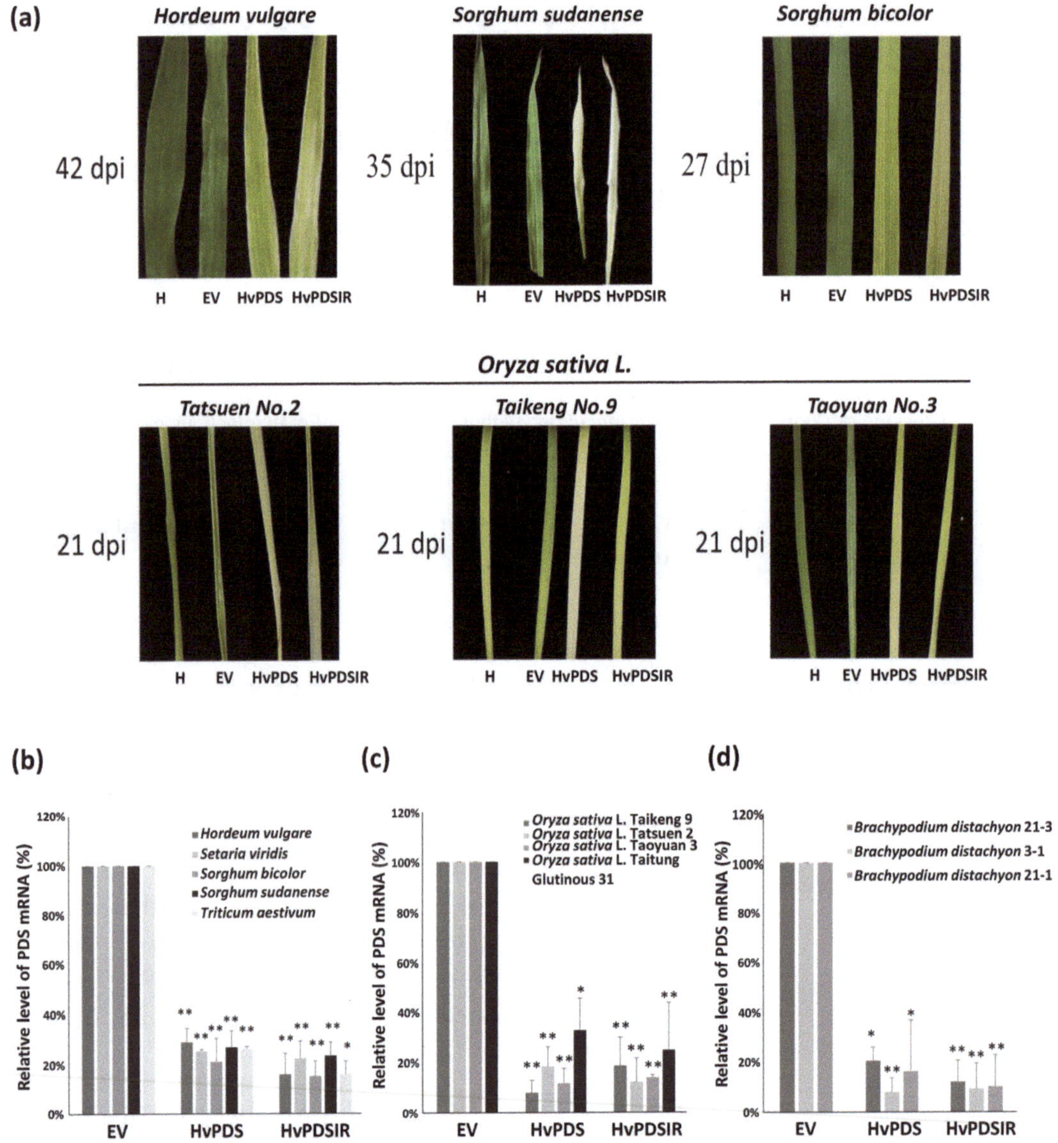

Fig. 3 *PDS* silencing efficiency of FoMV-VIGS vector in various Poaceae plants. (**a**) Phenotypes in *PDS*-silenced plants of barley (*Hordeum vulgare*), sudangrass (*Sorghum sudanense*), sorghum (*Sorghum bicolor*), and rice (*Oryza sativa L.*, Taikeng No. 9, Tatsuen No. 2, and Taoyuan No. 3) at 42, 35, 27, and 21 dpi, respectively. (**b–d**) Relative accumulation of *PDS* mRNA in control (EV) and *PDS* knockdown (HvPDS or HvPDSIR) plants measured by real-time RT-qPCR. Data are mean ± SD from three independent experiments relative to the control (EV). The mRNA level of *PDS* was normalized to that of *Actin*. $*$ and $**$ indicate statistically significant differences by Student's *t*-test ($*P < 0.05$; $**P < 0.01$)

(10 mM stock), 3 μL cDNA, and 10 μL SYBR green Master Mix in each PCR tubes. Carry out PCR according to the following program: 1 cycle of 3 min at 95 °C (enzyme activation); 40 cycles of 3 s at 95 °C (denaturation) and 20 s at 60 °C

(annealing/extension). For *PDS*-silenced plants, *PDS* is quantified by real-time RT-qPCR with primers (qPDSF and qPDSR) that are conserved in different *Poaceae* plants and specific to *PDS*.

4. Analyze the data by the $2^{-\Delta\Delta CT}$ method [27].
5. Normalize the gene of interest to the endogenous control gene, *actin*. The *actin* gene is amplified with primers (qActinF and qActinR) that can be used for various Poaceae plants. For *PDS*-silenced plants, *PDS* expression in FoMVPDS- and FoMVPDSIR-infected plants was decreased to 8–33% and 9–15% of that of control plants (infected with FoMV vector, pCFV) at 14 days postinoculation (Fig. 3b–d).

4 Notes

1. To induce *PDS* silencing in different plant species by using one construct, an HvPDS fragment located at nt1402-nt1843 of *H. vulgare phytoene desaturase* mRNA partial cds (GenBank: AY062039.1) was selected for cloning into the pCFV vector. This fragment shares 95%, 92%, and 90% sequence similarity with that of *T. aestivum* (FJ517553), *B. distachyon* (FJ913272), and *Oryza sativa* L. (AF049356), respectively. The high similarity (more than 88%) has the opportunity to trigger gene silencing in different plants [8]. For the HvPDSIR, an inverted repeat fragment of nt1090-nt1149 of HvPDS cDNA, shares 95%, 88%, and 89% sequence similarity with that of *TaPDS*, *BdPDS*, and *OsPDS*, respectively, was cloned into pCFV.
2. Because the region and length of the insert fragments used for silencing vary depending on the gene, optimization is required. A fragment of >300 bp is recommended for use for silencing, and if the fragment is <300 bp, the silencing efficiency would be reduced [28]. Multiple sequence alignments can be used to determine the optimal sequence for silencing. To avoid nonspecific targeting, a gene-specific region should be selected for primer design. If the complete sequence of the gene is available, its untranslated region should be used for silencing. To silence all or several members of a gene family, a conserved sequence with more than 88% similarity should be selected.
3. The Fo5137F sequence primer is located 237 bases upstream of the cloning site in the pCFV vector.
4. We recommend always infection with pCFHvPDS or pCFHvPDSIR as the positive control. Suppression of *PDS* expression will result in plants exhibiting a photo-bleached phenotype that is easily observed by the naked eye and is

quite different from viral symptoms. The inverted repeat (IR) structure produced by the virus always induces better efficiency of VIGS; we recommend pCFHvPDSIR as a positive control [29].

5. Among the tested Poaceae plants in this report, only the seeds of *O. sativa* and *S. bicolor* need pretreatment to break dormancy and improve seed growth. Soak *O. sativa* seeds and *S. bicolor* seeds in water at room temperature for 3 days. Cover seeds with water at a depth of approximately 15 mm, and expose them to 28 °C for 2 days prior to sowing.
6. The primers designed should be specific to the silenced gene and outside the gene fragment that was cloned into the pCFV plasmid. The sizes of amplified fragments are typically about 200–300 bp.

Acknowledgments

This work was supported by the Ministry of Science and Technology, Taiwan (MOST-107-2313-B-005-026), and the Advanced Plant Biotechnology Center from The Featured Areas Research Center Program within the framework of the Higher Education Sprout Project by the Ministry of Education (MOE).

References

1. Dommes AB, Gross T, Herbert DB et al (2019) Virus-induced gene silencing: empowering genetics in non-model organisms. J Exp Bot 70:757–770
2. Ding SW, Voinnet O (2007) Antiviral immunity directed by small RNAs. Cell 130:413–426
3. Szittya G, Burgyan J (2013) RNA interference-mediated intrinsic antiviral immunity in plants. Curr Top Microbiol Immunol 371:153–181
4. Yang Z, Li Y (2018) Dissection of RNAi-based antiviral immunity in plants. Curr Opin Virol 32:88–99
5. Vaistij FE, Jones L, Baulcombe DC (2002) Spreading of RNA targeting and DNA methylation in RNA silencing requires transcription of the target gene and a putative RNA-dependent RNA polymerase. Plant Cell 14:857–867
6. Vaistij FE, Jones L (2009) Compromised virus-induced gene silencing in RDR6-deficient plants. Plant Physiol 149:1399–13407
7. Burch-Smith TM, Anderson JC, Martin GB et al (2004) Applications and advantages of virus-induced gene silencing for gene function studies in plants. Plant J 39:734–746
8. Holzberg S, Brosio P, Gross C et al (2002) *Barley stripe mosaic virus*-induced gene silencing in a monocot plant. Plant J 30:315–327
9. Tai YS, Bragg J, Edwards MC (2005) Virus vector for gene silencing in wheat. Biotechniques 39:310–312
10. Pacak A, Geisler K, Jorgensen B et al (2010) Investigations of *Barley stripe mosaic virus* as a gene silencing vector in barley roots and in Brachypodium distachyon and oat. Plant Methods 6:26
11. Yuan C, Li C, Yan LJ et al (2011) A high throughput *Barley stripe mosaic virus* vector for virus induced gene silencing in monocots and dicots. PLoS One 6:e26468
12. Ding XS, Schneider WL, Chaluvadi SR et al (2006) Characterization of a *Brome mosaic virus* strain and its use as a vector for gene silencing in monocotyledonous hosts. Mol Plant-Microbe Interact 19:1229–1239
13. Purkayastha A, Mathur S, Verma V et al (2010) Virus-induced gene silencing in rice using a vector derived from a DNA virus. Planta 232:1531–1540

14. Purkayastha A, Sharma S, Dasgupta I (2013) Virus-induced gene silencing for rice using agroinoculation. Methods Mol Biol 975:33–45
15. Liou MR, Huang YW, Hu CC et al (2014) A dual gene-silencing vector system for monocot and dicot plants. Plant Biotechnol J 12:330–343
16. Wang R, Yang XX, Wang N et al (2016) An efficient virus-induced gene silencing vector for maize functional genomics research. Plant J 86:102–115
17. Liu N, Xie K, Jia Q et al (2016) *Foxtail mosaic virus*-induced gene silencing in monocot plants. Plant Physiol 171:1801–1807
18. Mei Y, Zhang CQ, Kernodle BM et al (2016) A *Foxtail mosaic virus* vector for virus-induced gene silencing in maize. Plant Physiol 171:760–772
19. Mei Y, Whitham SA (2018) Virus-induced gene silencing in maize with a *Foxtail mosaic virus* vector. Methods Mol Biol 1676:129–139
20. Zhang J, Yu DS, Zhang Y et al (2017) Vacuum and co-cultivation agroinfiltration of (germinated) seeds results in *Tobacco rattle virus* (TRV) mediated whole-plant virus-induced gene silencing (VIGS) in wheat and maize. Front Plant Sci 8:393
21. Yang J, Zhang TY, Liao QS et al (2018) *Chinese wheat mosaic virus*-induced gene silencing in monocots and dicots at low temperature. Front Plant Sci 9:1627
22. Paulsen AQ, Niblett CL (1977) Purification and properties of *Foxtail mosaic-virus*. Phytopathology 67:1346–1351
23. Short MN, Davies JW (1987) Host ranges, symptoms and amino-acid compositions of 8 potexviruses. Ann Appl Biol 110:213–219
24. Robertson NL, French R, Morris TJ (2000) The open reading frame 5A of *Foxtail mosaic virus* is expressed in vivo and is dispensable for systemic infection. Arch Virol 145:1685–1698
25. Bouton C, King RC, Chen HX et al (2018) *Foxtail mosaic virus*: a viral vector for protein expression in cereals. Plant Physiol 177:1352–1367
26. Chang CY (2013) Development and application of *Foxtail mosaic virus*-based vector system. Dissertation, National Chung Hsing University
27. Schmittgen DH, Livak JK (2008) Analyzing real-time PCR data by the comparative C_T method. Nat Protoc 3:1101–1108
28. Dinesh-Kumar SP, Anandalakshmi R, Marathe R et al (2003) Virus-induced gene silencing. Methods Mol Biol 236:287–294
29. Lacomme C, Hrubikova K, Hein I (2003) Enhancement of virus-induced gene silencing through viral-based production of inverted-repeats. Plant J 34:543–553

Chapter 3

Virus-Induced Gene Silencing (VIGS) in Chili Pepper (*Capsicum* spp.)

Magda Lisette Arce-Rodríguez and Neftalí Ochoa-Alejo

Abstract

Virus-induced gene silencing (VIGS) is a transcript suppression technique that enables the functional characterization of genes in recalcitrant transformation plants. This technique consists in cloning a short fragment of a gene of interest into a viral vector, such as TRV (*Tobacco rattle virus*), and this viral construction is used to agro-infiltrate the plant. VIGS induces posttranscriptional gene silencing (PTGS) that results in the specific sequence degradation of target RNAs. Here we describe a VIGS protocol using the Gateway-based TRV vector for the study of genes in chili pepper plants.

Key words *Capsicum*, Chili pepper, Gene silencing, VIGS, TRV

1 Introduction

The term virus-induced gene silencing (VIGS) was used for the first time to describe the resistance of plants against viruses [1]. Later on, VIGS term was used to refer to a technique that involves the use of a recombinant virus to silence the expression of endogenous genes to study their functions in plants [2, 3]. VIGS induces the posttranscriptional gene silencing (PTGS), which is an epigenetic phenomenon causing sequence-specific degradation of endogenous mRNAs resulting in the diminution of the expression levels of a target gene. This method consists in cloning a partial sequence of an endogenous gene into a viral vector. The viral construction is transferred into a strain of *Agrobacterium tumefaciens* (such as GV2260, GV3101, among others) to infect the plant. Several studies have shown high-efficiency gene silencing in *Capsicum* using VIGS [4–9]. In this protocol, we used *Tobacco rattle virus* (TRV)-derived vector to silence chili pepper genes. TRV virus has a bipartite genome of RNA; TRV1 encodes the proteins responsible for the replication and movement of the virus in the plant, while TRV2 encodes the viral coat protein (CP). TRV2 has been

Vincent Courdavault and Sébastien Besseau (eds.), *Virus-Induced Gene Silencing in Plants: Methods and Protocols*, Methods in Molecular Biology, vol. 2172, https://doi.org/10.1007/978-1-0716-0751-0_3,

modified to introduce either multiple cloning sites (MCS) or the Gateway system in order to clone the fragment of a specific gene to study its function [2, 3]. In general, *Capsicum* plants infected with TRV show a slight yellow streaked in the leaves [8] (Fig. 1A). The phytoene desaturase (*PDS*) gene is the positive control commonly used to test the efficiency of gene silencing in plants because a photobleaching phenotype is clearly visible at approximately 15 days post-infiltration (dpi) and persists until fruit development (Fig. 1B). Here, we describe the VIGS technique for chili pepper plants using the Gateway-based TRV protocol, which is highly efficient for gene function studies (Fig. 2).

2 Materials

2.1 Cloning into the pTRV2 Gateway Vector

1. Sterilized plastic material (tubes, tips, etc.).
2. Trizol reagent and RNA purification kit.
3. cDNA synthesis kit containing reverse transcriptase enzyme and Oligo(dT)$_{12-18}$ primer.
4. PCR reagents (Taq DNA polymerase recombinant, dNTP's).
5. PCR purification kit.
6. TA cloning Gateway entry vector (e.g., pCR8/GW/TOPO vector, Invitrogen).
7. TE buffer (10 mM Tris–HCl and 1 mM EDTA, pH 8.0): prepare separately Tris–HCl and EDTA solutions. Weigh 12.1 g of Tris base and dissolve in 80 ml distilled water. Adjust pH 8.0 using 1 M HCl and afore to 100 ml with distilled water. Sterilize by autoclave at 121 °C for 20 min. Weigh 46.52 g of EDTA, and dissolve in 200 ml distilled water. Adjust pH 8.0 using 10 M NaOH, and fill up to 250 ml with distilled water. Sterilize by autoclave at 121 °C for 20 min. Mix 1 ml of 1 M Tris–HCl and 0.2 ml of 0.5 M EDTA (pH 8.0) in 80 ml distilled water. Adjust pH 8.0 using 1 M HCl, and afore to 100 ml with distilled water. Sterilize in autoclave at 121 °C for 20 min.
8. 0.2 M $CaCl_2$: weigh 0.88 g of $CaCl_2$ and dissolve in 20 ml distilled water. Fill up to 30 ml of distilled water. Sterilize by autoclave at 121 °C for 20 min.
9. 2 mg/ml Proteinase K solution: mix 2.5 ml of 100% glycerol, 0.05 ml of 1 M Tris–HCl, 0.5 ml of 0.2 M $CaCl_2$, and 1.75 ml of distilled water. Add 10 mg of proteinase K in the solution and dissolve. Make aliquots and store at −20 °C.
10. Nitrocellulose membrane filter of 0.025 μm pore.
11. *Escherichia coli* TOP10 chemically competent cells and *E. coli* DH10B electrocompetent cells.

Fig. 1 TRV virus infection symptoms in chili pepper plant leaves and the photobleaching phenotype caused by the *PDS* gene silencing in leaves and chili pepper fruits. (**A**) Leaves of chili pepper plants showing a slight yellow streaked by TRV virus infection: (a) chili pepper plant infected with pTRV2 after 48 days post-infiltration (dpi). Red arrows show leaves with symptoms; (b) close-up of a leaf showing TRV infection symptoms. (**B**) Chili pepper plants and fruits exhibiting the characteristic photobleaching phenotype caused by the *PDS* gene silencing: (a) 1-week-old agroinfected chili pepper plant showing photobleaching after 12 dpi; (b) chili pepper plant showing photobleaching

12. M13 forward 5′-GTAAAACGACGGCCAG-3′ and M13 reverse 5′-CAGGAAACAGCTATGAC-3′.
13. Luria-Bertani (LB) agar and broth: weigh 10 g casein peptone, 5 g yeast extract, and 10 g NaCl. For semisolid medium weigh 16 g agar. Add 800 ml distilled water, and dissolve. Adjust pH 7.0 using NaOH 1 N. Adjust the final volume to 1 L with distilled water. Autoclave at 121 °C for 20 min.
14. Antibiotic stocks of 100 mg/ml spectinomycin and 50 mg/ml kanamycin in water. Sterilize by filtration through a 0.2 μm filter. Store at −20 °C.
15. Isolation plasmid kit.
16. LR Clonase II enzyme (Invitrogen).
17. *EcoRI* restriction enzyme.
18. pTRV2 Gateway vector (YL279): containing attR1 and attR2 recombination site, donated by the Arabidopsis Biological Resource Center (ABRC) of the Ohio State University, a gift from Dr. Savithramma Dinesh-Kumar.
19. Electroporation cuvettes 0.2 cm.
20. Equipment: Nanodrop, thermocycler, electrophoresis equipment, water baths, orbital shakers, centrifuges, electroporator, and 37 °C room.

2.2 Agro-infiltration

1. *A. tumefaciens* strain GV2260 (or any other preferred strain; e.g., GV3101, EHA105, and LB4404) electrocompetent cells.
2. YEB agar and broth: weigh 5 g beef extract, 1 g yeast extract, 5 g peptone, and 5 g sucrose. Semisolid medium contains 16 g agar. Add 800 ml distilled water and dissolve. Adjust pH 7.2 using NaOH 1 N. Adjust the final volume to 1 L with distilled water. Autoclave at 121 °C for 20 min.
3. 50 mg/ml Rifampicin stock solution in DMSO. Kept at −20 °C.
4. Wash buffer: weigh 2.03 g of $MgCl_2$ and dissolve in 1 L of distilled water. Sterilize in autoclave at 120 °C, 20 min; store at 4 °C.
5. 150 mM acetosyringone: weigh 0.2943 g of acetosyringone, and dissolve in 10 ml DMSO; store at −20 °C.
6. Infiltration medium: 10 mM $MgCl_2$, 10 mM 2-(4-morpholino)-ethane sulfonic acid (MES), and 150 μM acetosyringone.

Fig. 1 (continued) after 48 dpi; (c) leaf showing almost 100% of photobleaching; (d), (e), and (f), *PDS* gene silencing phenotype in chili pepper fruits of 40, 50, and 60 days post-anthesis (dpa), respectively. (Photos taken by Dr. María del Rocío Gómez-García and Dr. Magda Lisette Arce-Rodríguez)

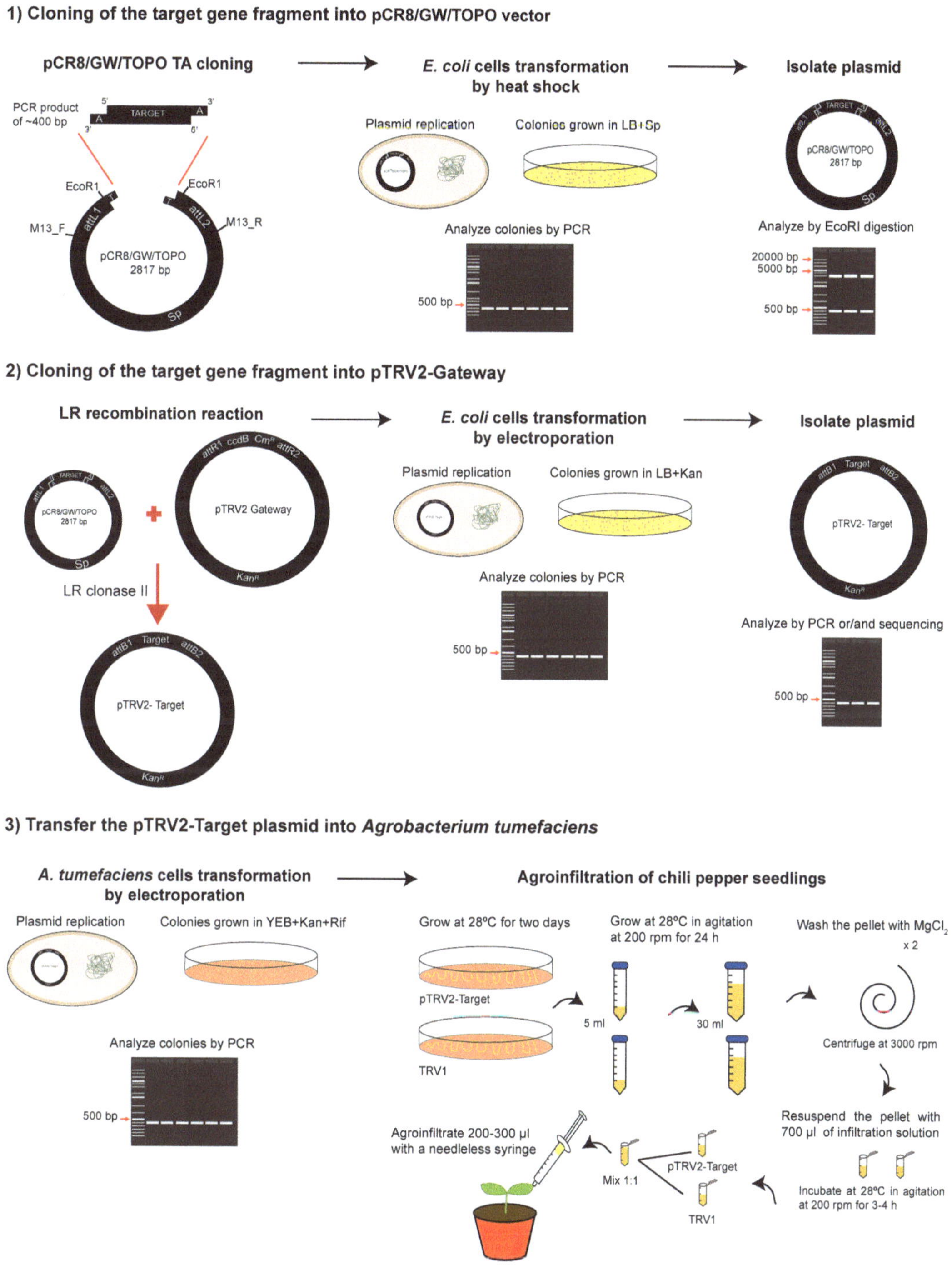

Fig. 2 Flowchart of the experimental steps for VIGS protocol using the Gateway-based TRV vector in chili pepper plants

Infiltration medium should be prepared at the same day as the agroinfiltration protocol. For 100 ml of infiltration solution, weight 0.1952 g of MES, and dissolve in 100 ml of sterilized 10 mM $MgCl_2$. Add 100 μl of 150 mM acetosyringone.

7. Silwet L-77 (100% copolymer of polyester and silicone, non-ionic surfactant).
8. pTRV2 forward vector sequence 5′-GGAGAATACGGTAGAA CG-3′ and reverse 5′-CAGTGCCGCTAGTAACCC-3′.
9. Equipment: 28 °C room, centrifuges, electroporator, and orbital shaker.
10. Bacterial growth material: 15- and 50-ml Falcon tubes, Petri dishes, spreader, toothpicks, and racks.

2.3 Plant Material and Growth Conditions

1. Chili pepper seedlings (1–4 weeks old).
2. Growth chamber with a 16 h photoperiod and 66% relative humidity at 26 °C (*see* **Note 1**).
3. Commercial solution fertilizer (N/P/K 30:20:10).

3 Methods

3.1 pTRV2 Target Gene Construction

3.1.1 Primers Design

1. Identify the coding region of the target gene.
2. Select a specific sequence of 150–500 nucleotides in the target gene.
3. Design the primers with specific properties: 40–60% in GC, 55–60 °C in annealing temperature, 18–30 bp in length, non--self-annealing, non-complementarity, non-hairpin formation, non-secondary structures, and non-dimer.

3.1.2 Cloning of the Target Gene Fragment into the pCR8/GW/TOPO Vector

1. Isolate total RNA using Trizol Reagent, and purify with a kit according to the manufacturer. Use a tissue sample where the target gene shows high expression level.
2. Check the quality of the RNA by 0.8% agarose gel electrophoresis, and quantify the sample at 260 nm using a NanoDrop. Synthesize cDNA from 1 μg of RNA using a reverse transcriptase following the manufacturer manual.
3. Determine the optimum annealing temperature of the primers by a PCR gradient using recombinant Taq DNA polymerase according to the protocol instructions. The amplification reaction is prepared by mixing 8.8 μl of sterilized distilled water, 1.25 μl of 10× PCR buffer, 0.5 μl of 50 mM $MgCl_2$, 1 μl of 40 ng/μl primer mix, 0.25 μl of 10 mM dNTP's, 0.5 μl of 100–200 ng/μl template cDNA, and 0.2 μl of Taq DNA polymerase (5 U/μl) at a 12.5 μl final volume. Perform the thermal cycling as follows: 94 °C for 5 min, 30 cycles (94 °C for

30 s, 45–65 °C for 30 s, and 72 °C for 1 min), and finally 72 °C for 7 min. Check 8 μl by 1% agarose gel electrophoresis to confirm the insert amplification. You should see a single band of the correct size.

4. Amplify the fragment of the target gene by PCR to clone it into the pCR8/GW/TOPO vector. The amplification reaction is prepared by mixing 17.6 μl of sterilized distilled water, 2.5 μl of 10× PCR buffer, 1 μl of 50 mM $MgCl_2$, 2 μl of 40 ng/μl primer mix, 0.5 μl of 10 mM dNTP's, 1 μl of 100–200 ng/μl template cDNA, and 0.4 μl of Taq DNA polymerase (5 U/μl) at a 25 μl final volume. Prepare four reactions to complete 100 μl or one reaction of 100 μl volume when applicable. Perform the thermal cycling as follows: 94 °C for 5 min, 30 cycles (94 °C for 30 s, specific annealing temperature for 30 s, and 72 °C for 1 min), and finally 72 °C for 7 min.
5. Analyze 5 μl of the purified PCR product by 1% agarose gel electrophoresis. You should visualize a single and discrete band of the expected size.
6. Proceed with a PCR product purification kit (*see* **Note 2**).
7. The PCR product is cloned into the pCR8/GW/TOPO vector. The mix consists of 1–4.5 μl of the purified fresh PCR product, 1 μl saline solution (1.2 M NaCl, 0.06 M $MgCl_2$), 0.5 μl pCR8/GW/TOPO vector, and water to a final volume of 6 μl. Mix with a vortex, and spin briefly. Incubate 20–60 min at room temperature (22–25 °C), and then put on ice (*see* **Note 3**).
8. Transform *E. coli* TOP10 chemically competent cells with 2 μl of the cloning reaction. Mix 2 μl cloning reaction with 50 μl cells into a vial, incubate on ice 20 for min, heat shock the cells at 42 °C for 1 min, immediately transfer the tube onto ice for 2 min, add 250 μl LB (or SOC) broth medium, and incubate with agitation (200 rpm) horizontally at 37 °C for 1 h. Spread 30 and 100 μl on an LB agar medium plate containing 100 μg/ml spectinomycin. Incubate overnight at 37 °C.
9. Analyze ten colonies by PCR with the specific gene primers or M13 universal primers. The mix reaction is the same as described before (Subheading 3.1.2, **step 4**), but the DNA template is the colony. Visualize 12 μl by 1% agarose gel electrophoresis.
10. Isolate the plasmid of ~3 positive colonies with a commercial isolation plasmid kit.
11. Check the liberation of the insert by *EcoRI* digestion. The restriction mix reaction consists of 1 μg plasmid, 3 μl 10× Buffer, 1 μl *EcoRI* (1 U/μl), and water to complete 30 μl. Incubate at 37 °C for 1–2 h, and inactivate the reaction at

65 °C for 20 min. Check 16 μl by 1% agarose gel electrophoresis to confirm the presence of the insert.

3.1.3 LR Recombination Reaction of the pCR8/GW/TOPO Target Gene with the pTRV2 Gateway Vector

1. Proceed with the LR Clonase II enzyme reaction following the instructions of the manufacturer manual with the modifications detailed below.
2. Mix 150–1000 ng of pCR8/GW/TOPO target gene, 150–500 ng of pTRV2 Gateway vector, and TE buffer (pH 8.0) to a final volume of 8 μl into 0.2 ml microcentrifuge tube at room temperature (*see* **Note 4**).
3. Add 1 μl of LR Clonase II enzyme and mix by vortexing and centrifuge briefly.
4. Incubate the reaction at 25 °C overnight (16 h).
5. Add 0.5 μl of the proteinase K solution, mix by vortexing, and spin briefly.
6. Incubate at 37 °C for 10 min to terminate the reaction.
7. Put on ice immediately and proceed with the bacteria transformation protocol.
8. Dialyze the recombination reaction. Put a nitrocellulose membrane filter of 0.025 μm pore over distilled water (shiny side up). Load the mixture on the filter, and incubate for 15–20 min. Recover the recombination reaction into a 0.2 ml microcentrifuge tube.
9. Mix 2 μl recombination reaction with 35 μl *E. coli* DH10B electrocompetent cells in a 0.2 ml microcentrifuge tube. Mix by vortexing and centrifuge briefly.
10. Put the mixture into a Gene Pulser cuvette carefully, avoiding air bubble formation. Electroporate at 2.5 kV.
11. In a laminar flow hood, add 450 μl of LB broth medium into the cuvette. Mix by pipetting, and transfer the culture to a 1.5 ml microcentrifuge tube. Incubate with agitation (200 rpm) horizontally at 37 °C for 1 h.
12. In a laminar flow hood, plate 100 μl on an LB agar medium plate containing 50 μg/ml kanamycin. Centrifuge the rest of cells at 14,000 × *g* for 2 min, remove the supernatant until 100 μl of total volume is left, and plate. Incubate overnight at 37 °C.
13. Analyze ten colonies by PCR with the specific-gene primers. The mix reaction is the same as described before (Subheading 3.1.2, **step 4**), but the DNA template in this case is the colony. Visualize 12 μl by 1% agarose gel electrophoresis.
14. Isolate the plasmid of ~3 positive colonies with a commercial isolation plasmid kit. Verify the presence of the insert by PCR amplification.

3.2 Chili Pepper Agroinfiltration

3.2.1 Transformation of Agrobacterium tumefaciens with the pTRV2 Target Gene Plasmid

1. Transform *A. tumefaciens* GV2260 electrocompetent cells with 2 μl of pTRV2 target gene plasmid in the same way as described before (Subheading 3.1.3, **steps 8–10**).
2. After electroporation, in a laminar flow hood, add 450 μl of YEB broth medium into the cuvette. Mix by pipetting and transfer the culture to a 1.5 ml microcentrifuge tube. Incubate with agitation (200 rpm) horizontally at 28 °C for 1 h. Plate 100 μl on a YEB semisolid medium plate containing 50 μg/ml of kanamycin and 50 μg/ml of rifampicin. Centrifuge the rest of cells at 14,000 × *g* for 2 min; remove the supernatant until 100 μl of total volume is left, and plate. Incubate overnight at 28 °C.
3. Analyze ten colonies by PCR with the specific gene primers. The mix reaction is the same as described previously (Subheading 3.1.2, **step 4**), but the DNA template is the colony. Optional: you could verify the PCR product by sequencing.

3.2.2 Germination and Growth of Chili Pepper Plants

1. Plant seeds in plastic trays containing a mixture of three parts of peat moss, one part of vermiculite, one part of perlite, one part of sludge, and two parts of forest soil.
2. Incubate in a growth chamber with a 16 h photoperiod and 66% relative humidity at 26 °C (*see* **Note 1**).
3. Transplant the chili pepper seedlings individually to 1.3 L plastic pots containing the same soil mixture.
4. Fertilize the plants every 2 weeks with 200 ml each with a commercial solution fertilizer at 3 g/L.

3.2.3 Agroinfiltration of the Chili Pepper Seedlings

1. Streak positive colonies of *A. tumefaciens* strain GV2260 containing either TRV1; empty pTRV2 or pTRV2 target gene on YEB agar medium with 50 μg/ml kanamycin and rifampicin. Incubate at 28 °C for 2 days.
2. Inoculate *Agrobacterium* culture in 5 ml of YEB liquid medium with 50 μg/ml kanamycin and rifampicin. Incubate at 28 °C for 1 day with agitation (200 rpm) in inclined position.
3. Inoculate the 5 ml of *A. tumefaciens* suspension into 30 ml of YEB liquid medium with 50 μg/ml kanamycin and rifampicin. Incubate at 28 °C for 1 day with agitation (200 rpm) in inclined position. We recommend an absorbance of $OD_{600} = 0.7–1.0$.
4. Pellet the cells by centrifuging at 1680 × *g* for 15 min. Discard the supernatant.
5. Resuspend the pellet with 30 ml of 10 mM $MgCl_2$ by vigorous manual agitation.
6. Centrifuge at 1680 × *g* for 15 min. Discard the supernatant.

7. Repeat **steps 5** and **6** once more.
8. Resuspend the cells in 0.7 ml of fresh infiltration solution by pipetting. Incubate with agitation (200 rpm) horizontally at 28 °C for 3–4 h.
9. Mix the *A. tumefaciens* suspensions containing pTRV2 target gene or empty pTRV2 with *A. tumefaciens* suspensions containing TRV1 in a 1:1 (V/V) proportion.
10. Add 0.5 μl of Silwet L-77 to facilitate the suspension penetration into the leaves.
11. Infiltrate by the abaxial zone cotyledons or youngest leaves of the seedlings (1–4 weeks old) with a 1 ml sterile needleless syringe (*see* **Note 5**).
12. Cover the plants from the light for 24 h immediately after the agroinfiltration. Keep the plants in a growth room or alternatively in a greenhouse.
13. Optional: agroinfiltrate the youngest leaves 1 month later in order to reinforce the infection.

3.3 Detection of TRV2 Transcripts in Leaves of Chili Pepper Plants

1. Collect leaves below and above the agroinfiltrated leaves and apical leaves from approximately 2-month-old plants. Select at least three biological replicates (three plants), preferentially showing TRV symptoms (Fig. 1A) (*see* **Note 6**).
2. Wash the leaves with distilled water, and freeze them immediately with liquid nitrogen. Store at −80 °C.
3. Isolate total RNA using Trizol reagent and purification kit according to the manual instructions.
4. Synthesize cDNA from 1 μg of RNA using reverse transcriptase following the manual instructions. Instead of oligo(dT), use specific reverse oligonucleotide of the pTRV2 vector sequence.
5. Amplify a fragment of the TRV2 by PCR to detect its presence. We use the oligonucleotides described in Subheading 2.2. The PCR mix consists of 8.8 μl sterilized distilled water, 1.25 μl 10× PCR buffer, 0.5 μl 50 mM $MgCl_2$, 1 μl of 40 ng/μl primer mix, 0.25 μl 10 mM dNTP's, 0.5 μl of 100–200 ng/μl template cDNA, and 0.2 μl of Taq DNA polymerase (5 U/μl) at a 12.5 μl final volume. The PCR conditions are as follows: 94 °C for 5 min, 30 cycles of 94 °C for 30 s, 57.5 °C for 30 s, and 72 °C for 1 min and finally 72 °C for 7 min. Visualize 8 μl PCR product on 1% agarose gel electrophoresis; the expected amplified band size is 654 bp (*see* **Note 7**).

4 Notes

1. Alternatively, a greenhouse can be used to keep plants; however, we recommend controlled conditions until the infection symptoms appear.
2. If you visualize more than two products, purify only the predicted amplicon from the agarose gel.
3. The ligation mixture of pCR8/GW/TOPO target gene can be kept at −20 °C overnight before bacteria transformation.
4. The LR Clonase II enzyme mix catalyzes in vitro recombination between the pCR8/GW/TOPO vector containing attL-flanked target gene and pTRV2 Gateway vector containing attR sites.
5. Approximately 80% of the leaf should be infiltrated. Approximately 100–300 μl of the bacterial suspension is needed to infiltrate each seedling.
6. Chili pepper plants agroinfiltrated with the pTRV2 vector do not always show a slight yellow streak in the leaves; therefore, plants that do not exhibit TRV virus symptoms could be infected and silenced. Check the silencing phenotype before discarding plants without virus symptoms.
7. The annealing temperature of pTRV2 primers is very specific (57.5 °C); probably you will need to standardize this in your own thermocycler equipment.

Acknowledgments

We thank Dr. María del Rocío Gómez-García and Dr. César Aza--González for their efforts in the standardization of the Gateway-based TRV VIGS protocol in chili pepper plants in our laboratory. This work was supported by a grant from Consejo Nacional de Ciencia y Tecnología (Conacyt), Mexico, project 1570.

References

1. Van Kammen A (1997) Virus-induced gene silencing in infected and transgenic plants. Trends Plant Sci 2:409–411
2. Ratcliff FG, Martín-Hernandez AM, Baulcombe DC (2001) Tobacco rattle virus as a vector for analysis of gene function by silencing. Plant J 25:237–245
3. Liu Y, Schiff M, Dinesh-Kumar SP (2002) Virus-induced gene silencing in tomato. Plant J 31:777–786
4. Chung E, Seong E, Kim YC et al (2004) A method of high frequency virus-induced gene silencing in chili pepper (*Capsicum annuum* L. Cv. Bukang). Mol Cells 17:377–380
5. Stewart C, Kang BC, Liu K et al (2005) *The Pun 1* gene for pungency in pepper encodes a putative acyltransferase. Plant J 42:675–688
6. Abraham-Juarez MR, Rocha-Granados MC, Lopez MG et al (2008) Virus-induced silencing of *Comt*, p*Amt* and *Kas* genes results in a

reduction of capsaicinoid accumulation in chili pepper fruits. Planta 227:681–695

7. Wang JE, Li DW, Gong ZH et al (2013) Optimization of virus-induced gene silencing in pepper (*Capsicum annuum* L.). Genet Mol Res 12:2492–2506
8. Arce-Rodríguez ML, Ochoa Alejo N (2015) Silencing *AT3* gene reduces the expression of p*Amt*, *BCAT*, *Kas*, and *Acl* genes involved in capsaicinoid biosynthesis in chili pepper fruits. Biol Plant 59:477–484
9. Arce-Rodríguez ML, Ochoa Alejo N (2017) An R2R3-MYB transcription factor regulates capsaicinoid biosynthesis. Plant Physiol 174:1359–1370

Chapter 4

Virus-Induced Gene Silencing in Diploid and Tetraploid Potato Species

Jinping Zhao, Haolang Jiang, Guanyu Wang, Zonghua Wang, Jingao Dong, and Junqi Song

Abstract

Potato is the world's fourth largest food crop and a vegetatively propagated model polyploid plant. To facilitate genomic studies in potato, here we describe detailed protocols to silence genes in both diploid potato *Solanum bulbocastanum* and tetraploid potato cultivars such as Maris Bard, Arran Pilot, Ancilla, and Serrana using tobacco rattle virus (TRV)- or potato virus X (PVX)-induced gene silencing (VIGS) system, respectively. The established VIGS system represents an efficient and powerful approach for functional analysis of genes involved in growth, development, metabolism, and responses to biotic and abiotic stresses in potato.

Key words Virus-induced gene silencing (VIGS), Diploid potato, Tetraploid potato, Tobacco rattle virus (TRV), Potato virus X (PVX), *Solanum bulbocastanum*, *Solanum tuberosum*

1 Introduction

Potato (*Solanum tuberosum*) is the fourth most important food crop and the most widely grown non-cereal crop worldwide primarily due to its high nutritional value, high energy production, and relatively low-input requirements [1]. As a vegetatively propagated polyploid crop, several features such as high levels of outcrossing rate, heterozygosity, and genetic diversity make potato an attractive dicotyledonous model plant for crop breeding and functional genomics. To date, genome sequences of three potato species have been completed, including the reference genome of the doubled monoploid *S. tuberosum* Group Phureja clone DM1-3, the wild diploid species *S. commersonii*, and the diploid inbred clone M6 of *S. chacoense* [2–4]. With the availability of genome sequences in potato, efficient and robust genetic tools need to be developed to determine the function of specific genes.

Vincent Courdavault and Sébastien Besseau (eds.), *Virus-Induced Gene Silencing in Plants: Methods and Protocols*, Methods in Molecular Biology, vol. 2172, https://doi.org/10.1007/978-1-0716-0751-0_4,

VIGS is an RNA-mediated posttranscriptional gene silencing approach and is exploited by plants as an adaptive defense mechanism against invading viruses [5, 6]. VIGS has been increasingly developed as a simple, rapid, and efficient genetic tool to knock down specific genes. Currently, a large collection of viruses have been modified and constructed as VIGS vectors in a wide variety of plant species including potato [7, 8]. Among these viruses, TRV-based vectors represent the most widely used VIGS system due to broad host range, aggressiveness, and mild viral symptoms of TRV [9, 10]. As a positive-sense and single-stranded RNA virus, TRV has a bipartite genome. RNA1 encodes two replicase proteins, a movement protein and a cysteine-rich protein, both of which are required for viral replication and systemic movement. RNA2 encodes a coat protein and two nonstructural proteins, which can be engineered to carry fragments of target genes [11]. Both RNA1 and RNA2 are necessary and sufficient for viral propagation and systemic spread [11, 12]. PVX belongs to the *Potexvirus* genus and has a positive-sense single-stranded RNA genome encoding an RNA-dependent RNA polymerase, TGBp1, TGBp2, TGBp3, and a coat protein [13]. The latter four proteins are essential for cell-to-cell and long-distance movement but are dispensable for replication [14–16]. PVX has a relatively broad host range including a number of solanaceous species [17] and has been developed as a VIGS vector [8, 10, 18]. However, many wild and cultivated potato species are not well amenable to VIGS. *Solanum bulbocastanum* is a wild diploid potato species native to central Mexico and provides promising sources of resistance to a variety of potato diseases including late blight, soft rot, potato cyst nematode, and aphids [19–21]. Three major resistance genes against late blight, *RB* (also known as *Rpi-blb1*), *Rpi-blb2*, and *Rpi-blb3*, were cloned from *S. bulbocastanum* [19, 22, 23]. Functional homologs of *RB* have been identified in other phylogenetically distant wild potato species such as *S. verrucosum* and *S. stoloniferum* [24, 25], suggesting that *RB* is of ancient origin and functionally conserved throughout the evolution of potato [26]. Therefore, *S. bulbocastanum* can potentially serve as a model potato species for functional analysis.

To facilitate functional genomic studies in potato, we have optimized the VIGS protocol in potato by surveying a collection of over 30 potato accessions including wild and cultivated potato species, examining different VIGS vectors, and evaluating various inoculation methods (agroinfiltration, sap inoculation, and graft inoculation). Our results showed that the endogenous phytoene desaturase gene (*PDS*) can be effectively silenced in diploid *S. bulbocastanum* using TRV-based vectors and the conventional agroinfiltration method, whereas the PVX-based system in combination with agroinfiltration and toothpick inoculation works more efficiently in tetraploid *S. tuberosum* cultivars such as Ancilla, Arran Pilot, Marius Bard, and Serrana. The protocols established here provide a powerful tool for functional analysis of genes in various biological processes in diploid and tetraploid potato species.

2 Materials

2.1 Plant Culture

1. In vitro potato plants: Ancilla, Arran Pilot, Marius Bard, and Serrana (https://www.ars-grin.gov/nr6/; National Research Support Program (NRSP)-6 Potato Genebank, Sturgeon Bay, WI, USA).
2. Diploid potato seeds: *Solanum bulbocastanum* (https://www.ars-grin.gov/nr6/; National Research Support Program (NRSP)-6 Potato Genebank, Sturgeon Bay, WI, USA).
3. Murashige and Skoog Basal Salt Mixture.
4. 1000× Gamborg's vitamin stock: 100 g/L myo-inositol, 1 g/L nicotinic acid, 1 g/L pyridoxine HCl, and 10 g/L thiamine.
5. Culture tubes (25× 150 mm).
6. Filter papers (20–24 cm in diameter).
7. Bleach (5.75% available chlorine).
8. 36.5–38% HCl.
9. Gellan Gum.
10. Sterile tweezers, lancets, and blades.
11. pH meter.
12. Burner or infrared sterilizer and laminar flow hood.
13. Ventilation hood.
14. Growth room conditions: 20–22°C, 16 h light/8 h dark photoperiod and light intensity 120 $\mu E/m^2/s$.
15. Soil (e.g., Metro-Mix 360, Sun Gro Horticulture, USA).
16. Pots, trays, and transparent tray domes.

2.2 Gene Cloning

1. RNA extraction kit.
2. Reverse transcription kit.
3. PCR reagents: DNA polymerase with dNTP.
4. Restriction enzymes: *Pst*I and *Sma*I.
5. T4 polymerase.
6. 10 mM dATP and 10 mM dTTP.
7. 1 M DTT.
8. Ethanol.
9. 25:24:1 (v/v/v) Phenol/chloroform/isopropanol.
10. 24:1 (v/v) Chloroform/isopropanol.
11. 3 M sodium acetate pH 4.0.

12. *E. coli* strains: DH5α for maintaining the destination constructs of ligation-independent cloning (LIC cloning); DB3.1 for maintaining the PVX-LIC and pTRV2e vectors (*see* **Note 1**).
13. PVX VIGS vector: PVX-LIC [27].
14. TRV VIGS vectors: pTRV1 (TRV genome 1) [28] and pTRV2e (TRV genome 2) [29, 30].
15. Cloning primers for gene of interest containing the LIC linker sequences (*see* **Note 2**).
16. LB liquid medium and LB plates containing antibiotics.
17. 50 mg/mL chloramphenicol stock solution in 90% ethanol, 50 mg/mL kanamycin stock solution in water, and 50 mg/mL rifampicin stock solution in DMSO.
18. PCR machine, shaker, vortex, and incubator.

2.3 Agrobacterium Infiltration

1. *Agrobacterium tumefaciens* strains GV3101.
2. Infiltration solution: 10 mM $MgCl_2$, 10 mM MES pH 5.6, and 200 μM acetosyringone.
3. 200 mM acetosyringone stock solution.
4. Needleless syringes (1 mL).
5. Toothpicks.
6. Spectrophotometer and pH meter.
7. Shaker, centrifuge, and incubator.

3 Methods

3.1 Generate in Vitro S. bulbocastanum Plants from Seeds

1. Prepare MS medium supplemented with 1 × Gamborg vitamins and gellan gum (3 g/L). Melt the medium, aliquot 10–15 mL into culture tubes, and autoclave for 20 min at 121°C and 15 psi.
2. Pour ~20 seeds of *S. bulbocastanum* into a 1.5 mL microcentrifuge tube sitting in a rack. Place the tube with lid open into a desiccator in a fume hood.
3. Place a 250 mL beaker inside the desiccator and add 100 mL of bleach (*see* **Note 3**).
4. Add 3 mL of HCl into the beaker containing bleach and immediately seal the desiccator with the lid.
5. After 4 h, open the desiccator in fume hood and seal the microcentrifuge tube.
6. Drop the seeds into the culture tubes containing MS medium with one seed in each tube.

7. Cover culture tube with caps and grow them in a tissue culture room at 22°C with a 16 h light/8 h dark photoperiod.

3.2 Propagate in Vitro Plants

1. Cut approximately 1-cm-long nodal stem segments from 4-week-old in vitro plants on two filter papers under sterile conditions.
2. Propagate plants by placing one nodal segment into each culture tube containing MS medium with the stem facing upward.
3. Cover tubes with caps, and grow in a tissue culture room at 22°C with a 16 h light/8 h dark photoperiod.
4. Propagate in vitro plants every 4 weeks.

3.3 Grow in Vitro Plants in Soil

1. Grow plants in a tissue culture room for 4 weeks.
2. Fill the soil into pots and soak the pots in a tray with water.
3. Remove MS medium from roots of in vitro plants with water, and transplant them into soil.
4. Grow plants in a growth room at 22°C with a 16 h light/8 h dark photoperiod and a light intensity of 120 μE/m^2/s.
5. Cover the tray with a transparent plastic dome to maintain moisture until new leaves emerge.

3.4 Construction of the VIGS Vectors

1. The sequences of target genes in potato are obtained through BLAST searches against the Spud DB Solanaceae Genomics Resource at http://solanaceae.plantbiology.msu.edu/.blast/blast.html or the Sol Genomics Network (SGN) at https://solgenomics.net/tools/blast/. The *N. benthamiana* phytoene desaturase (*NbPDS*, Niben101Scf01283g02002.1) and 4-hydroxy-3-methylbut-2-enyl diphosphate reductase (*NbIspH*, Niben101Scf10575g00008.1) genes are used as queries to identify homologous *StPDS* and *StIspH* genes in potato, respectively.
2. VIGS constructs are designed using the SGN VIGS tool at http://vigs.solgenomics.net/ via BLAST against the *S. tuberosum* ITAG v1 genome (*see* **Note 4**).
3. Primers are designed using Prime3Plus at https://primer3plus.com/cgi-bin/dev/primer3plus.cgi with default parameters.
4. LIC cloning primers are generated by adding the LIC1 and LIC2 linker sequences (*see* **Note 5**) to the reverse and forward primers designed using Primer3Plus, respectively.
5. To clone target gene sequences, total RNAs are isolated from 4-week-old in vitro plants with an RNA extraction kit, and cDNAs are synthesized using 1 μg of total RNAs as template with an RNA reverse transcription kit following manufacturer's instructions. The gene fragments are amplified in a total volume of 50 μL with a high-fidelity DNA polymerase.

6. To purify the PCR products, add equal volume of ddH_2O and 30% PEG8000/30 mM $MgCl_2$, vortex vigorously, centrifuge at 18,000 × *g* for 20 min, and remove the supernatant immediately.
7. Add 1 mL of 70% ethanol, vortex vigorously, and centrifuge at 18,000× *g* for 20 min. Immediately remove the supernatant after centrifugation, dry the pellet, and dissolve the pellet in 40 μL of ddH_2O.
8. To set up the T4 DNA polymerase reaction, mix on ice the PCR products (up to 200 ng) with 0.1 μL of T4 DNA polymerase (3 U/μL), 0.25 μL of dATP (100 mM), and 0.05 μL of DTT (1 mM), and add ddH_2O to 5 μL. Incubate the mixture at 37°C for 15–30 min, followed by 75°C for 20 min to inactivate the enzyme.
9. Digest the PVX-LIC and pTRV2e vectors with *Sma*I or *Pst*I, respectively.
10. To purify the linearized PVX-LIC or pTRV2e vectors, add equal volume of phenol/chloroform/isopropanol (25:24:1), vortex vigorously, centrifuge at 14,000× *g* for 10 min, and transfer the aqueous phase to a new centrifuge tube.
11. Add an equal volume of chloroform/isopropanol (24:1) to the supernatant, vortex vigorously, centrifuge at 14,000× *g* for 10 min, and transfer the supernatant to a new centrifuge tube.
12. Add 2.5× volume of ethanol and 1/10 volume of 3 M NaAc (pH 4.0) to the supernatant, invert to mix completely, incubate at −20°C for 15 min, centrifuge at 14,000× *g* for 10 min, and discard the aqueous phase.
13. Rinse the pellet with 1 mL of 75% ethanol, centrifuge at 14,000× *g* for 10 min, discard the aqueous phase, dry the pellet, and dissolve the digested plasmid with 100 μL of ddH_2O.
14. Treat the digested PVX-LIC or pTRV2e vectors with T4 DNA polymerase in the presence of 0.5 mM dTTP and 1 mM DTT at 37°C for 15–30 min to produce the 14-nt sticky 5′ ends. Inactivate T4 DNA polymerase at 75°C for 20 min.
15. To set up the LIC reaction, mix 5 μL of the T4 DNA polymerase-treated PVX-LIC or pTRV2e vector with 5 μL of the T4 DNA polymerase-treated PCR products. Incubate the mixture at 70°C for 5 min, and slowly cool down to 22°C with a ramp at 0.1°C/s. Incubate at 22°C for 30 min (*see* **Notes 6** and 7).
16. Transform 5 μL of the LIC reaction into the *E. coli* strain DH5α and spread cells on LB plates containing 50 μg/mL kanamycin.

17. Verify the VIGS constructs by colony PCR using gene-specific primers followed by sequencing.

3.5 TRV-Mediated VIGS in Diploid Potato

1. Transform the pTRV1 and pTRV2e constructs carrying target genes into *Agrobacterium* strain GV3101. Grow transformants on a LB plate containing 50 μg/mL kanamycin and 50 μg/mL rifampicin.
2. Analyze *Agrobacterium* transformants by colony PCR using gene-specific primers to confirm the presence of target genes in pTRV2e.
3. The positive colonies carrying the pTRV1 or pTRV2e constructs were each inoculated into 5 mL of liquid LB medium containing the same antibodies as described above. Grow the cultures at 28°C and 220 rpm for 16 h.
4. Centrifugate the cultures at 3400× *g* for 10 min at room temperature.
5. Resuspend the pellet with 5 mL of infiltration buffer and dilute to an OD_{600} of 1.0.
6. Mix the *Agrobacterium* cultures containing the pTRV1 vector or the pTRV2e construct carrying target genes at a 1:1 ratio.
7. Incubate the mixed cultures at room temperature for 4–6 h.
8. Transplant 4-week-old in vitro diploid potato plants from culture tubes into soil and cover with a dome 3 days before preparing *Agrobacterium* cultures. Plants that have been recovered in soil for 3 days are subjected to VIGS assay. Infiltrate the *Agrobacterium* cultures into the abaxial side of fully expanded leaves with 1-mL needleless syringe. Four leaves from each plant are selected for infiltration.
9. Grow the infiltrated plants in a growth chamber at 22°C with a 16 h light/8 h dark photoperiod and a light intensity of 120 μE/m^2/s.
10. Silencing phenotype usually appears in the newly developed leaves 2–4 weeks post inoculation (Fig. 1).

3.6 PVX-Mediated VIGS in Tetraploid Potato

1. Transform the PVX-LIC construct carrying the target gene into *Agrobacterium* strain GV3101. Grow transformants on a LB plate containing 50 μg/mL kanamycin and 50 μg/mL rifampicin.
2. Analyze the *Agrobacterium* transformants by colony PCR using gene-specific primers to confirm the presence of the target gene in PVX-LIC.
3. Streak the positive colonies carrying the PVX-LIC construct onto a LB plate with 200 μM acetosyringone and the same antibiotics as described above. Grow the cultures at 28°C for

Fig. 1 VIGS of the *IspH* gene in diploid potato species *S. bulbocastanum*. Photos were taken 4 weeks post inoculation. Scale bar represents 1 cm

16 h. The *Agrobacterium* cultures collected on a LB plate will be directly used to inoculate potato stems as described below.

4. The positive colonies carrying the PVX-LIC construct were inoculated into 5 mL of liquid LB medium containing the same antibodies as described above. Grow the cultures at 28°C and 220 rpm for 16 h.
5. Centrifugate the cultures at 3400× *g* for 10 min at room temperature.
6. Resuspend the pellet with approximately 5 mL of infiltration buffer, and dilute to an OD_{600} of 1.0.
7. Incubate the cultures at room temperature for 4–6 h.
8. Transplant 4-week-old in vitro diploid potato plants from culture tubes into soil and cover with a dome 3 days before preparing *Agrobacterium* cultures. Plants that have been recovered in soil for 3 days are subjected to VIGS assay. Infiltrate the *Agrobacterium* cultures carrying the PVX-LIC construct into the abaxial side of fully expanded leaves with 1-mL needleless syringe. Four leaves from each plant are selected for infiltration.
9. Immediately after infiltration, gently scratch the surface of stems from the lower part of the infiltrated potato plants by toothpick inoculation with the *Agrobacterium* cultures collected from a solid LB plate described in **step 3** (*see* **Note** 7).
10. Grow the infiltrated plants in a growth chamber at 22°C with a 16 h light/8 h dark photoperiod and a light intensity of 120 μE/m^2/s.
11. Silencing phenotype usually appears in the newly developed leaves 2–4 weeks post inoculation (Fig. 2) (*see* **Notes 8–10**).

Fig. 2 VIGS of the *PDS* gene in tetraploid potato species Maris Bard, Arran Pilot, Ancilla, and Serrana. Photos were taken 4 weeks post inoculation. Scale bar represents 1 cm

4 Notes

1. Both the pTRV2e and PVX-LIC are ligation-independent cloning (LIC) vectors that were constructed by inserting the LIC cassette containing *ccdB* into the TRV2 or PVX T-DNA vector by replacing the original multiple cloning site, respectively. The LIC VIGS vectors allow high-throughput cloning of target genes without the use of costly enzymes required for Gateway cloning. Both vectors are maintained and propagated in the *E. coli* strain DB3.1. pTRV1, pTRV2e, and PVX-LIC carry kanamycin resistance, while PVX-LIC carries additional chloramphenicol resistance.
2. For *StPDS*, the forward primer is 5′-CgACgACAAgACCgTAAGTTGGAGAAGTTAGTCGGAGTAC-3′, and the reverse primer 5′-gAggAgAagAgCCgTAAGCTAAAATAATAACCACGACGCT-3′. For *StIspH*, the forward primers is 5′-CgACgACAAgACCgTGGGCAGGTTGTATCAACAATTTGTA-3′, and the reverse primer 5′-gAggAgAagAgCCgTTTTCCGGAAGAATTTGACCAGAAG-3′. The sequences for LIC linkers are underlined (*see* **Note 5**).
3. Work with bleach and hydrochloric acid separately. Wear appropriate personal protective equipment including gloves and a lab coat to avoid spills onto skin and clothes. Always perform chlorine gas sterilization in a fume hood.
4. The DNA fragments selected for VIGS should be at least 100 bp in length but no longer than 1000 bp. If the DNA fragments are smaller than 100 bp or longer than 1000 bp, silencing efficiency may be reduced.
5. LIC1 linker sequence: 5′-CgACgACAAgACCgT-3′; LIC2 linker sequence: 5′-gAggAgAagAgCCgT-3′.
6. Incubation of the LIC reaction mixture at 4°C overnight increases the efficiency of LIC cloning.

7. Infiltration of the *Agrobacterium* culture together with simultaneous scratching of the stem surface on the lower part of the plant with toothpicks carrying the same culture greatly improves silencing efficiency in tetraploid potato species.
8. Silencing of *PDS* results in inhibition of the carotenoid synthesis, leading to a characteristic photobleached phenotype [8, 11]. Similarly, silencing of *IspH* impairs the synthesis of chlorophylls and carotenoids, causing the plants to display an albino phenotype [31].
9. VIGS phenotype typically takes 10–20 days to appear after infiltration [8, 11]. However, silencing efficiency and phenotype development are dependent on the sizes and sequences of target genes, plant growth conditions, growth stage of plants, and selection of VIGS vectors. It is recommended to include either the *PDS* or *IspH* gene as a positive control to monitor the progression of VIGS assays while silencing genes of interest.
10. pTRVR1 and pTRVR2e work better for silencing genes in the diploid potato species *S. bulbocastanum* using agroinfiltration alone, whereas PVX-LIC works more efficiently for tetraploid potato species using agroinfiltration in combination with stem inoculation.

Acknowledgments

We thank the NRSP-6 US Potato Genebank for providing the wild and cultivated potato species and Dr. Yule Liu from Tsinghua University for providing the TRV1 and TRV2e vectors. This work was supported by a startup fund from the Texas A&M AgriLife Research and the Hatch Project TEX0-1-9675 from USDA National Institute of Food and Agriculture to JS, the scholarship 201707877008 from China Scholarship Council to HJ, and the scholarship 201708130105 from China Scholarship Council to GW.

References

1. Barrell PJ, Meiyalaghan S, Jacobs JM et al (2013) Applications of biotechnology and genomics in potato improvement. Plant Biotechnol J 11(8):907–920
2. Leisner CP, Hamilton JP, Crisovan E et al (2018) Genome sequence of M6, a diploid inbred clone of the high-glycoalkaloid-producing tuber-bearing potato species *Solanum chacoense*, reveals residual heterozygosity. Plant J 94(3):562–570
3. Aversano R, Contaldi F, Ercolano MR et al (2015) The *Solanum commersonii* genome sequence provides insights into adaptation to

stress conditions and genome evolution of wild potato relatives. Plant Cell 27(4):954–968
4. The Potato Genome Sequencing Consortium, Xu X, Pan S et al (2011) Genome sequence and analysis of the tuber crop potato. Nature 475:189
5. Voinnet O (2001) RNA silencing as a plant immune system against viruses. Trends Genet 17(8):449–459
6. Baulcombe DC (1999) Fast forward genetics based on virus-induced gene silencing. Curr Opin Plant Biol 2(2):109–113
7. Brigneti G, Martín-Hernández AM, Jin H et al (2004) Virus-induced gene silencing in Solanum species. Plant J 39(2):264–272
8. Faivre-Rampant O, Gilroy EM, Hrubikova K et al (2004) Potato virus X-induced gene silencing in leaves and tubers of potato. Plant Physiol 134(4):1308–1316
9. Senthil-Kumar M, Mysore KS (2011) Virus-induced gene silencing can persist for more than 2 years and also be transmitted to progeny seedlings in *Nicotiana benthamiana* and tomato. Plant Biotechnol J 9(7):797–806
10. Dobnik D, Lazar A, Stare T et al (2016) *Solanum venturii*, a suitable model system for virus-induced gene silencing studies in potato reveals StMKK6 as an important player in plant immunity. Plant Methods 12(1):29
11. Ratcliff F, Martin-Hernandez AM, Baulcombe DC (2001) Technical Advance: Tobacco rattle virus as a vector for analysis of gene function by silencing. Plant J 25(2):237–245
12. MacFarlane SA, Vassilakos N, Brown DJ (1999) Similarities in the genome organization of tobacco rattle virus and pea early-browning virus isolates that are transmitted by the same vector nematode. J Gen Virol 80(1):273–276
13. Huisman MJ, Linthorst HJM, Bol JF et al (1988) The complete nucleotide sequence of Potato virus X and its homologies at the amino acid level with various plus-stranded RNA viruses. J Gen Virol 69(8):1789–1798
14. Beck DL, Guilford PJ, Voot DM et al (1991) Triple gene block proteins of white clover mosaic potexvirus are required for transport. Virology 183(2):695–702
15. Angell SM, Davies C, Baulcombe DC (1996) Cell-to-cell movement of Potato virus X is associated with a change in the size-exclusion limit of plasmodesmata in trichome cells of *Nicotiana clevelandii*. Virology 216 (1):197–201
16. Chapman S, Kavanagh T, Baulcombe D (1992) Potato virus X as a vector for gene expression in plants. Plant J 2(4):549–557
17. Brunt AA (1996) Plant viruses online: descriptions and lists from the VIDE database. University of Idaho, Moscow, ID
18. Ruiz MT, Voinnet O, Baulcombe DC (1998) Initiation and maintenance of virus-induced gene silencing. Plant Cell 10(6):937–946
19. Song J, Bradeen JM, Naess SK et al (2003) Gene *RB* cloned from *Solanum bulbocastanum* confers broad spectrum resistance to potato late blight. Proc Natl Acad Sci U S A 100 (16):9128–9133
20. Brown CR, Mojtahedi H, Zhang L-H et al (2009) Independent resistant reactions expressed in root and tuber of potato breeding lines with introgressed resistance to *Meloidogyne chitwoodi*. Phytopathology 99 (9):1085–1089
21. Hawkes JG (1990) The potato: evolution, biodiversity and genetic resources. Belhaven Press, London
22. Park T-H, Gros J, Sikkema A et al (2005) The late blight resistance locus Rpi-blb3 from *Solanum bulbocastanum* belongs to a major late blight R gene cluster on chromosome 4 of potato. Mol Plant-Microbe Interact 18 (7):722–729
23. van der Vossen EAG, Gros J, Sikkema A et al (2005) The Rpi-blb2 gene from *Solanum bulbocastanum* is an Mi-1 gene homolog conferring broad-spectrum late blight resistance in potato. Plant J 44(2):208–222
24. Liu Z, Halterman D (2006) Identification and characterization of RB-orthologous genes from the late blight resistant wild potato species *Solanum verrucosum*. Physiol Mol Plant Pathol 69(4–6):230–239
25. Vleeshouwers VG, Rietman H, Krenek P et al (2008) Effector genomics accelerates discovery and functional profiling of potato disease resistance and *Phytophthora infestans* avirulence genes. PLoS One 3(8):e2875
26. van der Vossen E, Sikkema A, Hekkert B et al (2003) An ancient R gene from the wild potato species *Solanum bulbocastanum* confers broad-spectrum resistance to *Phytophthora infestans* in cultivated potato and tomato. Plant J 36 (6):867–882
27. Zhao J, Liu Q, Hu P et al (2016) An efficient Potato virus X-based microRNA silencing in *Nicotiana benthamiana*. Sci Rep 6:20573
28. Liu Y, Schiff M, Marathe R et al (2002) Tobacco Rar1, EDS1 and NPR1/NIM1 like genes are required for N-mediated resistance to tobacco mosaic virus. Plant J 30 (4):415–429

29. Sha A, Zhao J, Yin K et al (2014) Virus-based microRNA silencing in plants. Plant Physiol 164(1):36–47
30. Zhao J, Liu Y (2016) Virus-based MicroRNA silencing. Bio-protocol 6(2):e1714
31. Page JE, Hause G, Raschke M et al (2004) Functional analysis of the final steps of the 1-deoxy-D-xylulose 5-phosphate (DXP) pathway to isoprenoids in plants using virus-induced gene silencing. Plant Physiol 134 (4):1401–1413

Chapter 5

Virus-Induced Gene Silencing (VIGS) in Cassava Using Geminivirus Agroclones

Syed Shan-e-Ali Zaidi, Kumar Vasudevan, Ezequiel Matias Lentz, and Hervé Vanderschuren

Abstract

Virus-induced gene silencing (VIGS) is an efficient, low-cost, and rapid functional validation tool for candidate genes in planta. The VIGS approach is particularly suitable to perform reverse genetics studies in crop species. Here we present a detailed method to perform VIGS in cassava, from target gene fragment to agroinoculation and VIGS quantitation.

Key words VIGS, Cassava, *African cassava mosaic virus*, Geminiviruses

1 Introduction

Validation of gene functions in plants can sometimes be a challenging task for non-model crop species due to their recalcitrance to genetic transformation and the absence of well-characterized T-DNA insertional mutant collections. The increasing availability of omics tools to study crop species [1] enables the rapid identification of candidate genes associated with improved crop traits. The development of fast reverse genetics tools such as virus-induced gene silencing (VIGS) can be instrumental to rapidly and efficiently decipher gene functions in crop species.

Cassava (*Manihot esculenta* Crantz) is mainly grown in tropical regions in South America, sub-Saharan Africa, India, and Southeast Asia. Cassava provides staple food to an estimated 800 million people worldwide [2]. Cassava ranks among the most important source of food in developing countries with maize, rice, and wheat. However, cassava production is severely constrained by two viral diseases in sub-Saharan Africa and the Indian subcontinent: cassava mosaic disease (CMD) and cassava brown streak disease (CBSD) [3]. Because cassava breeding is a lengthy and tedious process,

Vincent Courdavault and Sébastien Besseau (eds.), *Virus-Induced Gene Silencing in Plants: Methods and Protocols*, Methods in Molecular Biology, vol. 2172, https://doi.org/10.1007/978-1-0716-0751-0_5,

cassava biotechnology has emerged as an important tool to perform fundamental studies as well as to develop cassava varieties with improved performance [1, 4–8]. Despite progress in the establishment of protocols for the genetic transformation of cassava [9–15], production of transgenic cassava has remained a lengthy process with low transformation efficiency in several important cassava genotypes. Therefore, VIGS is emerging as an efficient, low-cost, and rapid functional validation tool for candidate genes in cassava [15].

The first VIGS in cassava was established with a modified infectious clone of *African cassava mosaic virus* Cameroon isolate (ACMV-[CM]). The infectious clone was introduced in cassava plants by particle bombardment, and ACMV-[CM] infection led the sequence-specific silencing of leaf-expressed genes whose partial sequences were included in the modified ACMV-[CM] isolate [16]. Recently, we developed a modified agroclone of the *African cassava mosaic virus* Nigeria isolate (ACMV-[NOg]) [17] for efficient and high-throughput VIGS in cassava leaves and roots [15]. The unmodified isolate is contained in two separate pCAMBIA plasmids (DNA-A and DNA-B) and can be used as a positive control in infection and VIGS assays [15]. The modified DNA-A used as VIGS vector contains a partially deleted *AV1* gene and a small multiple cloning site, allowing easy insertion of DNA fragments according to the gene-silencing strategy ([15] and Fig. 1). In this chapter we describe a step-by-step guide for performing ACMV-mediated VIGS analysis in cassava (Fig. 2).

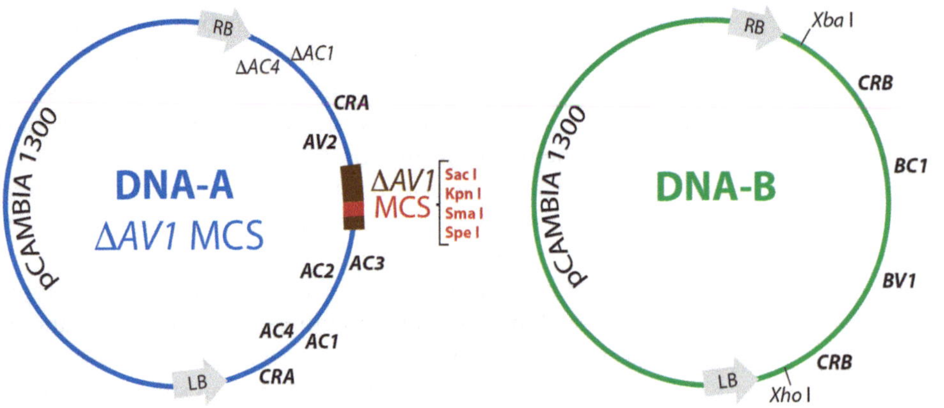

Fig. 1 Schematic representation of VIGS ACMV-[NOg] agroclones. Viral coding sequences (AC1, AC2, AC3, AC4, AV1, AV2, BC1, BV1), viral common regions A and B (CRA, CRB), left and right border (LB, RB), and multiple cloning site (MCS)

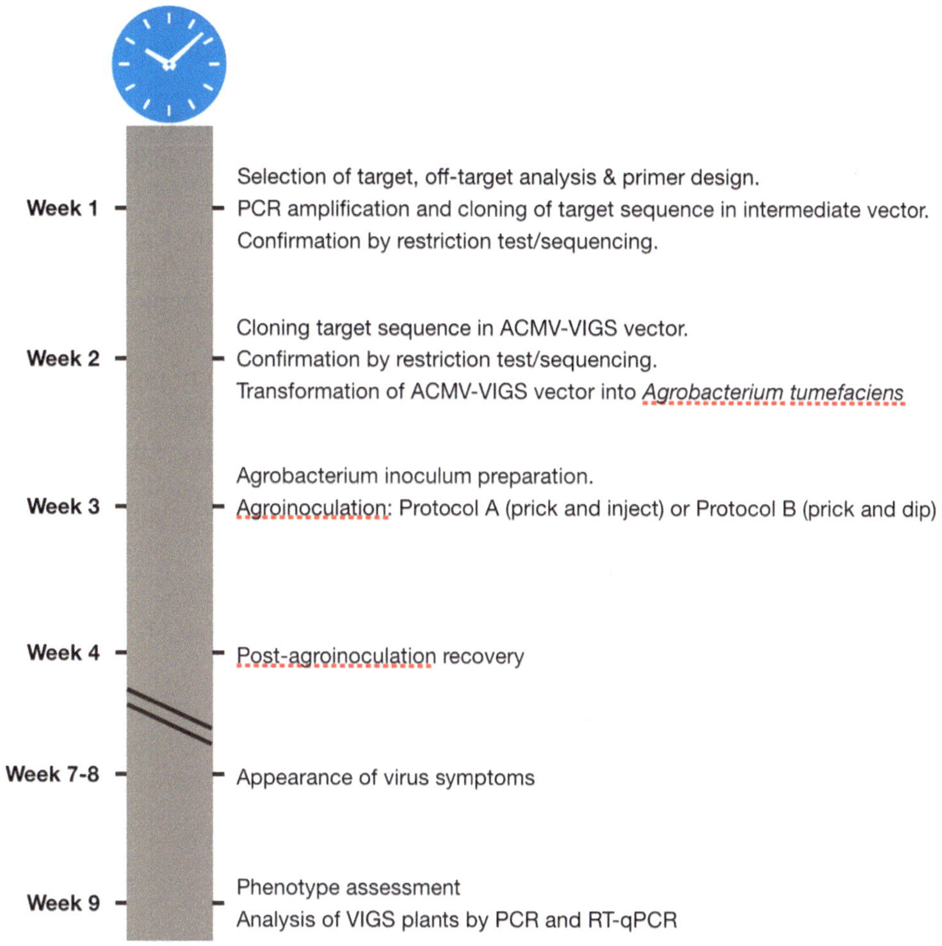

Fig. 2 A timeline of VIGS assay in cassava

2 Materials

Use ultrapure water (double-distilled and deionized) for preparing all solutions. Prepare and store all reagents at room temperature, unless indicated otherwise. Strictly follow all biosafety and waste disposal regulations. Materials for RNA extraction should be autoclaved or baked at 180 °C for 4 h.

2.1 Plant Material and Bacterial Strains

1. Shoot cultures of cassava cultivar (cv.) 60444 originally obtained from the International Institute of Tropical Agriculture (IITA, Ibadan, Nigeria) and maintained in the in vitro collection.
2. Agrobacterium strain AGL1.
3. *E. coli* competent cells (DH5α or other regular cloning strain).

2.2 Media and Antibiotics

1. CBM (basic shoot culture medium): 1 × MS salts with vitamins, 2 μM $CuSO_4$, 2% (m/v) sucrose, 0.3% (m/v) Gelrite, pH 5.8, autoclaved.
2. YEB (yeast extract broth) solid medium: 1 g/L yeast extract, 5 g/L beef extract, 5 g/L peptone, 5 g/L sucrose, 1.5% (m/v) agar, pH 7.2, autoclaved.
3. YEB (yeast extract broth) liquid medium: 1 g/L yeast extract, 5 g/L beef extract, 5 g/L peptone, 5 g/L sucrose, pH 7.2, autoclaved. Supplemented with $MgSO_4$ sterile solution (2 mM final concentration).
4. LB (Luria-Bertani) medium: 1% (m/v) tryptone, 0.5% (m/v) yeast extract, 1% (m/v) NaCl, pH 7.0 ± 0.2.
5. Antibiotics: 100 g/L ampicillin solution, 50 g/L rifampicin solution, 100 g/L carbenicillin solution, and 50 g/L kanamycin solution.

2.3 Plasticware and Other Consumables

1. Pots with soil, trays, and transparent lids.
2. 1 mL disposable cuvettes for electroporation.
3. 1.5 and 2 mL plastic tubes.
4. Insulin syringe 1 mL (0.33 mm/29 G/12.7 mm).
5. Needles 21 G.
6. Pipettes 25 mL, sterile.
7. Sterile polypropylene 15 and 50 mL tubes.
8. Sterile plastic Petri dishes, 90 mm.
9. RNAse-free tips.
10. qPCR plates.

2.4 Equipment

1. Autoclave.
2. Controlled environment chamber/room (28 °C, 16-h day length, and 50% humidity).
3. Controlled environment chamber/room (28 °C, 16-h day length, and >80% humidity).
4. Greenhouse: 28 °C, 16-h day length, 22 klx, 50% humidity.
5. Fridge (4 °C) and freezer (−20 and −80 °C).
6. Electroporator and electroporation cuvette.
7. Incubator shaker (28 °C).
8. Inoculation loops.
9. Laminar flow hood.
10. Mortar and pestle.
11. pH meter.
12. Precision balance.

13. Refrigerated centrifuge for 50 mL tubes.
14. Spatula, scalpel, and forceps.
15. Spectrophotometer (e.g., NanoDrop).
16. Thermal cycler.
17. Vortexer.
18. Electrophoresis unit.
19. UV transilluminator.
20. Safety mask/goggles (UV protection).
21. Real-time PCR system.

2.5 Reagents

1. cDNA Synthesis Kit including oligo dT.
2. 2× SYBR Master Mix.
3. High-fidelity proofreading DNA polymerase.
4. Agarose, 1× TAE, Bet.
5. Gel purification DNA kit.
6. Blunt sub-cloning vector kit.
7. 2× PCR mix including Taq polymerase and dNTPs.
8. Plasmid purification kit.
9. Restriction enzymes: KpnI and SpeI or SacI and SmaI.
10. T4 DNA ligase.
11. ACMV DNA-A vector, ACMV DNA-B vector, and ACMV-VIGS vector (modified ASMV DNA-A vector contains a partially deleted *AV1* gene and a small multiple cloning site) (Fig. 1).
12. ACMV-VIGS vector MCS reverse primer 5-′-TGTCATTAGAGCTGCTGATCATGT-3′.
13. 100 mM Acetosyringone.
14. Liquid nitrogen.
15. Extraction buffer: 50 mM Tris–HCl, pH 9, 10 mM EDTA, 2% (m/v) SDS, 100 mM LiCl, and 10 μg/mL proteinase K.
16. 25:24:1 (v/v/v) phenol/chloroform/isoamyl alcohol.
17. 1 M acetic acid.
18. 100% (v/v) And 70% (v/v) ethanol.
19. TE buffer: 10 mM Tris–HCl, pH 8, and 1 mM EDTA.
20. Reference gene PP2A (Manes.09G039900) primers for qPCR: PP2A-cDNA fw 5′-TGCAAGGCTCACACTTTCATC-3′ and PP2A-cDNA rv 5′-CTGAGCGTAAAGCAGGGAAG-3′.
21. Primer pair to detect ACMV-DNA-A: ACMV_DNA-A_F 5′-GGGAAAATCACGATTACGAACAATC-3′ and ACMV_DNA-A_R 5′-GTATACACGCCATTCATTGCTTG-3′.

22. Primer pair to detect ACMV-DNA-B: ACMV_DNA-B_F 5-′-CTATCGAATGACAGCCTGTGCTA-3′ and ACMV_DNA-B_R 5′-CCCCTGGAAGGATACGTCAC-3′.

3 Methods

3.1 Preparing Cassava Plantlets

1. Maintain cassava cv. 60444 plantlets in vitro in CBM under the growth conditions 28 °C, 16-h day length and 50% humidity (*see* **Note 1**).
2. Monitor the plantlets for optimal growth; when roots are well developed (normally 4–6 weeks), transfer the in vitro cassava plantlets to soil under the growth conditions 28 °C, 16-h-day length (*see* **Note 2**).
3. Place the plants in a tray, and cover with transparent lid to maintain high humidity for 1 week (*see* **Note 3**).
4. Acclimatize the plantlets in soil for 3–5 weeks before using them for VIGS experiment (*see* **Note 3**).

3.2 Cloning of Target Gene into the ACMV-VIGS Vector

3.2.1 Cloning of Target Gene in Intermediate Vector

1. Choose a specific target gene segment of ~500 bp (*see* **Note 4**). Blast the chosen target fragment against the cassava genome to ensure the selection of specific sequence and to avoid any off-targets.
2. Design gene-specific primers containing KpnI and SpeI restriction sites (or SacI and SmaI) as overhangs in the forward and reverse primers, respectively (*see* **Note 5**).
3. Amplify the target gene segment using high-fidelity proofreading DNA polymerase. Use 50 ng of template, 200 μM of dNTPs, 0.1–0.5 μM of primers, and 3 U of polymerase in a final volume of 50 μL. PCR conditions: Initial denaturation at 98 °C for 30 s; 30 cycles of denaturation at 98 °C for 10 s, annealing at 50–72 °C (depending on Tm value of your primer pair) for 30 s, extension at 72 °C for 1 min, and final extension at 72 °C for 2 min.
4. Load 5 μL of the PCR reaction on agarose gel and perform electrophoresis. After confirming the presence of the PCR product with the expected size, run in a preparative agarose gel the remaining 45 μL using new TAE buffer. Working on a UV or blue light LED transilluminator, excise the PCR band from the agarose gel, purify the DNA fragment with the specific kit, and quantify the final preparation.
5. Clone the eluted DNA amplicon in a blunt sub-cloning vector (*see* **Note 6**).
6. Set up a ligation reaction as per cloning kit instructions.

7. Add 5–10 μL of the ligation mixture to 50 μL of *E. coli* competent cells, and proceed with heat shock at 42 °C for 50 s, and immediately place the tubes on ice for 2 min.
8. Add 0.5 mL of cold LB broth (without antibiotics), and incubate the tubes in a shaker at 37 °C for at least 1 h.
9. Spread an aliquot of 200 μL on a LB plate containing 100 mg/L of ampicillin (or the respective antibiotic based on the vector that is being used), and incubate overnight in dark at 37 °C.
10. Prepare colony PCR reaction: 7.5 μL of 2× PCR mix, 0.25 μL of each primer (primer pair provided with the cloning vector, 10 μM), and 7 μL of nuclease-free water. Pick a single colony with sterile toothpick or 10 μL tip, dip into the PCR tube, and mix.
11. Set up PCR: Initial denaturation at 95 °C for 5 min; 25 cycles denaturation at 95 °C for 30 s, annealing at 58 °C for 30 s, extension at 72 °C for 1 min, and final extension at 72 °C for 5 min.
12. Proceed with electrophoresis and select positive colonies.
13. Culture one or two positive clones in LB broth medium containing appropriate antibiotics; overnight at 37 °C in a shaking incubator (200 rpm).
14. Extract plasmid from overnight bacterial culture using a plasmid purification kit (*see* **Note 7**).
15. Confirm the cloned DNA segment by restriction analysis and sequencing.

3.2.2 Cloning of Target Gene in ACMV-VIGS Vector

1. Digest 1 μg of the intermediate vector carrying the target gene with KpnI and SpeI (or SacI and SmaI) (for insert).
2. In parallel digest 1 μg of the ACMV-VIGS vector using the same enzymes.
3. Run the digested samples on a 1% (m/v) agarose gel electrophoresis, and dissect the insert-band and backbone-band in separate tubes, and proceed with gel elution using a gel purification DNA kit.
4. Quantify the concentration of insert and backbone, and set up a ligation reaction with a T4 DNA ligase using a 5:1 to 20:1 ratio of insert to vector. Incubate under room temperature for 1 h.
5. Follow the same procedure from **steps** 7 to **11** in Subheading 3.2.1 using LB plates with kanamycin 50 mg/L for "**step 9**."
6. Perform colony PCR as described in Subheading 3.2.1, using the target gene cloning primer as forward primer and ACMV-VIGS MCS reverse primer.

7. Confirm cloning of the selected DNA fragment into the ACMV-VIGS vector by restriction analysis or sequencing using ACMV-VIGS MCS reverse primer.

3.3 Transformation of ACMV-VIGS Vector into *A. tumefaciens* by Electroporation

3.3.1 Preparation of Competent *Agrobacterium* Cells

1. Streak *Agrobacterium tumefaciens* strain AGL-1 (*see* **Note 8**) from a glycerol stock onto YEB + 100 mg/L carbenicillin + 50 mg/L rifampicin plates, and incubate at 28 °C for 2 days in the dark.
2. Pick a colony using a sterile inoculation loop, and inoculate 5 mL of YEB + 100 mg/L carbenicillin + 50 mg/L rifampicin in a 15 mL sterile disposable Falcon tube. Grow overnight in an incubator shaker at 28 °C and 200 rpm.
3. Take 0.5 mL from the overnight culture to inoculate 25 mL of YEB + 100 mg/L carbenicillin + 50 mg/L rifampicin, in sterile 250 mL flasks.
4. Grow in an incubator shaker at 28 °C and 200 rpm until the OD_{600} of the culture reaches 0.7–1.0, which is typically 24 h.
5. Transfer the culture to sterile 50 mL conical Falcon tube, and pellet the cells by centrifugation at 4 °C for 20 min at 3000 × *g*. Discard the supernatant.
6. For all the following steps, *Agrobacterium* must be kept on ice.
7. Wash in the same culture volume of 10% (v/v) cold glycerol, and centrifuge at 4 °C for 20 min at 3000 × *g*. Discard the supernatant.
8. Repeat the washing step and discard the supernatant.
9. Resuspend the cells in 10% (v/v) of cold glycerol using 0.02× of the original culture volume (e.g., 200 μL if the original culture was in 10 mL of medium).
10. Make 100 μL aliquots, and use for electroporation, or freeze in liquid nitrogen, and store at −80 °C.

3.3.2 Transformation of Competent *Agrobacterium* Cells via Electroporation

1. Put bacterial cells, plasmid DNA, and electroporation cuvette on ice.
2. Add 100–200 ng of plasmid DNA to the bacterial cells (max volume 10 μL).
3. Transfer the solution containing bacteria and plasmid to the cuvette.
4. Set the Gene Pulser to 2.5 kV, 200 Ω, and 25 μFD.
5. Run the Gene Pulser and as soon as the pulse is finished; add 400 μL LB to the cuvette immediately.
6. Place the electroporated bacteria in a 2 mL tube, and incubate in a shaker at 28 °C for 1–2 h.

7. Spread an aliquot of 200 μL on a LB plate with 100 mg/L carbenicillin, 50 mg/L rifampicin, and 50 mg/L kanamycin.
8. Incubate in the dark at 28 °C for 48 h.

3.4 Preparation of the Agrobacterium Solution for Agroinoculation

Proceed with the following steps in parallel for *Agrobacterium* carrying ACMV DNA-A, ACMV-VIGS, or ACMV DNA-B vectors in separate tubes/flasks.

1. To make the starter cultures, inoculate 50 mL sterile tubes containing 5 mL of YEB with 100 mg/L carbenicillin, 50 mg/L rifampicin, and 50 mg/L kanamycin with *Agrobacterium* colonies transformed with the agroclones separately, and grow them for 24–36 h at 28 °C and 150 rpm until the $OD_{600\ nm}$ reaches values >1.
2. Add 2 mL of the starter cultures to 1 L Erlenmeyer flasks containing 100 mL of YEB with 100 mg/L carbenicillin, 50 mg/L rifampicin, and 50 mg/L kanamycin.
3. Grow cultures for 24–36 h at 28 °C and 150 rpm until they reach an $OD_{600\ nm}$ of 1.5–2.
4. Transfer the culture into two sterile 50 mL tubes, and centrifuge the agrobacteria suspensions at 4000 × *g* for 20 min at room temperature.
5. After centrifugation, carefully resuspend the bacterial pellets in 25 mL of sterile deionized water (in each tube) at room temperature, and centrifuge again at 4000 × *g* for 20 min.
6. Carefully resuspend the washed bacterial pellets in 10–20 mL of LB, and adjust the final value of $OD_{600\ nm}$ at 4 with LB.
7. Mix an equal volume of *Agrobacterium* carrying the ACMV DNA-A vector or ACMV-VIGS vector and *Agrobacterium* carrying the ACMV DNA-B vector (*see* **Note 10**).
8. Add acetosyringone in the bacterial suspensions to a final concentration of 200 μM.
9. Incubate at room temperature for 3 h on a shaker at 100 rpm.

3.5 Agroinoculation of Plants

3.5.1 Protocol A (for Small Number of Plants)

1. Take 1 mL insulin syringe (0.33 mm/29 G/12.7 mm), and carefully inject the cassava plantlet with agrobacteria suspension at least three times near the axillary buds without damaging the meristems.
2. Use the same syringe to gently puncture the stem five to seven times.
3. Keep the inoculated plantlets covered, and incubate at 20–24 °C in the dark overnight.
4. Transfer the inoculated plantlets to the greenhouse, and keep the trays covered with transparent lid for 2–4 days (*see* **Notes 3** and **11**).

3.5.2 Protocol B (High-Throughput Option) (See Note 12)

1. Remove the leaves from 4–5 in. upper part of the stem of cassava plantlets.
2. Use a 21 G needle, previously soaked in agrobacteria suspension, to puncture the 4–5 in. upper part of the stem at least three times.
3. Dip the plantlets into a 50 mL tube containing the *Agrobacterium* suspension for 30 s.
4. Allow the plantlets to recover for 4 days under high humidity conditions (28 °C, 16-h day length, >80% humidity).
5. Transfer the plantlets to greenhouse (*see* **Notes 2** and **10**).

3.6 Analysis of VIGS in Plants by PCR and Phenotyping

The ACMV-VIGS infection can be detected by performing a multiplex PCR using primer pairs for ACMV DNA-A and DNA-B in the same PCR reaction. The expected product size for DNA-A is 1 kb and for DNA-B is 0.5 kb.

3.6.1 PCR Detection of ACMV-VIGS Infection

1. Use 50 ng of template, 200 μM of dNTPs, 0.25 μL of each primer (10 μM) (primer pair for DNA-A and DNA-B), and 1 U of polymerase and nuclease-free water to a final volume of 20 μL. As a positive control for PCR, use DNA-A and DNA-B plasmids (1:1 ratio) as template.
2. Set up PCR conditions: Initial denaturation at 98 °C for 30 s; 35 cycles of denaturation at 98 °C for 10 s, annealing at 58 °C for 30 s, and extension at 72 °C for 1 min; and final extension at 72 °C for 5 min.
3. Observe the PCR product by agarose gel electrophoresis. The presence of double DNA bands at 1 and 0.5 kb confirms the presence of ACMV infection.

3.6.2 Scoring the Agroinoculated Plants for the Silencing Phenotype

1. Observe the agroinoculated plants for development of cassava mosaic disease symptoms in emerging leaves during 16 weeks for each plant. In case the VIGS phenotype is not expected to be detected visually, the highest silencing could be found in those leaves showing the strongest mosaic symptoms [15].
2. Score the plants using the following 0–5 scale. (0) No silencing or VIGS symptoms; (1) faint mosaic; (2) yellow mosaic, malformation, and 5% size reduction; (3) severe mosaic, distortion, and reduced size; (4) severe mosaic, severe distortion, and up to 50% size reduction; (5) leaf reduced to veins and 50–80% size reduction [5, 17–19].

3.7 Analysis of VIGS in Plants by RT-qPCR

3.7.1 Extraction of Leaf RNA [20] and cDNA Synthesis

1. For each biological replicate, select three symptomatic intermediate leaves, and pool them.
2. Grind to a fine powder 2–3 g of leaves in a mortar with liquid nitrogen.
3. Transfer the ground leaf tissue to a 50 mL polypropylene tube containing 10 mL of extraction buffer.
4. Vortex vigorously for 10 s and incubate in a water bath at 45 °C for 15 min.
5. Centrifuge at 5000 × *g* for 15 min, and transfer the supernatant to a new polypropylene tube.
6. Add 1 volume of phenol/chloroform/isoamyl alcohol (25:24:1), vortex, and centrifuge at 5000 × *g* for 15 min.
7. Transfer the aqueous phase to a new tube, and add 0.7 volume of ethanol followed by 0.2 original volume of 1 M acetic acid.
8. Incubate 1 h at −80 °C.
9. Centrifuge at 10,000 × *g* for 20 min at 4 °C and discard the supernatant.
10. Add 10 mL of 70% (v/v) ethanol and centrifuge 10 min. Discard the supernatant.
11. Repeat the washing step with 70% (v/v) ethanol, and then make an additional wash with 100% (v/v) ethanol.
12. Dry the pellet by leaving the open tube in a laminar flow, and resuspend it in 1 mL of TE buffer.
13. Perform a quality check, and quantify all the samples with a spectrophotometer or a NanoDrop system.
14. Treat RNA samples with DNase I, and convert aliquots to cDNA with a kit according to the manufacturers' instructions.
15. Keep RNA, DNAse I-treated RNA, and cDNA-converted samples at −80 °C.

3.7.2 Primer Design for qPCR

1. Design a primer pair for qPCR for the gene(s) of interest, following these criteria; (1) product size, 150 bp; (2) amplicon obtained from the 3′ end of the target mRNA; (3) primers designed to amplify across an intron junction; (4) specific for the gene of interest; and (5) all other important features such us self-complementarity and melting temperature can be automatically checked by using specific software, such us Primer Express.
2. Determine primer efficiency by making several dilutions of the cDNA, with four replicates for each dilution, and include a water control that should give a CT value ≥35. Free online software can be used for the calculation of primer efficiency, such as those offered by Merck and Thermo Fisher Scientific.

Primer efficiency values of 1.8–2.0 are expected, calculated from results with a correlation coefficient of at least 0.99. Values above 2.0 indicate unspecific amplification, incorrect dilutions, or inhibition of the PCR.

3. The constitutive gene *PP2A* (Manes.09G039900) have been already validated to be used as reference gene for RT-qPCR in cassava [21] (PP2A-cDNA fw and PP2A-cDNA rv primers).

3.7.3 Quantification of VIGS by qPCR

1. Assemble qPCR reactions for the target gene and PP2A with 5–50 ng of cDNA, 1 μM of primers, and a 2× SYBR Green master mix in a final volume of 20 μL. Include at least two technical replicates per sample. Include in parallel for each cDNA sample a total RNA DNAse I-treated sample as blank. Obtain Ct values from the qPCR machine, and import the data into a spreadsheet.
2. Analyze first each RNA sample for DNA contamination, and exclude those samples with positive amplification.
3. Calculate relative expression ratio using the equation $\mathrm{RE} = E_{\mathrm{target}}^{(\mathrm{CTcontrol} - \mathrm{CTtarget})}/E_{\mathrm{PP2A}}^{(\mathrm{CTcontrol} - \mathrm{CTreference})}$ (Pfaffl method [22]), where E_{target} is the primer efficiency determined in Subheading 3.7.2 and E_{PP2A} is the primer efficiency of the constitutive *PP2A* gene [21].
4. Analyze significant differences in gene expression levels with two-tailed *T* tests.

4 Notes

1. This protocol can be used with ACMV susceptible cassava genotypes. Here, we used the model cassava cv. 60444, which is popular in the cassava research community because of its amenability for efficient genetic transformation [9, 23]. The cv. 60444 is highly susceptible to ACMV [17, 24].
2. Thoroughly wash the roots of plantlets before transferring them to soil to avoid fungal growth and contamination.
3. Regularly remove dead and infected leaves from the agroinoculated plants to avoid fungal infections.
4. The ACMV-VIGS vector was constructed by replacing a fragment of 459 bp of AV1 gene (coding for capsid protein) with a 30 bp multiple cloning site (MCS) containing four restriction sites, i.e., SacI, KpnI, SmaI, and SpeI. Thus, the target gene segment of ~500 bp is advisable for full infectivity of the ACMV-VIGS clone.
5. Choose any two enzyme combinations that are suitable for your target gene sequence.

6. Alternatively, digest the purified PCR product directly, and clone it into the ACMV-VIGS vector. Intermediate cloning in a TA-cloning vector is recommended to facilitate the ligation in the final ACMV-VIGS vector.
7. Consider preparing a glycerol stock to store the bacterial clones for future use.
8. We recommend using *Agrobacterium* strain AGL-1 for high infection rate. However, other *Agrobacterium* strains like LBA4404 and GV3101 can also be used.
9. In case YEB is not available, simple LB broth (Luria-Bertani [25]) can also be used.
10. For inoculation of each agroclone, equal volume of agrobacterium carrying DNA-A (or ACMV-VIGS vector) and DNA-B is required.
11. Discard all the agroinoculated plants only after autoclaving at the recommended conditions.
12. Both agroinoculation protocols lead to infection rates that are above 80% (Protocol A) and 60% (Protocol B) in cassava. Protocol A is suitable for a smaller number of plants (20–30) as it is more time-consuming than the Protocol B.

Acknowledgments

The authors acknowledge financial support from the Belgian FNRS (S.S.Z.; grants M.i.S.F.4515.17 and V4 /705 to H.V.) and LEAP Agri Grant 288 (to H.V.) and the Swiss Agency for Development and Cooperation, Government of Switzerland (K.V.; Cassava Network grant (Indo-Swiss Collaboration in Biotechnology) to H.V.).

References

1. Vanderschuren H, Nyaboga E, Poon JS et al (2014) Large-scale proteomics of the cassava storage root and identification of a target gene to reduce postharvest deterioration. Plant Cell 26:1913–1924
2. FAO (2013) Save and grow: cassava. A guide to sustainable production intensification. ISBN: 978-92-5-107641-5, 140pp
3. Rey C, Vanderschuren H (2017) Cassava mosaic and brown streak diseases: current perspectives and beyond. Annu Rev Virol 4:429–452
4. Vanderschuren H (2012) Strengthening African R&D through effective transfer of tropical crop biotech to African institutions. Nat Biotechnol 30:1170–1172
5. Mehta D, Sturchler A, Anjanappa RB et al (2019) Linking CRISPR-Cas9 interference in cassava to the evolution of editing-resistant geminiviruses. Genome Biol 20:80
6. Beyene G, Chauhan RD, Wagaba H et al (2016) Loss of CMD2-mediated resistance to cassava mosaic disease in plants regenerated through somatic embryogenesis. Mol Plant Pathol 17:1095–1110
7. Bull SE, Seung D, Chanez C et al (2018) Accelerated ex situ breeding of GBSS- and PTST1-edited cassava for modified starch. Sci Adv 4:eaat6086
8. Rosas-Diaz T, Zhang D, Fan P et al (2018) A virus-targeted plant receptor-like kinase promotes cell-to-cell spread of RNAi. Proc Natl Acad Sci U S A 115:1388–1393

9. Bull SE, Owiti JA, Niklaus M et al (2009) Agrobacterium-mediated transformation of friable embryogenic calli and regeneration of transgenic cassava. Nat Protoc 4:1845–1854
10. Zainuddin IM, Schlegel K, Gruissem W et al (2012) Robust transformation procedure for the production of transgenic farmer-preferred cassava landraces. Plant Methods 8:24
11. Niklaus M, Gruissem W, Vanderschuren H (2011) Efficient transformation and regeneration of transgenic cassava using the neomycin phosphotransferase gene as aminoglycoside resistance marker gene. GM Crops 2:193–200
12. Nyaboga E, Njiru J, Nguu E et al (2013) Unlocking the potential of tropical root crop biotechnology in east Africa by establishing a genetic transformation platform for local farmer-preferred cassava cultivars. Front Plant Sci 4:526
13. Chetty CC, Rossin CB, Gruissem W et al (2013) Empowering biotechnology in southern Africa: establishment of a robust transformation platform for the production of transgenic industry-preferred cassava. New Biotechnol 30:136–143
14. Mehta D, Hirsch-Hoffmann M, Were M et al (2018) A new full-length circular DNA sequencing method for viral-sized genomes reveals that RNAi transgenic plants provoke a shift in geminivirus populations in the field. Nucleic Acids Res 47(2):e9
15. Lentz EM, Kuon JE, Alder A et al (2018) Cassava geminivirus agroclones for virus-induced gene silencing in cassava leaves and roots. Plant Methods 14:73
16. Fofana IB, Sangare A, Collier R et al (2004) A geminivirus-induced gene silencing system for gene function validation in cassava. Plant Mol Biol 56:613–624
17. Vanderschuren H, Alder A, Zhang P et al (2009) Dose-dependent RNAi-mediated geminivirus resistance in the tropical root crop cassava. Plant Mol Biol 70:265–272
18. Zhang P, Vanderschuren H, Futterer J et al (2005) Resistance to cassava mosaic disease in transgenic cassava expressing antisense RNAs targeting virus replication genes. Plant Biotechnol J 3:385–397
19. Fauquet C, Fargette D (1990) African cassava mosaic-virus— etiology, epidemiology, and control. Plant Dis 74(6):404–411
20. Soni R, Murray JA (1994) Isolation of intact DNA and RNA from plant tissues. Anal Biochem 218:474–476
21. Moreno I, Gruissem W, Vanderschuren H (2011) Reference genes for reliable potyvirus quantitation in cassava and analysis of Cassava brown streak virus load in host varieties. J Virol Methods 177:49–54
22. Pfaffl MW (2001) A new mathematical model for relative quantification in real-time RT-PCR. Nucleic Acids Res 29:e45
23. Liu J, Zheng Q, Ma Q et al (2011) Cassava genetic transformation and its application in breeding. J Integr Plant Biol 53:552–569
24. Vanderschuren H, Akbergenov R, Pooggin M et al (2007) Transgenic cassava resistance to African cassava mosaic virus is enhanced by viral DNA-A bidirectional promoter-derived siRNAs. Plant Mol Biol 64:549–557
25. Bertani G (1951) Studies on lysogenesis. I. The mode of phage liberation by lysogenic Escherichia coli. J Bacteriol 62(3):293–300

Chapter 6

Virus-Induced Gene Silencing of Cell Wall Genes in Flax (*Linum usitatissimum*)

Maxime Chantreau and Godfrey Neutelings

Abstract

Plants have developed defense mechanisms against viruses by using an RNA silencing-based process, which has many common features with the endogenous RNA silencing pathway used for regulating the level of transcripts derived from developmental genes. In the virus-induced gene silencing (VIGS) method, it is possible to take advantage of this mechanism by inserting a plant nucleic fragment within the viral genome to knock down the corresponding gene. This tool has been used in many species as a fast and easy reverse genetics technique in order to gain information on the role of genes with poorly understood functions. Here we describe in detail two *Agrobacterium*-mediated infection protocols in flax, based on a whole plant vacuum infiltration and a leaf syringe infiltration that systemically impact the transcript levels in the stem.

Key words VIGS, Flax, Gene silencing, Agroinfiltration, *Tobacco rattle virus*

1 Introduction

Since the recent development of high-throughput sequencing technologies, a very large amount of data concerning the structure and contents of plant genomes are now available. To better understand the role of the identified genes, it is necessary to develop new functional genomic approaches or to adapt the existing techniques that have been efficiently used on model plants, to non-model plants. VIGS utilizes the natural ability of plants to destroy a viral RNA by using their endogenous RNAi machinery. It is therefore possible to exploit this mechanism to specifically target a gene or a family of genes with homologous sequences, if a fragment of the corresponding plant transcript is present in the viral RNA [1]. When using the very popular *Tobacco rattle virus* (TRV), the genome is cloned in two separate vectors, one carrying the viral replication and movement functions, while the other one hosts a gene encoding the coat protein and a cloning site for the plant target gene [2]. Both can be separately introduced in the plant by

Vincent Courdavault and Sébastien Besseau (eds.), *Virus-Induced Gene Silencing in Plants: Methods and Protocols*, Methods in Molecular Biology, vol. 2172, https://doi.org/10.1007/978-1-0716-0751-0_6,

Agrobacterium tumefaciens coinfection or by other alternative methods.

Flax (*Linum usitatissimum*) is used as a model to study the cell wall of the long bast fiber cells located beneath the epidermis in the stem cortex [3]. These fibers are characterized by a very thick secondary cell wall (CWII) containing high amounts of crystalline cellulose but low lignin contents in contrast to the heavily lignified cell walls typically found in the xylem tissue [4]. In this species, functional genomic approaches have been performed successfully by producing stable transgenic lines after *Agrobacterium tumefaciens* transformation. Unfortunately, the production of such lines in flax remains time-consuming and difficult to carry out on several genes at the same time [5]. To circumvent these limits, we have adapted the VIGS approach initially developed on *Nicotiana* species by using the same viral vectors on flax [6]. Indeed this species was previously shown to be susceptible to the *Tobacco rattle virus* (TRV) although plant contaminations were only rarely reported in the field [7].

The transfer of the VIGS approach on a new species requires the use of a positive control to check if the plant is indeed receptive to the viral genome. In general a vector containing a fragment of a phytoene desaturase (*PDS*) gene is used. The disruption of this gene causes a developmental arrest of the chloroplasts at the proplastid stage and the inhibition of many genes in the carotenoid, chlorophyll, and gibberellic acid biosynthesis pathways [8]. The plants develop an albino phenotype meaning that there is no need to proceed to molecular characterization of the transformed plants, since the silencing can be easily visualized. The *PDS* orthologs are well conserved among plants, so it is easy to design degenerated primers for species that have not yet been fully sequenced [9]. In the case of flax, the genome sequence was obtained and annotated [10], so the reference of the *PDS* gene is given in the materials chapter below. The use of this control allowed us to evaluate the efficiency of two different transformation protocols. The *Agrobacterium tumefaciens* strains containing both vectors were either introduced in the plant by vacuum infiltration consisting of soaking the whole plant in the bacterial suspension or by leaf syringe infiltration.

Here, we also propose the use of a cellulose synthase cDNA fragment which originated from the *LusCESA1_A* gene [11] to target the three paralogous gene families, *LusCESA1*, *LusCESA3*, and *LusCESA6*, involved in the biosynthesis of the primary cell wall. The silencing of these genes is responsible for the reduction of plant growth and perturbation of leaf morphology. The use of *LusCESA* results in a more systemic effect of the silencing and a less harsh metabolic perturbation compared to *LusPDS*, showing that *LusCESA* can be successfully used as a positive control when performing VIGS on flax.

2 Materials

2.1 Plant Growth

1. Seeds of *Linum usitatissimum* cultivar Barbara.
2. Pots filled with potting soil.
3. Culture room/chamber maintained at 20 °C with 16 h light and 8 h dark cycle using artificial light.

2.2 RNA Extraction

1. Liquid nitrogen.
2. Mortar and pestle heated in a 200 °C oven for 8 h.
3. RNase-free microcentrifuge tubes and tips.
4. RNA purification reagent (e.g., TRI Reagent).
5. Chloroform.
6. Isopropanol.
7. 70% Ethanol.
8. RNase-free water.
9. NanoDrop.
10. cDNA synthesis kit including oligo dT.

2.3 Plasmid Recombination

1. Thermocycler.
2. High-Fidelity DNA Polymerase Kit.
3. For *LusPDS* gene Lus10021967 primers:

   LusPDS-F1: 5′-GGATCCTCGAGTGACTACTGAGGTGTTC-3′ and

   LusPDS-R1: 5′-GAGCTCTCCGGTCAAACCATATGTGAAC-3′.

   For *CESA* gene Lus10018902 primers:

   LusCESA-F1: 5′-GAAGACGAGTTCAGCTATCAAG-3′ and

   LusCESA-R1: 5′-AGCATCGTCAGCCATTTGGAG-3′.
4. TA Cloning Vector Kit.
5. *pTRV1* (CD3-1039) and *pTRV2* (CD3-1040) available from TAIR (https://www.arabidopsis.org).
6. 1× TAE: 0.04 M Tris–HCl, pH 8.0, 5 mM sodium acetate, and 1 mM EDTA.
7. Agarose, GelRed, and electrophoresis equipment.
8. Plasmid purification kit and PCR cleanup kit.
9. *BamH*I and *Sac*I restriction enzymes with buffers.
10. T4 DNA ligase kit.
11. Chemocompetent TOP10 *E. coli* cells (F-*mcr*A Δ(*mrr-hsd*RMS-*mcr*BC) φ80*lac*ZΔM15 Δ*lac*X74 *rec*A1 *ara*D139 *gal*U *gal*KΔ(*ara-leu*)7697 *rps*L (StrR) *end*A1 *nup*G).

2.4 Bacterial Transformation

1. *Agrobacterium tumefaciens* C58C1 pGV2260 strain.
2. LB medium: 10 g/L Bacto Tryptone, 5 g/L yeast extract, and 0.17 M NaCl. For solid media, 1.5% agar is added.
3. Supplemented LB medium: LB medium containing 10 mM MES and 20 μM acetosyringone.
4. SOC medium: 0.5% yeast extract, 2% tryptone, 10 mM NaCl, 2.5 mM KCl, 10 mM $MgCl_2$, 10 mM $MgSO_4$, and 20 mM Glucose.
5. Antibiotic independent stock solutions: 100 mg/mL rifampicin, 25 mg/mL ampicillin, and 50 mg/mL kanamycin.
6. 50% Glycerol (v/v) solution.
7. Electroporator and 2 mm electroporation cuvettes.
8. 17 mm × 95 mm Round-bottom culture tubes.

2.5 Plant Transformation

1. Agroinfiltration medium: 10 mM $MgCl_2$, 10 mM MES, and 150 μM acetosyringone.
2. 1 M MES solution.
3. 1% Silwet-L77 solution.
4. Vacuum bell jar and vacuum pump.
5. 1 mL syringe.

3 Methods

3.1 Plant Growth

1. Grow flax plants (*Linum usitatissimum* cultivar Barbara; *see* **Note 1**) in a greenhouse under 16 h/20 °C day and 8 h/18 °C night conditions. A total of 12 seeds should be placed in 10 × 10 cm pots filled with potting soil.
2. When the plants are 15 days old, proceed to the VIGS transformation (Fig. 1a).

3.2 RNA Extraction and cDNA Synthesis (See Note 2)

1. Grind 100 mg of frozen flax stems in liquid nitrogen using a mortar and a pestle.
2. Transfer the powder to a microtube, add 1 mL of RNA extraction reagent, and vortex for 10 min.
3. After 5 min incubation at room temperature, centrifuge for 10 min at 12,000 × *g*, 4 °C.
4. Add 200 μL chloroform to the supernatant and mix for 15 s.
5. Incubate for 15 min at room temperature and centrifuge for 15 min at 12,000 × *g*, 4 °C.
6. Add 500 μL isopropanol to the supernatant.
7. Incubate for 10 min at room temperature and centrifuge for 10 min at 12,000 × *g*, 4 °C.

Fig. 1 *LusPDS* gene inactivation in young flax plants. (**a**) 15-Day-old plants before infiltration. (**b**) Leaf albino phenotype observed on 28-day-old plants. (**c**) Representative leaf and stem phenotypes of inactivated and control plants

8. Wash the pellet with 1 mL ethanol 70% (v/v) and centrifuge for 5 min at 7500 × *g*, 4 °C.
9. Dry the pellet on the bench and dissolve it with 30 μL RNase-free water.
10. Quantify the RNA sample using a NanoDrop (*see* **Note 3**).
11. 1 μg Of total RNA was reverse-transcribed using a cDNA synthesis kit including oligo dT according to the manufacturer's instructions.

3.3 Plasmid Recombination

1. Proceed to a 50 μL PCR amplification reaction with the *LusPDS* (*see* **Note 4**) and *LusCESA* (*see* **Note 5**)-specific primers using a High-Fidelity DNA polymerase and 1 μL of stem cDNA as template. The amplification parameters are 30 s 98 °C denaturation step followed by 35 cycles with 10-s denaturation at 98 °C, 30-s annealing at 56 °C, and 30-s extension at 72 °C. Finish with a 2 min step at 72 °C.
2. Check if the fragments were correctly amplified in a 1.5% agarose gel dissolved in 1× TAE supplemented with GelRed. Compare the size of the fragments with those of a ladder.

3. Clone the fragments in a TA cloning plasmid according to the manufacturer's instructions.
4. Extract the plasmids with a plasmid purification kit.
5. Digest 1 μg of the obtained plasmids with *BamHI* and *SacI*.
6. Cut the gel containing the *LusPDS* and *LusCESA* digested fragments, and purify with a gel and PCR cleanup kit.
7. Digest 1 μg of the *pTRV2* plasmid with *BamHI* and *SacI*.
8. Proceed to the ligation of the purified fragments with the linearized *pTRV2* plasmid by using T4 DNA ligase. The molar vector/insert ratio should be 1:3.
9. Transform TOP10 cells to multiply the recombinant plasmids.
10. Grow on selective solid LB medium containing 50 μg/mL kanamycin.
11. Screen the colonies by PCR using the above protocol. Replace the cDNA volume by water, and suspend a small fraction of a bacterial colony collected on a pipet tip directly in the mix.
12. Extract the plasmids as in **step 4**.

3.4 Plasmid Integration into Agrobacterium tumefaciens

1. Grow the C58C1 pGV2260 strain on solid LB supplemented with 100 μg/mL rifampicin, 25 μg/mL ampicillin, for 2 days at 28 °C.
2. Pick a single colony and grow for 16 h in 10 mL LB with antibiotics at 28 °C, 200 rpm.
3. Inoculate 100 mL of the same medium under the same conditions with 1 mL of this culture until the OD600 reaches 0.5.
4. Incubate the cultures for 10 min on ice.
5. Centrifuge at 4000 × *g* for 15 min, 4 °C.
6. Resuspend the bacteria in 25 mL ice-cold water.
7. Centrifuge at 4000 × *g* for 15 min, 4 °C.
8. Resuspend the bacteria in 15 mL ice-cold water.
9. Centrifuge at 4000 × *g* for 15 min, 4 °C.
10. Resuspend the bacteria in 5 mL glycerol 10%.
11. Centrifuge at 4000 × *g* for 15 min, 4 °C.
12. Resuspend the bacteria in 0.2 mL glycerol 10%.
13. Freeze 40 μL aliquots in liquid nitrogen.
14. Add 1 μL of plasmid solution at 50 ng/μL to 40 μL competent *Agrobacterium* cells.
15. Transfer to a 2 mm electroporation cuvette previously chilled on ice. Electroporate using the following conditions: 2.0 kV, 200 Ω, and 25 μF. The typical time constant should be about 5 ms.

16. Add 960 μL of SOC, mix gently, and transfer to a 17 mm × 95 mm round-bottom culture tube.
17. Grow for 2.5 h at 28 °C, 200 rpm.
18. Spread 100 and 200 μL on full selective solid LB (100 μg/mL rifampicin, 25 μg/mL ampicillin, 50 μg/mL kanamycin), and grow for at least 48 h until single colonies appear.

3.5 Preparation of the Bacteria for Agroinfiltration

1. Inoculate a single colony of *Agrobacterium* containing *pTRV1* and the recombinant *pTRV2* plasmids in 5 mL full selective LB for 16 h at 28 °C, 200 rpm.
2. Dilute the culture to 1/10 with the same medium supplemented with 10 mM MES and 20 μM acetosyringone, and grow as previously done.
3. Centrifuge the cultures at room temperature for 15 min at 4000 × *g*, and resuspend the pellets with the agroinfiltration medium in order to reach OD600 of 0.7.
4. Leave the suspensions on the bench at room temperature for 4–6 h, and mix the *pTRV1* and recombinant *pTRV2* cultures at ratio 1:1.

3.6 Plant Transformation by Vacuum Infiltration

1. Remove the 15-day-old plants from the pots, and discard the remaining soil on the roots.
2. Transfer 50 mL of the *Agrobacterium* suspension supplemented by 0.05% Silwet to a 250 mL beaker, and immerge up to four plantlets. The same suspension can be used up to five times.
3. Place the beaker in a vacuum bell jar.
4. Apply vacuum for a 5-min period and then quickly open the valve to let the air in.
5. Drain the plants on paper towels.
6. Transplant in individual pots.

3.7 Plant Transformation by Syringe Infiltration

1. Inject the bacterial mix in the two bottom leaves of 15-day-old plants using a 1 mL syringe.
2. Cut the apex of the plants immediately after agroinfiltration (*see* **Note 6**).
3. For the *LusPDS* and *LusCESA* constructs, the silencing was effective in the stem and leaves 13 days after infiltration (*see* **Notes 7** and **8**) as illustrated in the Figs. 1 and 2.

Fig. 2 Effect of *LusCESA* silencing by VIGS. (**a**) Phenotype of wild-type and *LuCESA*-silenced plants. (**b**) Curly leaf phenotype on the silenced plants

4 Notes

1. The performance of transformation by *Agrobacterium* and regeneration of plants is highly dependent on flax cultivars. When stable transformation experiments are performed, the highest number of genetically modified plants is usually obtained for oil cultivars. We have previously tested VIGS with the *LusPDS* construct on the oleane oil cultivar, but no albino phenotype was observed. When the fiber cultivar Diane was tested, the percentage of plants carrying white leaves was lower compared to Barbara.
2. Phenol-containing solutions are easy to use and provide high amounts of RNA from most flax organs. When choosing this extraction protocol, it is important to use personal protective equipment and to work under a fume hood. An alternative method is the use of a RNA purification kit [12] which also provides high-quality RNA.
3. The use of a microvolume spectrophotometer is generally sufficient to quantify and control the purity of RNA used for cDNA synthesis.
4. The gene references are those of phytozome (https://phytozome.jgi.doe.gov). The two flax genes with the highest percentage of identity with the *Nicotiana PDS* sequence are

Table 1
Comparison of the *LusPDS* gene inactivation frequency (% of plants with silenced leaves) by vacuum and syringe infiltration

Duration of infiltration (min)	Number of plants	Inactivation frequency (%)
Vacuum infiltration		
2	39	71.79
5	38	65.79
Syringe infiltration		
–	39	84.62

Lus10021967 and Lus10041260. They are both 98.8% identical, but the second sequence has an additional stretch of 228 nucleotides.

5. The sequence of the *LusCESA* fragment used in this work has between 69% and 100% identity with the *LusCESA* genes involved in primary cell wall biosynthesis [11]. This ensures an effect on at least 11 genes essential for the cell wall construction in growing cells.
6. In order to increase the success rate of the silencing, it is absolutely essential to remove the apex of the plants immediately after the infiltration. If not, only few leaves and stem fragments will be receptive to the viral vectors.
7. This may vary depending on the construct. It may be useful to check the accumulation of the corresponding transcripts by semiquantitative RT-PCR or qRT-PCR.
8. In our hands, the performance of syringe infiltration was higher compared to the vacuum infiltration method (Table 1), but this may vary depending on the targeted gene but also on the cultivar. It may be useful to try both methods when designing your experiments on flax gene silencing.

References

1. Baulcombe DC (1999) Fast forward genetics based on virus-induced gene silencing. Curr Opin Plant Biol 2:109–113. https://doi.org/10.1016/S1369-5266(99)80022-3
2. Burch-Smith TM, Anderson JC, Martin GB, Dinesh-Kumar SP (2004) Applications and advantages of virus-induced gene silencing for gene function studies in plants. Plant J 39:734–746. https://doi.org/10.1111/j.1365-313X.2004.02158.x
3. Huis R, Morreel K, Fliniaux O et al (2012) Natural hypolignification is associated with extensive oligolignol accumulation in flax stems. Plant Physiol 158:1893–1915. https://doi.org/10.1104/pp.111.192328
4. Day A, Ruel K, Neutelings G et al (2005) Lignification in the flax stem: evidence for an unusual lignin in bast fibers. Planta 222:234–245. https://doi.org/10.1007/s00425-005-1537-1

5. Caillot S, Rosiau E, Laplace C, Thomasset B (2009) Influence of light intensity and selection scheme on regeneration time of transgenic flax plants. Plant Cell Rep 28:359–371. https://doi.org/10.1007/s00299-008-0638-2
6. Ratcliff F, Martin-Hernandez AM, Baulcombe DC (2001) Technical Advance. Tobacco rattle virus as a vector for analysis of gene function by silencing. Plant J 25:237–245. https://doi.org/10.1046/j.0960-7412.2000.00942.x
7. Brunt AA (Alan A), International CAB (1996) Viruses of plants: descriptions and lists from the VIDE database/edited by Alan Brunt ... [et al.]. CAB International, Wallingford, Oxon, UK
8. Qin G, Gu H, Ma L et al (2007) Disruption of phytoene desaturase gene results in albino and dwarf phenotypes in Arabidopsis by impairing chlorophyll, carotenoid, and gibberellin biosynthesis. Cell Res 17:471–482. https://doi.org/10.1038/cr.2007.40
9. Yamagishi M, Masuta C, Suzuki M, Netsu O (2015) Peanut stunt virus-induced gene silencing in white lupin (*Lupinus albus*). Plant Biotechnol 32:181–191. https://doi.org/10.5511/plantbiotechnology.15.0521a
10. Wang Z, Hobson N, Galindo L et al (2012) The genome of flax (*Linum usitatissimum*) assembled de novo from short shotgun sequence reads. Plant J 72:461–473. https://doi.org/10.1111/j.1365-313X.2012.05093.x
11. Chantreau M, Chabbert B, Billiard S et al (2015) Functional analyses of cellulose synthase genes in flax (*Linum usitatissimum*) by virus-induced gene silencing. Plant Biotechnol J 13:1312–1324. https://doi.org/10.1111/pbi.12350
12. Galindo-González L, Deyholos MK (2016) RNA-seq transcriptome response of flax (*Linum usitatissimum* L.) to the pathogenic fungus *Fusarium oxysporum* f. sp. lini. Front Plant Sci 7:1–22. https://doi.org/10.3389/fpls.2016.01766

Chapter 7

Virus-Induced Gene Silencing to Investigate Alkaloid Biosynthesis in Opium Poppy

Rongji Chen, Xue Chen, Jillian M. Hagel, and Peter J. Facchini

Abstract

Virus-induced gene silencing (VIGS) enables the targeted silencing of genes in opium poppy (*Papaver somniferum*) and has been used extensively to determine or support the physiological functions of benzylisoquinoline alkaloid biosynthetic enzymes. Here we describe detailed protocols involved in the application of VIGS to investigate BIA metabolism in opium poppy.

Key words *Papaver somniferum*, Benzylisoquinoline alkaloid, Biochemical genomics, Plant-specialized metabolism, Plant metabolic engineering

1 Introduction

Virus-induced gene silencing (VIGS) has been extensively used in a wide range of plant species for the transient suppression of gene expression in order to study the function of target genes. Compared with gene-silencing techniques mediated by stable genetic transformation, VIGS provides a time-efficient alternative, especially in plant species that are recalcitrant to the regeneration of intact transgenic plants. VIGS uses an intrinsic plant defense mechanism, mediated by the short interfering RNA (siRNA)/RNase complex, which responds to intracellular viral proliferation and extracellular viral movement [1]. Host gene silencing is achieved via posttranscriptional RNA degradation resulting from recognition and targeting of a specific region of the host mRNA by the siRNA/RNase complex, subsequent to infection of a plant cell by a recombinant virus harboring a corresponding host gene fragment. Despite a small number of plant DNA virus-derived VIGS systems, most VIGS systems originate from positive-strand RNA viruses, such as *Tobacco mosaic virus* (TMV), *Tobacco rattle virus* (TRV), and *Potato virus X* (PVX) [2–4]. The TRV-based VIGS system not only has been used successfully in model plants such as *Arabidopsis*

Vincent Courdavault and Sébastien Besseau (eds.), *Virus-Induced Gene Silencing in Plants: Methods and Protocols*, Methods in Molecular Biology, vol. 2172, https://doi.org/10.1007/978-1-0716-0751-0_7,

and tobacco but has also been adapted to non-model plant species including several basal eudicot species, such as *Aquilegia vulgaris* (common columbine), *Eschscholzia californica* (California poppy), *Papaver somniferum* (opium poppy), and *Thalictrum* spp. (meadow rue) [5–8]. In the TRV system, two binary T-DNA vectors were engineered to independently carry half of the bipartite viral genome [3]. Both vectors are required for plant infection: the pTRV1 vector contains the cDNA of viral RNA1 encoding proteins involved in viral replication and movement, whereas the pTRV2 vector carries the cDNA clone of viral RNA2 encoding the coat protein required for virion formation, as well as a multiple cloning site for insertion of a selected partial sequence of the targeted host gene. Both genome RNAs are under the control of the duplicated *Cauliflower mosaic virus* (CaMV) 35S promoter for strong and constitutive expression. *Agrobacterium*-mediated T-DNA transformation is used to deliver the TRV components into a plant cell, which initiates viral infection that spreads systematically throughout the plant.

Opium poppy is an ancient medicinal plant, which remains the sole commercial source of the narcotic analgesics morphine and codeine and has been comprehensively investigated by our group in the context of benzylisoquinoline alkaloid (BIA) biosynthesis. In addition to morphine and codeine, other pharmacologically important BIAs produced by opium poppy include thebaine, the precursor to several semisynthetic opiates such as oxycodone and naloxone, the cough suppressant and potential anticancer drug noscapine, and the vasodilator papaverine. For the functional characterization of BIA biosynthetic pathways in opium poppy, VIGS provides essential in planta evidence in support of in vitro enzyme assays and other lines of evidence [9–11]. In a typical VIGS experiment, the transcript level of a target gene will be determined by qRT-PCR in positively infiltrated plants, followed by BIA metabolic profiling using liquid chromatography–mass spectrometry (LC–MS; Fig. 1). Several factors can affect the BIA profile in VIGS-treated plants. Depending on the suppression level of the target gene, the chemical stability of the relevant pathway intermediates and their interactions with the associated catalytic enzymes (e.g., product inhibition), and other complicated yet often unknown metabolic regulation mechanisms, downregulation of a metabolic pathway gene will normally lead to an increased accumulation of upstream metabolic intermediates and/or a reduced level of downstream products. In addition, VIGS can also be used as a tool to investigate the function of uncharacterized gene candidates, as modulations in the BIA profile of VIGS-treated plants provide valuable clues to guide further functional studies of selected candidate genes [12].

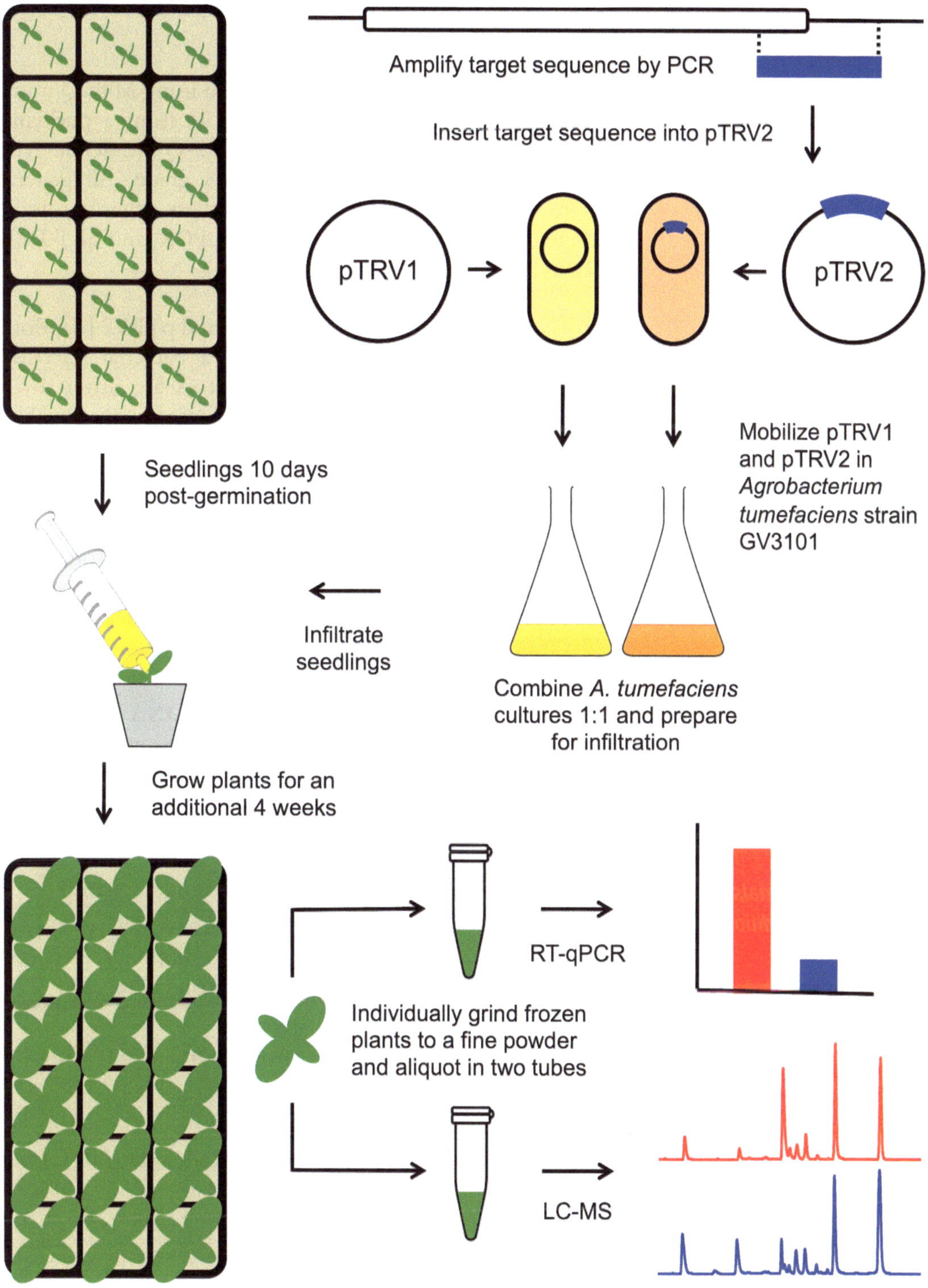

Fig. 1 VIGS workflow in opium poppy using *Tobacco rattle virus* (TRV) vectors. Ten-day-old seedlings are infiltrated with a mixed culture of *A. tumefaciens* harboring pTRV1 and pTRV2, the latter containing a fragment of the target gene. Subsequently, total RNA and metabolite extracts from approximately 6-week-old plants are subjected to RT-qPCR and HPLC analysis, respectively

2 Materials

All solutions for analytic purposes are prepared using Milli-Q water with a sensitivity of 18.7 MΩ. cm at 25 °C from a purifying deionized water system (e.g., Millipore Milli-Q Academic A10 System). Reagents are analytical or LC-MS grade. All general reagents are stored at room temperature unless otherwise indicated. Reagents and kits for cDNA synthesis, PCR, and qRT-PCR are stored at −20 °C.

2.1 Vectors

The system consists of binary vectors pTRV1 and pTRV2 [3]. Both carry a kanamycin-resistant gene used as a bacterial selection marker. The viral genome on the two vectors are driven by a duplicated CaMV 35S promoter and terminated with nopaline synthase gene.

1. pTRV1: this vector contains genes coding for replicase, movement, and cysteine-rich proteins.
2. pTRV2: the vector harbors genes encoding a coat protein and two nonstructural proteins. It has a multiple cloning site (MCS) for inserting the target gene/genes. pTRV2 is used to build various pTRV2 constructs containing fragments of the target gene/genes.
3. pTRV2-PapsPDS′: PapsPDS′ is a 436-bp cDNA fragment of *P. somniferum* phytoene desaturase (PapsPDS) gene [5]. This construct is served as a positive control through observing photo-bleaching phenotypes for leaves and floral organs due to silencing of the *PapsPDS* gene.

2.2 Plant Materials and Growth Conditions

1. Seeds of *P. somniferum* cultivar Bea's Choice (*see* **Note 1**).
2. Potting soil mixture: The mixture consists of 32% (v/v) peat moss, 14% (v/v) Turface MVP turf conditioner (calcined non-swelling illite clay), 27% (v/v) vermiculite, and 27% (v/v) perlite.
3. Pots (8 × 8 × 11 cm), pot trays (6 × 3 pots), and transparent tray lids.
4. Plant fertilizer (N-P-K = 14-14-14).
5. Growth chamber with a photoperiod of 16 h light (20 °C) and 8 h dark (18 °C).

2.3 Transformation of A. tumefaciens

1. *A. tumefaciens* strain GV3101 [13].
2. LB liquid medium: weigh out 10 g of tryptone, 5 g of yeast extract, and 10 g of NaCl in a 1 L glass bottle. Add 900 mL of distilled water and stir until fully dissolved. Adjust the pH to 7.0 with 1 N NaOH. Bring up to 1 L with distilled water. Sterilize by autoclaving. For LB solid medium, add 15 g of agar to 1 L of LB liquid medium, and mix before autoclaving.

3. 20 mg/mL Fentamicin stock solution: dissolve gentamycin sulphate (stored at −20 °C) in sterile Milli-Q water. Sterilize by passing through a 0.22 μm syringe filter. Keep the stock solution at −20 °C for up to 1 year.
4. 50 mg/mL Kanamycin stock solution: dissolve kanamycin in sterile water. Sterilize by passing through a 0.22 μm syringe filter. Keep the stock solution at −20 °C for up to 1 year.
5. Sterile 10% (v/v) glycerol.
6. High-speed centrifuge.
7. Incubating orbital shaker.
8. Liquid nitrogen.
9. Freezer (−80 °C).
10. Electroporation system and electroporation cuvettes (e.g., 0.2 cm–gap).
11. Bacterial cell spreaders.

2.4 Agroinfiltration

1. 1 M MES buffer pH 5.6: dissolve 21.3 g of [2-(N-morpholino) ethanesulfonic acid] monohydrate in 100 mL water. Adjust the pH to 5.6 with 10 N NaOH. Sterilize by passing through a 0.22 μm filter (*see* **Note 2**).
2. 1 M $MgCl_2$ solution: dissolve 20.3 g of [magnesium chloride] hexahydrate in 100 mL water. Sterilize by autoclaving.
3. 100 mM Acetosyringone (3′,5′-Dimethoxy-4-′-hydeoxyacetophenone): dissolve 19.62 mg acetosyringone (stored at −20 °C) in 985 μL of dimethyl sulfoxide (DMSO). Prepare 100 mM acetosyringone fresh just before use.
4. LB induction medium: to 100 mL of LB liquid medium, add 100 μL of 50 mg/mL kanamycin stock solution, 1 mL of 1 M MES buffer (pH 5.6), 1 mL of 1 M $MgCl_2$, and 20 μL of 100 mM acetosyringone (prepared fresh).
5. Infiltration solution: for 100 mL, add 1 mL of 1 M MES (pH 5.6), 1 mL of 1 M $MgCl_2$, and 200 μL of 100 mM acetosyringone (prepared fresh just before use). Bring up to 100 mL with sterile Milli-Q water. Prepare fresh infiltration solution.
6. Syringes (5 mL).

2.5 Preparation of Plant Tissue Samples

1. 1.7 mL Microcentrifuge tubes.
2. Sterile 2.0 mL SafeSeal microcentrifuge tubes.
3. Mortars and pestles.
4. Spatulas.
5. Liquid nitrogen.
6. Freezer (−80 °C).

2.6 RNA Isolation and Quantification

1. DEPC-treated water: add 1 mL of DEPC (diethyl pyrocarbonate) to 1 L of Milli-Q water. Incubate at 37 °C for 12 h. Sterilize by autoclaving.
2. RNA extraction buffer: 2% (w/v) CTAB, 1% (w/v) PVP (MW 40,000), 100 mM Tris–HCl pH 8.0, 20 mM EDTA, 1.4 M NaCl, and 2% ß-mercaptoethanol. Prepare the RNA extraction buffer without ß-mercaptoethanol. Sterilize by autoclaving. Add ß-mercaptoethanol to a final concentration of 2% just before use.
3. 10 M LiCl: prepare with DEPC-treated water. Sterilize by autoclaving.
4. 24:1 (v/v) Chloroform/isoamyl alcohol.
5. Dry ice.
6. Mixer mill and adapter sets.
7. Water bath.
8. Vortex mixer.
9. Refrigerated microcentrifuge.
10. Vacuum concentrator.
11. NanoDrop spectrophotometer.
12. Freezer (−80 °C).
13. Sterile 1.7 mL microcentrifuge tubes, 200 and 1000 μL tips (*see* **Note 3**).
14. 50× TAE stock solution: dissolve 242 g of Tris base in distilled water, and add 57.1 mL of glacial acetic acid and 100 mL of 0.5 M EDTA (pH 8.0). Bring up to 1 L with distilled water.
15. 10 mg/mL Ethidium bromide solution: dissolve ethidium bromide in sterile Milli-Q water.
16. 1% Agarose-TAE gel: weigh out 1 g of agarose into a flask, and add 100 mL of 1× TAE. Heat until agarose is completely dissolved. Cool down to 60 °C, and add 3 μL of 10 mg/mL ethidium bromide. Mix and pour into a gel tray.
17. 6× Agarose gel loading dye: 60% glycerol, 10 mM Tris–HCl pH 7.6, 60 mM EDTA, 0.03% bromophenol blue, and 0.03% xylene cyanol FF. Dissolve bromophenol blue and xylene cyanol FF in Tris–HCl (pH 7.6) and water. Add glycerol at last.
18. Gel electrophoresis apparatus.

2.7 cDNA Synthesis

1. Thin wall 0.2 mL PCR tubes.
2. cDNA synthesis kit.
3. PCR thermal cycler.

2.8 PCR Screening

1. Primers flanking the fragment cloned in pTRV2:

   pTRV2 forward 5′-GGTCAAGGTACGTAGTAGAG-3′.

   pTRV2 reverse 5′-CGAGAATGTCAATCTCGTAGG-3′.
2. Thin-wall 0.2 mL PCR tubes.
3. PCR reagents: 10 mM dNTPs, 10× Taq buffer, and Taq DNA polymerase (5 U/μL).
4. PCR thermal cycler.
5. Gel electrophoresis apparatus and solutions.

2.9 qRT-PCR

1. Gene-specific real-time PCR primers designed to exclude the sequence cloned in pTRV2.
2. Primers for endogenous reference genes:

   GAPDH_forward: 5′-CTCATTTGAAGGGTGGAGC-3′.

   GAPDH_reverse: 5′-GTCATTGCGTGGACAGTGG-3′.

   Ubiquitin_forward: 5′-CCATTTGGTGCTTCGTCTAC-3′.

   Ubiquitin_reverse: 5′-CAAGCCATAGCTGAAACACC-3′.
3. 2× SYBR Green Master mix.
4. 96-Well reaction plates and adhesive film kit adapted for qPCR.
5. Real-time PCR thermal cycler.

2.10 Alkaloid Extraction and LC-MS/MS Analysis

1. Freeze-dryer.
2. 50% Methanol/50% acetonitrile (v/v) mixture.
3. Orbital shaker.
4. Microcentrifuge.
5. Vials for LC MS/MS.
6. Solvent A [10 mM ammonium acetate (pH 5.5), 5% (v/v) acetonitrile]: to prepare 1 L of solvent A, dissolve 770.8 mg of ammonium acetate in 950 mL of LC-MS-grade water, and adjust the pH to 5.5 with concentrated acetic acid. Add 50 mL of 100% acetonitrile, and mix the solution thoroughly. Filter the solution through a 0.22 μm membrane filter.
7. Solvent B (100% acetonitrile).
8. HPLC coupled to a triple-quadrupole MS.
9. HPLC C18 column.
10. Compressed nitrogen gas.

3 Methods

3.1 Building of pTRV2 Constructs with Target Gene Fragments

1. To silence a unique gene in the genome, select a specific region of the target gene that minimizes suppression of the other homologous genes. To silence multiple members simultaneously for a gene family in the genome, choose a conserved region of the target gene that maximizes suppression of members of the gene family. Alternatively, select a short (longer than 70 bp) very specific region for each member of the gene family, and build a chimeric construct for silencing multiple members simultaneously for a gene family in the genome [14].
2. Synthesize or amplify by PCR the selected regions of the target gene/genes, and clone the target sequence into pTRV2.
3. Purify pTRV2 constructs, and dissolve the purified pTRV2 constructs in sterile Milli-Q water to a final concentration of 80 ng/μL, and store at −20 °C until use.

3.2 Preparation of Opium Poppy Seedlings for Agroinfiltration

1. Fill growth pots with potting soil mixture.
2. Saturate the potted soil mixture with tap water.
3. Assemble 18 filled pots in a pot tray, cover the tray with a transparent lid, and keep the pots at room temperature for 1–2 days.
4. Plant 15–20 seeds in each pot with the moist soil mixture (*see* **Note 4**).
5. Water the pots (with sown seeds) gently and carefully.
6. Germinate the seeds in a growth chamber. Cotyledons should start to emerge from the soil 4–5 days after seeding.
7. Continue to grow the seedlings in the same growth chamber for about 10 days to the 2-true leaf stage for agroinfiltration (*see* **Note 5**).
8. For agroinfiltration, select five seedlings at the 2-true leaf stage for each pot.

3.3 Preparation of Electrocompetent A. tumefaciens *GV3101 [15, 16]*

1. Streak out strain GV3101 from a frozen glycerol stock onto a LB plate with 40 μg/mL gentamicin, and incubate the plate at 28 °C for 2–3 days to form single colonies.
2. Inoculate a single colony from the plate into 20 mL of LB broth with 40 μg/mL gentamycin, and incubate the culture overnight at 28 °C with shaking at 200 rpm. Start this culture early in the morning.
3. Subculture 5 mL of the overnight-grown culture into 500 mL of LB broth without antibiotics early in the morning.

4. Incubate the 500 mL culture at 28 °C with shaking at 200 rpm until $OD_{600} = 0.6$–0.8. It will take about 7 h to reach this cell density.
5. Transfer the culture from the flask to centrifuge bottles and chill on ice for 20 min.
6. Centrifuge at 4000 × *g* for 10 min at 4 °C (*see* **Note 6**).
7. Discard supernatant, and resuspend cells thoroughly in 10–20 mL of sterile ice-cold Milli-Q water.
8. Add sterile ice-cold Milli-Q water to a final volume of 200 mL and mix thoroughly.
9. Repeat **steps 6–8** to wash the cells twice.
10. Repeat **step 6** to pellet the cells, and resuspend the cells in 5 mL of sterile ice-cold 10% glycerol.
11. Dispense the cells in 50 μL aliquots, freeze the aliquots in liquid nitrogen, and store them at −80 °C (*see* **Note 7**).

3.4 Transformation of A. tumefaciens GV3101 with pTRV Vectors

1. Cool an electroporation cuvette, and thaw an aliquot (50 μL) of competent *Agrobacterium tumefaciens* cells on ice.
2. Once the competent cells are thawed, add 1 μL of vector at 80 ng/μL (pTRV1, pTRV2, pTRV2 construct, or pTRV2-PapsPDS′) to the cells, and mix the suspension gently with a pipette.
3. Place the suspension at the bottom of the cuvette between the aluminum plates, and let it sit on ice for 5 min.
4. Turn on electroporator, and set the pulse controller resistance at 200 Ω for the low range and 500 Ω for the high range, capacitance at 25 μF, and voltage at 2.5 kV.
5. Insert the cuvette into the cuvette slide. Push the slide to make the cuvette contact the chamber electrodes.
6. Press both pulse buttons simultaneously and hold until a beep tone sounds. Release both pulse buttons when the tone sounds. The time constant should be less than 1 s.
7. Immediately, add 1 mL of LB broth without antibiotic to the cuvette, and transfer the suspension to a 1.7 mL sterile microcentrifuge tube.
8. Incubate the suspension with slow shaking (50 rpm) at 28 °C for 1 h.
9. Plate out a portion of the cell culture to LB-agar plates containing 50 μg/mL kanamycin in different volumes (*see* **Note 8**).
10. Incubate the plates at 28 °C for 2–3 days to obtain single colonies (*see* **Note 9**).

3.5 Agroinfiltration of Opium Poppy Plants

It takes about 10 days to get plants ready after the cotyledons emerge from the soil. To prepare plants for infiltration, grow opium poppy seedlings to the 2-true leaf stage, and water the plants in the morning on the infiltration day. To perform the infiltration with the plants, it takes 3 days to complete all procedures including preparing *Agrobacterium* strains and conducting the infiltration. A typical infiltration experiment consists of pTRV1-containing GV3101 coinfection strain, pTRV2 construct-containing GV3101 test strain, empty vector pTRV2-containing GV3101 negative control, strain and pTRV2 PapsPDS′-containing GV3101 positive control strain. The following preparation for a test/control strain is sufficient to infiltrate more than 100 seedlings. Kanamycin is used at the final concentration of 50 μg/mL when addition is indicated.

1. Select single colonies, and streak out *A. tumefaciens* carrying pTRV1, pTRV2 constructs, pTRV2, and pTRV2 PapsPDS' on LB plates containing kanamycin, and incubate the plates at 28 °C for 2–3 days to form single colonies.
2. Initiate starter cultures at noon on day 1 by inoculating a single colony into 2 mL of LB broth containing kanamycin. Grow the cultures with shaking (200 rpm) at 28 °C overnight (*see* **Note 10**).
3. Subculture 0.5 mL of each starter culture into 100 mL of LB induction medium containing kanamycin at noon on day 2. Grow the cultures with shaking (200 rpm) at 28 °C overnight.
4. Harvest the bacteria in the morning (~10 am) on day 3 by centrifuging the cultures at 3000 × *g* for 15 min at room temperature.
5. Discard the media, and resuspend the pellet in freshly prepared infiltration solution to an OD_{600} of 18 for each bacterial strain (*see* **Note 11**).
6. Mix pTRV1-containing GV3101 coinfection strain with a test/control strain at a ratio of 1:1(v/v) (*see* **Note 12**).
7. Incubate the cell mixture on bench at room temperature for 2–3 h before infiltration.
8. Infiltrate the cell mixture manually into the apical meristem of the 2-true leaf seedlings using a 5-mL needleless syringe (*see* **Note 13**).
9. Place pots with the infiltrated plants in a tray, and cover the tray with a transparent lid to minimize moisture loss from the infiltrated plants. Do not water the infiltrated plants for 24 h post-infiltration.

10. Grow the infiltrated plants in the same growth chamber for 30–32 days. Water the plants to keep appropriate moisture. Apply fertilizer to the plants 14 days post-infiltration.
11. Harvest whole young plants including roots and clean the soil with tap water. Wrap the plant samples with aluminum foil, and immediately place the samples in an insulated foam container filled with dry ice.
12. Store the samples at −80 °C until use.

3.6 Preparation of Samples for RNA and Alkaloid Extraction

1. Weigh a clean 1.7 mL microcentrifuge tube and a clean sterile 2.0 mL SafeSeal microcentrifuge tube. Record and label their weights on the tubes.
2. Freeze a mortar and pestle by adding liquid nitrogen to the mortar with pestle.
3. Grind a young plant to fine powder under liquid nitrogen using the frozen mortar and pestle.
4. Freeze two microcentrifuge tubes and a metal spatula in liquid nitrogen.
5. Using the frozen spatula, transfer approximately 150 mg of the ground powder to the 2.0 mL SafeSeal microcentrifuge. This sample will be used for RNA isolation.
6. Using the frozen spatula, transfer 100–200 mg of the ground powder to the 1.7 mL microcentrifuge tube. This sample is for alkaloid extraction.
7. Store the samples at −80 °C until use (*see* **Note 14**).

3.7 Determination of Gene-Silencing Efficiency

3.7.1 RNA Isolation and Quantification [17, 18]

To determine the degree of gene silencing in VIGS-treated plants, it is essential to isolate high-quality total RNA. Many methods and commercial extraction kits are available. The following method used in our lab consistently produces high-quality RNA for VIGS analysis.

1. Chill the mixer mill adapter set (grinding block) and grinding beads at −80 °C for 30 min.
2. Take out the chilled adapter set and grinding beads and put them on dry ice.
3. Take out the samples stored at −80 °C (from Subheading 3.6, **step** 7), and place the samples in the tube holder of the adapter set.
4. To each sample tube, add two chilled grinding beads. Chill the adapter set with the samples and beads again at −80 °C for 20 min.
5. Prepare the RNA extraction buffer by adding ß-mercaptoethanol to the RNA extraction buffer without

ß-mercaptoethanol to a final concentration of 2% (*see* **Note 15**).

6. Prewarm the RNA extraction buffer by incubating in a 65 °C water bath.
7. Take out the adapter set loaded with samples (stored at −80 °C), and install the set on the clamps of the mixer mill.
8. Grind the samples for 1 min at the oscillation frequency of 30 Hz (1/s).
9. Immediately, put the ground samples on dry ice to keep them frozen.
10. To each ground sample (~150 mg) in the tube, add 600 μL of the prewarmed RNA extraction buffer. Incubate the sample with beads at 65 °C for 10 min with occasional vertex.
11. Take the sample out from the 65 °C water bath, and add 600 μL of chloroform/isoamyl alcohol (24:1, v/v). Invert the tube vigorously and vortex.
12. Spin the sample at 17,000 × *g* for 10 min at 4 °C. Recover approximately 600 μL of the supernatant using a 200 μL pipette, and transfer the supernatant to a sterile 1.7 mL microcentrifuge tube (*see* **Note 16**).
13. To the supernatant in the tube, add 600 μL of chloroform/isoamyl alcohol (24:1, v/v). Invert the tube vigorously and vortex.
14. Spin the sample at 17,000 × *g* for 10 min at 4 °C and recover the supernatant.
15. Add 0.4 volumes of 10 M LiCl to the supernatant, and mix the sample well (*see* **Note 17**).
16. Incubate the sample at 4 °C for 1 h.
17. Spin the sample at 17,000 × *g* for 20 min at 4 °C, and then carefully remove all the supernatants.
18. Wash the pellet (RNA) three times with cold 70% ethanol. For each wash, spin the sample at 17,000 × *g* for 10 min at 4 °C, and remove all the supernatants.
19. Dry the pellet (RNA) in Speed Vac Vacuum Concentrator without heating for about 10 min (*see* **Note 18**).
20. Resuspend the RNA pellet in 50 μL of sterile Milli-Q water to dissolve RNA (*see* **Note 19**).
21. Use 2 μL of the RNA preparation to estimate the RNA concentration/quantity on a NanoDrop spectrophotometer. The recommended RNA concentration is above 500 ng/μL. Assess RNA purity by determining the ratio of A260/A280 and the ratio of A260/A230. Both ratios should be very close to, or greater than, 2.0.

22. Check RNA integrity by running the RNA samples on a 1% agarose TAE gel and examining the ribosomal RNA (rRNA) bands. Detecting bands at high molecular weight indicates DNA contamination (*see* **Note 20**).

3.7.2 cDNA Synthesis from Total RNA Preparation

1. Synthesize cDNA from 1 μg of total RNA preparation using a reverse transcription kit including a balanced ratio of Oligo (dT)s and random primers kit following the manufacturer's instructions.
2. Use a PCR thermal cycler to conduct the reaction. Start the reaction by incubating the mixture for 10 min at 25 °C, followed by 50 min synthesis at 42 °C. Terminate the reaction by heating at 85 °C for 5 min, and then cool down to 4 °C.
3. Finalize the cDNA preparation by adding 90 μL of sterile Milli-Q H_2O to 10 μL reaction mixture.
4. Store the cDNA preparation at −20 °C for use.

3.7.3 PCR Screening for Samples Containing pTRV2-Derived Transcripts

1. Set up PCR reaction in a total volume of 10 μL. It contains 0.2 μL of 10 mM dNTP, 1 μL of 10× Taq buffer, 0.1 μL of 10 μM pTRV2 forward primer, 0.1 μL of 10 μM pTRV2 reverse primer, 0.08 μL of Taq DNA polymerase (5 U/μL), 8 μL of cDNA preparation, and 0.52 μL of H_2O.
2. Perform PCR by incubating the reaction for 2 min at 94 °C followed by 30 cycles of amplification (20 s at 94 °C followed by 30 s at 52 °C and then 1 min at 72 °C) and 5 min at 72 °C for final extension. At the end of PCR, the products are cooled down to 4 °C.
3. Run the PCR products on 1.7% agarose TAE gels for selecting the desired cDNA samples that show the expected PCR amplicon size.

3.7.4 Evaluation of the Gene Silencing by qRT-PCR

Gene silencing leads to reduced gene expression by degrading the target gene transcripts. Quantitative real-time PCR (qRT-PCR) is used to analyze the relative transcript abundance for a target gene. For qRT-PCR, follow the principles of MIQE guidelines [19].

1. Follow good guidelines to design qRT-PCR primers [20].
2. GAPDH and ubiquitin are used as stable endogenous reference genes [11, 21].
3. Use sterile Milli-Q H_2O to dilute the cDNA preparation with the dilution ratio of 1:1 (*see* **Note 21**).
4. Set up qRT-PCR reaction in a total volume of 10 μL. It contains 5 μL of a 2× SYBR Green Master Mix, 0.5 μL of 10 μM forward primer, 0.5 μL of 10 μM reverse primer, and 4 μL of the cDNA preparation (1:1 dilution). Do sufficient technical replicates for each sample.

5. PCR amplification is initiated by 2 min incubation at 50 °C followed by 2 min incubation at 95 °C and then 40 cycles of amplification (1 s at 95 °C followed by 30 s at 60 °C). The melt curve is established by 1 s at 95 °C, 30 s at 60 °C, and then slowly increasing to 95 °C at the rate of 0.15 °C/S.
6. Evaluate the degree of gene silencing by comparing the relative transcript abundance in the VIGS-treated plants with that in the control plants by using the comparative C_T method ($\Delta\Delta C_T$).

3.8 LC-MS/MS Alkaloid Analysis

3.8.1 Alkaloid Extraction and Sample Preparation for LC-MS/MS Analysis

1. Turn on the freeze dryer and the freezer inside the freeze-dryer, and wait until the temperature is below −50 °C.
2. Take the samples (in a cardboard box) out of the −80 °C freezer (from Subheading 3.6, **step** 7), and snap freeze the samples by pouring liquid nitrogen into the box.
3. Open the lid for each tube and place the samples into the freeze-dryer chamber.
4. Turn on the freeze-dryer's vacuum pump. Close all valves on the chamber, and seal the chamber completely.
5. Dry the samples for 48 h below −60 °C.
6. Take the dried sample out of the freeze-dryer, and immediately weigh the samples (tube + dried sample).
7. To each 1 mg of a dried sample, add 40 μL of methanol/acetonitrile (50%/50%), and then vortex the samples.
8. Incubate the samples on a shaker to extract alkaloids at room temperature at 200 rpm for 4 h.
9. Store the alkaloid extracts at −20 °C until use.
10. Take the samples out of the −20 °C freezer, and centrifuge the samples at 21,000 × *g* at 4 °C for 40 min.
11. Recover the supernatants (alkaloid extracts), and use methanol/acetonitrile (50%/50%) mixture to dilute the alkaloid extracts with the dilution factor of 1:50 (*see* **Note 22**).
12. Use 10 μL of each 1:50 dilution alkaloid extract for LC-MS/MS analysis.

3.8.2 LC-MS/MS Analysis of Alkaloid Extracts [22]

Alkaloid extracts are analyzed using an HPLC coupled to a triple-quadrupole MS. We recommend the following procedures and conditions to perform analyses:

1. 10 μL of each sample (1:50 dilution) is injected.
2. Analytes are eluted on a gradient consisting of Solvent A and Solvent B at a flow rate of 0.6 mL/min. Solvent B is increased from 0% to 60% (v/v) over 8 min, reaches 99% at 10 min, and remains constant until 14 min. Subsequently, Solvent B is

reduced to 0% over 0.1 min and remains 0% for 3 min to re-equilibrate the column.

3. The mass analyzer operates in positive ion mode via electrospray ionization (ESI) source. Conditions for the source and interface are as follows: gas temperature 350 °C, gas flow rate 10 L/min, nebulizer gas pressure 50 psi, capillary voltage 4000 V, and fragmentor voltage 110 V.
4. Analyses are performed in full-scan mode, with quadrupole 1 and 2 set to radio frequency and the third quadrupole scanning from 200 to 700 *m/z*. Alkaloids are identified in extracted ion chromatograms (EIC) based on a comparison of retention times and collision-induced dissociation (CID) spectra with those of authentic standards. For CID spectra, collision energy of 25 eV is applied to quadrupole 2. Quadrupole 3 detects the resulting fragments by scanning the product ions from 40 to 700 *m/z*.
5. Multiple reaction monitoring (MRM) and ESI^+-MS/MS are used to analyze minor alkaloids due to the high sensitivity and selectivity of this method (*see* **Note 23**).

4 Notes

1. *P. somniferum* cv. Bea's Choice is a garden variety. It grows well in small pots and has high effectiveness and efficiency for the stable genetic transformation.
2. 1 M MES buffer (pH 5.6) stock is used to prepare LB induction medium containing kanamycin and infiltration solution. It is important to adjust the pH to 5.6 for 1 M MES buffer due to that acetosyringone will be inactivated at the pH 7.0.
3. We recommend using the RNase, DNase, and pyrogen-free microcentrifuge tubes and tips.
4. Adjust the number of seeds planted for each pot based on the actual germination rate.
5. *P. somniferum* seedlings have a narrow window for highly efficient VIGS. It is important to select correct stage seedlings. Determine when seedlings are ready for agroinfiltration by leaf stage and leaf morphology, not by the days after germination.
6. All operations are performed on ice or under 4 °C from this step onward.
7. Electrocompetent *A. tumefaciens* GV3101 cells can be stored at −80 °C for up to 1 year.
8. The purpose of plating out in different volumes is to obtain well-separated single colonies on plates. Plating 20 and 180 μL separately on to two plates usually gives good results.

9. Incubation at 28 °C for more than 3 days may form satellite colonies close to transformants.
10. Grow a start culture of pTRV1-containing GV3101 coinfection strain in equal culture volume for each test or control strain.
11. Do not overdilute the cells because the cell suspension actually injected into a seedling is *extremely* small in volume.
12. For example, if you have 10 mL of pTRV1-containing GV3101 coinfection strain ($OD_{600} = 18$) and 7.5 mL of empty vector pTRV2-containing GV3101 negative control strain ($OD_{600} = 18$), simply add 7.5 mL of pTRV1-containing GV3101 coinfection strain ($OD_{600} = 18$) to the 7.5 mL of pTRV2-containing GV3101 negative control strain ($OD_{600} = 18$) to get 15 mL of the cell mixture. Usually, 15 mL of the cell mixture is sufficient for infiltration of more than 100 seedlings.
13. To do infiltration with a 5-mL needleless syringe, pull the plunger to take in the cell mixture, then press the tip of the syringe against the apical meristem and leaves while simultaneously applying gentle counterpressure to the other side, and push the plunger gently to inject the cell mixture.
14. It is recommended to isolate RNA as soon as possible. In our lab, we usually isolate RNA in 2–3 h after grinding the samples.
15. A simple operation formula is to add 100 μL of β-mercaptoethanol to 5 mL of RNA extraction buffer or 200 μL of β-mercaptoethanol to 10 mL of RNA extraction buffer.
16. Do not disturb the interface during recovering the supernatant.
17. The final LiCl concentration should be about 2.9 M. LiCl efficiently precipitates RNA, but not DNA, protein, or carbohydrate. LiCl-precipitated RNA samples do not require further purification for use in cDNA synthesis reactions.
18. Do not overdry RNA samples. It is difficult to dissolve the overdried RNA samples in water.
19. After adding Milli-Q water to the dried RNA samples, let the samples stand on ice for 20–30 min, and then gently mix the samples, and spin briefly.
20. Formaldehyde gels are not required for this assessment. For good-quality RNA preparations, the intensity of the upper 28S rRNA band should be about twice that of the lower 18S rRNA band. In addition, the two rRNA bands should be crisp and tight. If the two rRNA bands have similar intensity, RNA degradation may have occurred. Smearing below the 18S

rRNA band implicates a poor integrity for the RNA preparation.

21. Adjust the dilution ratio to an appropriate ratio according to actual situations.
22. Appropriately diluting samples helps produce good signals and avoid column overload. Concentrations can range from picomolar (pM) to nanomolar (nM) for the full-scan mode and from femtomolar (fM) to nM for the MRM mode.
23. The MRM mode is powerful due to its high sensitivity and selectivity. In this mode, a precursor ion of interest is preselected in the first quadrupole. The ion is then fragmented in the second quadrupole (Q2). Finally, fragments (product ions) generated in Q2 are analyzed in the third quadrupole.

Acknowledgments

This work was supported by funds awarded through the Industrial Research Assistance Program (IRAP; Project 86155) operated by the National Research Council of Canada to Epimeron Inc.

References

1. Lange M, Yellina AL, Orashakova S et al (2013) Virus-induced gene silencing (VIGS) in plants: an overview of target species and the virus-derived vector systems. In: Becker A (ed) Virus-induced gene silencing, Methods in molecular biology (Methods and protocols), vol 975. Humana Press, Totowa, NJ
2. Kumagai MH, Donson J, Della-Cioppa G et al (1995) Cytoplasmic inhibition of carotenoid biosynthesis with virus-derived RNA. Proc Natl Acad Sci U S A 92:1679–1683
3. Liu Y, Schiff M, Marathe R et al (2002) Tobacco Rar1, EDS1 and NPR1/NIM1 like genes are required for N-mediated resistance to tobacco mosaic virus. Plant J 30:415–429
4. Lu R, Malcuit I, Moffett P et al (2003) High throughput virus-induced gene silencing implicates heat shock protein 90 in plant disease resistance. EMBO J 22:5690–5699
5. Hileman LC, Drea S, Martino G et al (2005) Virus-induced gene silencing is an effective tool for assaying gene function in the basal eudicot species *Papaver somniferum* (opium poppy). Plant J 44:334–341
6. Kramer EM, Holappa L, Gould B et al (2007) Elaboration of B gene function to include the identity of novel floral organs in the lower eudicot *Aquilegia* (Ranunculaceae). Plant Cell 19:756–766
7. Wege S, Scholz A, Gleissberg S et al (2007) Highly efficient virus-induced gene silencing (VIGS) in California poppy (*Eschscholzia californica*): an evaluation of VIGS as a strategy to obtain functional data from non-model plants. Ann Bot 100:641–649
8. Di Stilio VS, Kumar RA, Oddone AM et al (2010) Virus-induced gene silencing as a tool for comparative functional studies in *Thalictrum*. PLoS One 5:e12064
9. Chen X, Hagel JM, Chang L et al (2018) A pathogenesis-related 10 protein catalyzes the final step in thebaine biosynthesis. Nat Chem Biol 14:738–743
10. Farrow SC, Hagel JM, Beaudoin GAW et al (2015) Stereochemical inversion of (*S*)-reticuline by a cytochrome P450 fusion in opium poppy. Nat Chem Biol 11:728–732
11. Dang TTT, Chen X, Facchini PJ (2015) Acetylation serves as a protective group in noscapine biosynthesis in opium poppy. Nat Chem Biol 11:104–106
12. Winzer T, Gazda V, He Z et al (2012) A *Papaver somniferum* 10-gene cluster for synthesis of the anticancer alkaloid noscapine. Science 336:1704–1708

13. Van Larebeke N, Engler G, Holsters M et al (1974) Large plasmid in *Agrobacterium tumefaciens* essential for crown gall-inducing ability. Nature 252:169–170
14. Zhou B, Zeng L (2017) Elucidating the role of highly homologous *Nicotiana benthamiana* ubiquitin E2 gene family members in plant immunity through an improved virus-induced gene silencing approach. Plant Methods 13:59
15. Shen WJ, Forde BG (1989) Efficient transformation of *Agrobacterium* spp. by high voltage electroporation. Nucleic Acids Res 17:8385
16. Mersereau M, Pazour GJ, Das A (1990) Efficient transformation of *Agrobacterium tumefaciens* by electroporation. Gene 90:149–151
17. Meisel L, Fonseca B, González S et al (2005) A rapid and efficient method for purifying high quality total RNA from peaches (*Prunus persica*) for functional genomics analyses. Biol Res 38:83–88
18. Fleige S, Pfaffl MW (2006) RNA integrity and the effect on the real-time qRT-PCR performance. Mol Asp Med 27:126–139
19. Bustin SA, Benes V, Garson JA et al (2009) The MIQE guidelines: minimum information for publication of quantitative real-time PCR experiments. Clin Chem 55:611–622
20. Bustina S, Huggettb J (2017) qPCR primer design revisited. Biomol Detect Quantif 14:19–28
21. Wijekoon CP, Facchini PJ (2012) Systematic knockdown of morphine pathway enzymes in opium poppy using virus-induced gene silencing. Plant J 69:1052–6311
22. Dang TT, Onoyovwi A, Farrow SC et al (2012) Biochemical genomics for gene discovery in benzylisoquinoline alkaloid biosynthesis in opium poppy and related species. Methods Enzymol 515:231–266

Chapter 8

A Biolistic-Mediated Virus-Induced Gene Silencing in Apocynaceae to Map Biosynthetic Pathways of Alkaloids

Pamela Lemos Cruz, María Isabel Restrepo, Thomas Dugé de Bernonville, Audrey Oudin, Thibaut Munsch, Arnaud Lanoue, Sébastien Besseau, Lucia Atehortùa, Nathalie Giglioli-Guivarc'h, Nicolas Papon, Marc Clastre, Inês Carqueijeiro, and Vincent Courdavault

Abstract

Monoterpene indole alkaloids (MIAs) are specialized metabolites synthesized in many plants of the Apocynaceae family including *Catharanthus roseus* and *Rauvolfia* sp. MIAs are part of the chemical arsenal that plants evolved to face pet and herbivore attacks, and their high biological activities also confer pharmaceutical properties exploited in human pharmacopeia. Developing robust and straightforward tools to elucidate each step of MIA biosynthetic pathways thus constitutes a prerequisite to the understanding of Apocynaceae defense mechanisms and to the exploitation of MIA cytotoxicity through their production by metabolic engineering. While protocols of virus-induced gene silencing (VIGS) based on *Agrobacterium*-based transformation have emerged, the recalcitrance of Apocynaceae to this type of transformation prompted us to develop an universal procedure of VIGS vector inoculation. Such procedure relies on the delivery of the transforming plasmids through a particle bombardment performed using a biolistic device and offers the possibility to overcome host specificity to silence genes in any plant species. Using silencing of geissoschizine oxidase as an example, we described the main steps of this biolistic mediated VIGS in *C. roseus* and *R. tetraphylla*.

Key words VIGS, Silencing, Apocynaceae, Alkaloids, Gene validation

1 Introduction

From the analgesic morphine produced by *Opium poppy* to the antiasthma ephedrine extracted from *Ephedra* spp., the bioactive compounds produced by plants are countless. The capacity of plants to architecture and decorate molecules goes beyond our own imagination such as in Apocynaceae producing the valuable monoterpene indole alkaloids (MIAs). Remarkable plants of this family first include the Madagascar periwinkle (*Catharanthus roseus*), source of vinblastine and vincristine, two MIAs that

Vincent Courdavault and Sébastien Besseau (eds.), *Virus-Induced Gene Silencing in Plants: Methods and Protocols*, Methods in Molecular Biology, vol. 2172, https://doi.org/10.1007/978-1-0716-0751-0_8,

constitute the first natural drugs used in chemotherapy, with special emphases on the most severe types of cancer such as Hodgkin's lymphoma and lung and brain cancer. It also encompasses *Rauwolfia* sp. such as *R. serpentina* and *R. tetraphylla*, known as the devil pepper, which both synthesize the antiarrhythmic ajamaline accumulated in stem barks and roots but frequently substituted by the semisynthetic propyl derivative—prajmaline—due to its low availability [1, 2].

Over the last five decades, unraveling the metabolic pathways of MIAs in those plants constituted an outstanding scientific topic that challenged numerous researchers of the specialized metabolism field. Besides deciphering the role of specialized metabolisms in plant/environment interactions, it allowed understanding how plants recruited enzymes from primary metabolisms to synthesize secondary metabolites. Nowadays, such elucidations find a renewed interest with the last developments in synthetic biology allowing us to consider other "biological factories" for the production of these valuables molecules, including genetically modified yeast or bacteria [3–5]. Developing new approaches to produce compounds of interest constitutes a concrete alternative to the classical supply relying on extraction from plants notably for MIA production. At midterms, it may provide an answer to the potential supply problems that may come from natural resources endangering or climate changes, as illustrated with the anti-malaria artemisinin, previously extracted from *Artemisia annua*, which can now be obtained through the synthesis of its precursor artemisinic acid by both types of engineered microbes [3]. However, conferring to other organisms the capacity to produce compounds of interest on demand relies on the transfer of the ad hoc producing machineries and thus on the identification of each gene of the corresponding biosynthetic pathways.

Undeniably, development of OMIC's approaches and especially RNA-seq constitutes a turning point in biosynthetic gene identification by providing access to the sequences of all the genes simultaneously expressed in the different cells/organs of plants. Gathering gene sequences co-expressed in organs accumulating valuable molecules gives a potential access to candidate genes involved in their biosynthesis and in turn rises the demand on characterization tools to understand the role of these genes. Although mutant plants generated by chemical mutagenesis, T-DNA tagging [6–9], or RNA interference [10, 11] gave potent results, the laborious and time-consuming task of mutant population characterization [12] promotes the development of new strategies. In this context, transiently silencing genes through virus-induced gene silencing (VIGS) rises a major breakthrough, conferring the capacity of evaluating the consequences of gene downregulation in short time and thus quickly elucidating the function of numerous candidate genes [13].

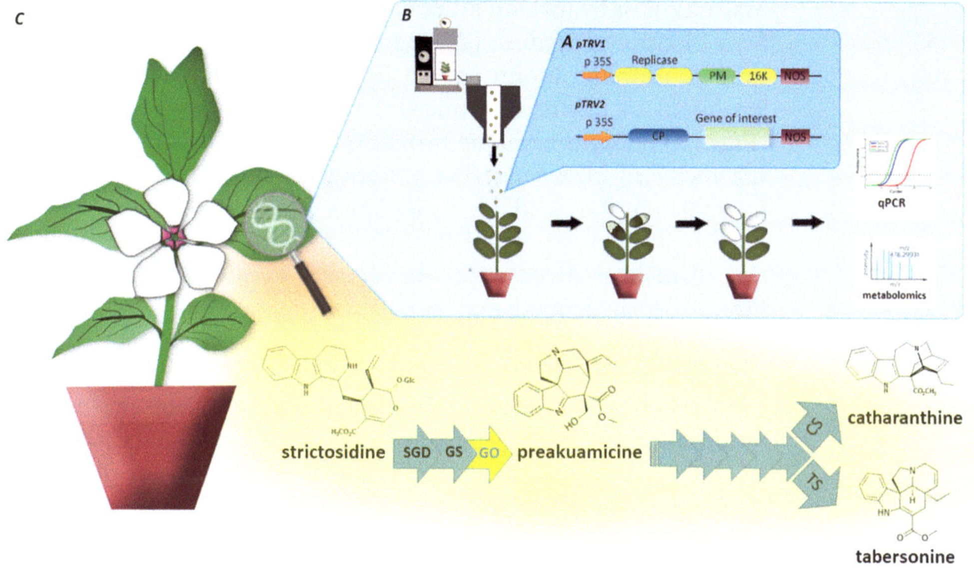

Fig. 1 Flowchart of TRV-mediated VIGS of the gene geissoschizine oxidase in *C. roseus*. From the viral vectors (**a**), the biolistic delivery (**b**) to the metabolic pathway (**c**)

From a technical point of view, VIGS deeds the capacity of plants' defense mechanisms to degrade double-stranded RNAs formed during invading virus replication [14]. It thus belongs to posttranscriptional gene-silencing approaches [14, 15]. Taking advantages of this process, VIGS involves the cloning of the viral component(s) in (binary) vectors to allow delivery in plants, under the control of a strong promoter such as CaMV35S for initial constitutive expression (Fig. 1a). A small fragment of the target plant gene (200–500 bp for instance) is inserted in the virus genome causing the degradation of the corresponding plant mRNA through virus RNA targeting by plant defense mechanisms [16–18]. Due to virus host specificities, different VIGS vectors based on distinct virus types have been developed over years expending the spectrum of plants potentially subjected to VIGS. Currently, VIGS vectors based on *Tobacco rattle virus* (TRV) are by far the most widely used. This is notably due to the high capacity of tissue infection including meristems with weak side effects [16, 19] and to the large array of TRV host plants ranging from *Arabidopsis* [20], *Solanaceae* sp. [17, 21], or more recalcitrant species such as plants from the Apocynaceae family [22, 23].

Delivery of the infectious material (from plasmids, RNA, to reformed viral particles) represents a critical step in the initiation of the VIGS procedure. Depending on virus and assayed plants, numerous techniques have been developed. For example, viral vectors can thus be transfected in plant cells mechanically [24], inoculated by *Agrobacterium* through vacuum/syringe leaf infiltration or by making holes with toothpick [13, 25] or delivered by a biolistic procedure [22, 23, 26]. Albeit more expensive and time-consuming, biolistic-mediated VIGS that involves the bombardment of plants with microparticles coated with viral vectors (Fig. 1b) presents a unique benefit relying on overcoming the host specificity of *Agrobacterium* resulting in the capacity to deliver transforming DNA/RNA in all plants. Besides being highly reproducible, it does not suffer from the pronounced defense reaction that plants may deploy after *Agrobacterium* infection [27]. Therefore, the biolistic-mediated inoculation of viral vectors constitutes a potent approach to perform VIGS in all plant species once efficient viruses identified for the targeted plant.

In the present report, we describe an efficient protocol of biolistic-mediated VIGS initially developed for the medicinal plants of the Apocynaceae family including *C. roseus*, *R. tetraphylla*, or *R. serpentina* and that can be transposed to many plants following minor adaptations. To describe protocol conduct, we illustrated how using this method we were able to uncover the missing enzymes catalyzing the ultimate steps to tabersonine and catharanthine biosynthesis, a story with more than 60 years that reveals the 31 enzymatic reactions needed to complete the anticancer MIA production [28, 29]. Through silencing of geissoschizine oxidase—GO (Fig. 1c)—in *C. roseus* leaves and using phytoene desaturase (PDS) as a reporter gene to visualize VIGS efficiency by leaf photobleaching, we highlighted the crucial steps of the whole silencing process from viral vector delivery up to analyses of gene silencing consequences.

2 Materials

2.1 Plant Materials

1. Seeds of *Catharanthus roseus* (Little Bright Eye) and *Rauwolfia tetraphylla*.
2. Greenhouse under controlled 16-h light/8-h dark cycle with white fluorescent light (intensity of 70 μmol/m^2/s).

2.2 Cloning in pTRV2

1. Liquid nitrogen.
2. Reverse transcriptase kit.
3. Micropipette set.
4. Sterile filter pipette tips.
5. PCR tubes.
6. Sterilized Milli-Q water.
7. 10 mM dNTP Mix.
8. Specific targeted gene oligonucleotides for PCR:

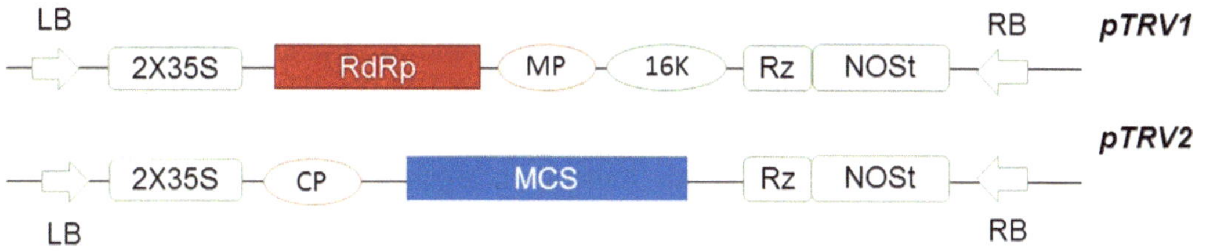

Fig. 2 *Tobacco rattle virus* (TRV)-based VIGS vectors. TRV cDNA clones were placed between duplicated 35S promoter (2×35S) and nopaline synthase terminator (NOSt) in a T-DNA vector. *RdRp* RNA-dependent RNA polymerase, *16K* 16 kDa cysteine-rich protein, *MP* movement protein, *CP* coat protein, *LB and RB* left and right borders of T-DNA, *Rz* self-cleaving ribozyme, *MCS* multiple cloning sites

   (a) PDS *C. roseus* forward: 5′-AGGTTTGGGGGTTTGTGT-3′; reverse: 5′-TACGCCTTGCTTTCTCATCC-3′.
   (b) PDS *R. tetraphylla* forward: 5′-AGGTTTGGGGGTTTGTGT-3′; reverse: 5′-TACGCCTTGCTTTCTCATCC-3′.
   (c) GO *C. roseus* forward: 5′-CTGAGAGGATCCTACAGTATGGCCCGA-3′; reverse: 5′-CTGAGAGGATC-CATCGTTAACAAGATGAGGAACCAAT-3′.

9. High-fidelity DNA polymerase.
10. Thermal cycler.
11. 50× TAE Buffer: 2 M Tris, 1 M glacial acetic acid, 50 mM EDTA, and pH 8.0.
12. 1% agarose gel: 1 g of agarose in 100 mL of 1× TAE buffer and 2 μg/mL ethidium bromide.
13. DNA electrophoresis equipment.
14. 6× Loading buffer: 0.25% (w/w) bromophenol blue, 40% (w/v) sucrose, and 0.25% (w/v) xylene-cyanol.
15. Spectrophotometer.
16. Gel extraction and PCR cleanup kit.
17. RNA purification kit.
18. TRV vectors, namely, pTRV1 and pTRV2-MCS, encoding the two genomic components of TRV (Fig. 2) (Source: Arabidopsis Biological Resource Center—ABRC; http://www.arabidopsis.org).
19. Restriction enzymes and buffers (including *BamHI*).
20. T4 DNA ligase and buffer.
21. *Escherichia coli* strain TOP10 chemically competent.
22. Luria-Bertani Broth: 10 g tryptone, 5 g yeast extract, and 10 g NaCl. Combine reagents, and dissolve in 800 mL deionized water, adjust pH to 7 using 1 M NaOH, and finally adjust volume to 1 L.

23. Luria-Bertani agar: 10 g tryptone, 5 g yeast extract, and 10 g NaCl. Combine reagents, and dissolve into 800 mL deionized water; adjust pH to 7 using 1 M NaOH. Add 20 g of agar and finally adjust volume to 1 L.
24. 50 mg/mL Kanamycin stock solution in sterilized Milli-Q water.
25. Incubator shaker at 37 °C, 200 rpm.
26. Centrifuge.

2.3 Particles Bombardment Procedure

1. Gold particles 1 μm.
2. 0.1 M Spermidine stock solution (*see* **Note 1**).
3. 2.5 M $CaCl_2$ stock solution.
4. Vortex.
5. Absolute ethanol (100%) and ethanol 70%.
6. Sonic bath.
7. Set of micro- and macro-carriers for Bio-Rad PDS1000/He delivery system Bio-Rad (1100-psi rupture disks).
8. Bio-Rad PDS1000/He delivery system Bio-Rad device.
9. Box for plants with acclimatization control.

2.4 qPCR Analysis

1. Quantitative PCR instrument.
2. Microcentrifuge.
3. qPCR SYBR Green Mix.
4. cDNA synthesis kit (reverse transcriptase).
5. RNA purification kit.
6. Oligonucleotides for qPCR:
   (a) qPDS *C. roseus* forward 5′-ATGCCCGTTGTTGATCA-TATTCGAT-3′; reverse 5′-ATCTCCTTTGATTGCTG ACCCATTA-3′.
   (b) qGO *C. roseus* forward: 5′-GCTGAGTTTATGTTGG CTGCTATGTT-3′; reverse: 5′-ATAGTTGGCAAAG ACAGACTAATCGT-3′.
   (c) qRSP9 *C. roseus* forward 5′-TTGAGCCGTATCAG AAATGC-3′; reverse 5′-CCCTCATCAAGCAGACCA TA-3′.
   (d) qPDS *R. tetraphylla* forward 5′-GATCGAGTAACTG ATGAGGTATTCA-3′; reverse 5′-AGCATGGTTCC AAAATGGCCTTTTTA-3′.
   (e) qActin *R. tetraphylla* forward 5′-TGACCTTGAAG-TACCCTATTGAGC-3′; reverse 5′-TGTACGACCAC TGGCATACAGAG-3′.
7. Sterile filter pipette tips.

8. Sterile 1.5 mL screw-top microcentrifuge tubes.
9. PCR tubes or 96 well plates with sealing films.
10. Sterilized Milli-Q water.
11. Microcentrifuge.

2.5 Alkaloid Analysis

1. Mixer mill.
2. Methanol supplemented with 0.1% of formic acid.
3. 90:10 (v/v) of water/acetonitrile (0.1% formic acid).
4. UPLC chromatography system coupled to a SQD mass spectrometer equipped with an electrospray ionization (ESI) source controlled by MassLynx 4.1 software (Waters, Milford, MA) with a C18 column (150 × 2.1 mm, i.d. 1.8 μm) at a stabilized flow rate of 0.4 ml/min at 55 °C.
5. The following multistep gradient was used: acetonitrile (B) versus water (A) system containing 0.1% (v/v) formic acid, at a flow rate of 0.1 mL/min, with 10:90:0.1 to 60:40:0.1 over 18 min.
6. QuanLynx™ software (Waters, UK).

3 Methods

3.1 Plant Germination and Growth Condition Pre-transformation

1. Seeds of *C. roseus* and *R. tetraphylla* are germinated at 28 °C in loam substrate, in greenhouse.
2. Once cotyledons are fully developed, plantlets are transferred to individual pots and cultivated in the same conditions until transformation.
3. For the VIGS procedure, the exact developmental stage of plantlets depends on plant species. For *C. roseus* and *R. tetraphylla*, the plants must have the first pair of leaves full developed and the second pair just emerged as previously accurated for both species [22, 23] (*see* Fig. 3; **Note 2**).

3.2 cDNA Synthesis of Target Genes

1. The total RNA from *C. roseus* and *R. tetraphylla* was isolated from leaf tissue using commercial kit following the instructions of the manufacturer.
2. Quantify total RNA with a spectrophotometer at 260/280 nm (*see* **Note 3**).
3. Analyze the integrity of the RNA by migration on 1% agarose gel.
4. The cDNA synthesis was performed with a cDNA synthesis kit from 1 μg total RNA primed with random hexamers (5 μM), following the manufacturer's instructions (*see* **Note 4**).

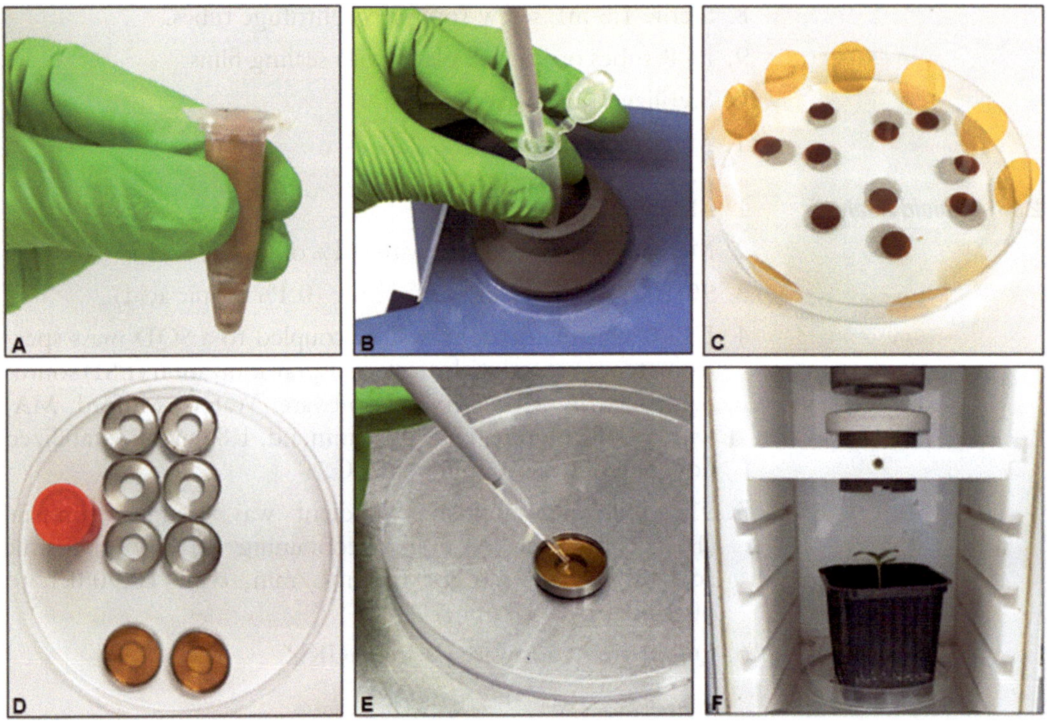

Fig. 3 Pre-bombardment procedure. (**a**) Glycerol-stocked gold particles. (**b**) Coat plasmid constructs onto gold particles. (**c**) Set of macro-carriers, microcarriers, and stopping screens from Bio-Rad used on the transformations. (**d**) Macro-carriers placed on the O-rings. (**e**) Gold-coated particles in 100% ethanol suspension being placed in the macro-carriers. (**f**) *C. roseus* plantlet placed in the Bio-Rad delivery system before transformation (stopping-screen-to-target distance of 9 cm)

5. We considered the sequence (3′-end fragments) of *C. roseus* PDS gene (GenBank accession #JQ655739), the ortholog *R. tetraphylla* PDS gene (similar to rsa_locus_3770_iso_7_-len_2441_ver_2), and GO (CYP71D1V1 GenBank accession #JN61301) as genes that will be silenced (*see* **Note 5**).
6. Amplify DNA fragments using high-fidelity DNA polymerase and the gene-specific oligonucleotides, incorporating the *BamH*I restriction site at both cDNA extremities (*see* **Note 6**).
7. The PCR products must be analyzed by agarose gel electrophoresis to ensure the correct amplification.
8. Purify the PCR product using a PCR cleanup kit.
9. Digest each amplified DNA fragment with *BamH*I.
10. Purify the product of the digestion using a PCR cleanup kit.

3.3 Silencing VIGS Constructs

1. Linearize the pTRV2 vector (around 1 μg) by digestion with the proper restriction enzymes, in this case, *BamH*I.
2. Purify the digestion product using a PCR cleanup kit.

3. Ligate the digested DNA fragments (around 100 ng) into the vector pTRV2 (around 100 ng) to create the final transformation vector (*see* **Note 7**).
4. Transform *E. coli* (strain TOP10) chemically competent cells with the product ligation following a standard heat shock protocol.
5. Select positive clones on Luria-Bertani agar media containing 50 μg/mL kanamycin through standard colony PCR method.
6. Extract the vector using a plasmid extraction kit. Supercoiled plasmids used for particle bombardment should be isolated from *E. coli* desirable yield and quality of the DNA, approx. 1 μg/μL (*see* **Note 8**).
7. Confirm the presence of the insert and its orientation by restriction digestion followed by DNA sequencing.

3.4 Particles Bombardment Procedure

3.4.1 Particle Preparations

1. 30 mg of 1 μm gold particles are weighted in a glass tube followed by dry heat at 180 °C for 8 h (*see* **Note 9**).
2. From this point on, all manipulations should be performed in a flow hood under sterile conditions. Add 1 mL of fresh 70% ethanol, and wash the gold particles for 5 min trough vortex and sonication in an ultrasonication bath. After 5 min, transfer the gold particles to a 1.5 mL sterile microcentrifuge tube (*see* **Note 10**).
3. Pellet the golden particles by centrifugation at 16,000 × *g* for 5 s.
4. Without disturbing the pellet, remove the supernatant.
5. Slowly add 1 mL of sterilized Milli-Q water to wash the pellet. Vortex, and sonicate for 2 min. Pellet the golden particles by centrifugation at 16,000 × *g* for 5 s, and remove the supernatant. Repeat this step two more times. Take side of a P200 pipette to eliminate the remaining water.
6. After washing the golden particles, resuspend in 500 μL of sterilized 50% glycerol (v/v) (*see* **Note 11**).

3.4.2 Coating of Plasmid Constructs onto Golden Particles

1. For each plant to be transformed, prepare a mix of 1 μg of plasmid constructs (pTRV1 and pTRV2—empty vector or harboring a target gene).
2. Add to a sterilized microcentrifuge tube 10 μL of glycerol stocked gold particles (*see* **Note 12**).
3. Mix 5 μL of 0.1 M spermidine with the gold particles while vortexing. Incubate for 3 min at the bench. Keep vortexing for short pulses during the reaction period (Fig. 3a, b).

4. Add the appropriate amount of mixed plasmids to the particles while vortexing. Wait for 3 min, and keep vortexing for short pulses during the reaction period (*see* **Note 13**).
5. Add 6 μL of 2.5 M $CaCl_2$ to the mix while vortexing to optimize the coating, and mix for additional 15 min with a constant speed vortex (1000 rpm).
6. Pellet the gold-coated particles at 16,000 × *g* for 2 s, and remove the supernatant.
7. Wash the pellet with 500 μL of freshly prepared 70% ethanol without disturbing the pellet. Repeat washing using 500 μL of 100% ethanol. Carefully, remove all the remaining liquid near the pellet.
8. Resuspend the gold-coated particles in 10 μL of 100% ethanol with the help of a vortex.
9. Distribute 10 μL of the particles onto the macro-carrier, and let it dry before starting the transformation (Fig. 3c–e).

*3.4.3 Particles Bombardment Procedure (See **Note 14**)*

1. Bombardment of *C. roseus* and *R. tetraphylla* plantlets are performed with the Bio-Rad PDS1000/He delivery system following the manufacturer's recommendations, using 1100-psi rupture disks and under a vacuum pressure of 27.5 in. of Hg.
2. A single parameter of the transformation conditions changes between the two varieties of plants due to their different size, *R. tetraphylla* being around twofold taller than *C. roseus.* Therefore, to maintain a similar distance between top leaves and stopping screen, the stopping-screen-to-target distance was adjusted at 9 and 6 cm, for *C. roseus* and *R. tetraphylla*, respectively.
3. A single bombardment per individual potted plantlet was performed.
4. Transformed plants were kept in greenhouse in a wet atmosphere during 2 days post-bombardment (dpb) followed by progressive reacclimation (Fig. 3f).
5. Plants are cultivated in a greenhouse until appearance of the phenotype.

3.5 Post-transformation Treatments and Analysis

The photobleaching of leaves caused by PDS silencing in the control plants is used to evaluate VIGS efficiency assuming that silencing of the target gene evolves similarly (Fig. 4). For *C. roseus*, leaf photo-bleaching typically begins around 7–10 days after bombarding, while it reaches completion after 21–25 days in neo-formed leaves.

Fig. 4 *C. roseus* and *R. tetraphylla* plantlets transformation. (**a**) *C. roseus* and *R. tetraphylla* plantlets pretransformation presenting one pair of fully expanded leaves. (**b**) *C. roseus* plantlet 20 days after bombardment (dbp) with pTRV2-PDS and the leaves from the wild-type plantlet; plantlet transformed with an empty vector (pTRV2-EV) or with the silencing PDS construct (pTRV2-PDS), all at 20 dpb. (**c**) *R. tetraphylla* plantlet 20 dbp with pTRV2-PDS and the leaves from the wild-type plantlet; plantlet transformed with an empty vector (pTRV2-EV) or with the silencing PDS construct (pTRV2-PDS), all at 20 dpb

3.5.1 Harvesting of Silenced Leaves

1. Collect the first two leaf pairs to emerge following inoculation.
2. In advance, prepare Safe-Lock sterilized microcentrifuge tubes containing three glass beads (*see* **Note 15**).
3. Weight the tubes.

4. Collect the two young leaves, one destined to RNA extraction for testing gene silencing by qPCR and the other one designed for MIA extraction. Flash-freeze the samples (*see* **Note 16**).
5. Store the samples at −80 °C upon use.

3.5.2 Quantitative PCR Analyses

1. Isolate total RNA from the silenced leaves collected.
2. Quantify total RNA with a spectrophotometer at 260/280 nm (*see* **Note 17**).
3. Analyze the integrity of the RNA by migration on 1% agarose gel.
4. Perform cDNA synthesis using 1 μg of total RNA using the reverse transcriptase kit.
5. Before starting the qPCR reactions, place all needed components on ice, vortex briefly each one, and spin down during 2 s to collect the contents at the bottom of the tube.
6. Prepare a master mix according to the protocol of the chosen SYBR green qPCR kit. Always add 10% volume to allow for pipetting error (*see* **Note 18**). As an example, a single reaction is contained in a 15 μL final volume, 6 μL of diluted template cDNA (18×), 0.5 μM forward, and reverse primers designed for the targeted gene and 2× SYBR green Mix.
7. To validate the efficiency of target gene silencing (GO), use in separate reactions primers designed for the targeted gene and primers for housekeeping gene (RPS9 for *C. roseus* or Actin for *R. tetraphylla*) (*see* **Note 19**).
8. Transfer the appropriated volume of master mix to each well on the plate or PCR tube.
9. Cover the PCR plate or the PCR tubes and centrifuge it briefly.
10. Program the qPCR protocol: 95 °C for 7 min (Initial denaturation), 40 cycles at 95 °C for 10 s (denaturation), and 60 °C for 40 s (annealing, extension, and read fluorescence). Hold at 4 °C is optional.
11. The results of the qPCR are normalized in comparison with a housekeeping gene using ΔΔCt method.

3.5.3 Alkaloids Extraction Protocol (See Note 20)

1. After 24 h at −80 °C, lyophilize the samples dedicated to MIA extraction for 24–48 h.
2. Grind the samples with mixer mill during 3 min at the maximal frequency.
3. Weight the tubes and determine the precise dry weight mass of each sample.
4. Add 1 mL of methanol: 0.1% formic acid solution.
5. Vortex during 1 h at 24 °C.

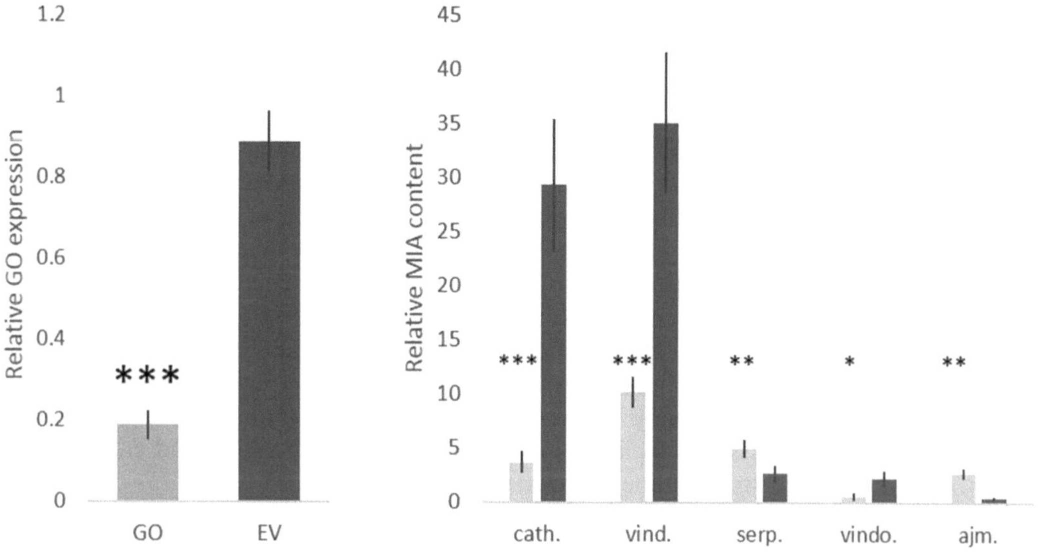

Fig. 5 Biolistic approach using TRV vectors successfully silenced GO in *C. roseus* leaves. (**a**) Relative expression of GO in leaves of pTRV2-EV (grey bar) and pTRV2-GO (white bar)-transformed plants, measured by qPCR. (**b**) Relative mean alkaloid content in leaves of pTRV2-EV (grey bar) and pTRV2-GO (white bar) transformed plants. (**c**) Cath, catharanthine; vind, vindoline; serp, serpentine; vindo, vindorosine; ajma, ajmalicine. Data represent mean ± SE of three technical replicates performed on five plants transformed with pTRV2-EV or pTRV2-GO. Asterisks denote statistical significance ($^{*}P < 0.05$; $^{**}P < 0.01$; $^{***}P < 0.001$, Student's *t*-test)

6. Spin down the tubes at 15,000 × *g* for 15 min; collect the supernatant (around 500 μL).
7. Perform a 1:10 (v/v) dilution of the supernatant with the water: acetonitrile solution.
8. Spin down the tubes at 15,000 × *g* for 15 min and collect 50 μL. Load the samples in the proper sampler tubes.
9. The analysis of the alkaloid content was carried out by UPLC-MS.
10. The following multistep gradient was used: acetonitrile (B) versus water (A) system containing 0.1% (v/v) formic acid, at a flow rate of 0.1 mL/min, with 10:90:0.1 to 60:40:0.1 over 18 min.
11. For results described in Fig. 5, MS experiments were carried out in positive mode in the selected ion-monitoring mode using *m/z* 337 for catharanthine ($[M-H]^-$, RT = 11.6 min), *m/z* 457 for vindoline ($[M-H]^-$, RT = 15.4 min), *m/z* 349 for serpentine ($[M+H]^+$, RT = 13.01 min), *m/z* 427 for vindorosine ($[M+H]^+$, RT = 15.03 min), and *m/z* 353 for ajmalicine ($[M+H]^+$, RT = 11.7 min). Relative quantification was performed by correcting peak areas by sample masses dry weight.

3.6 Conclusion

The results described in Fig. 5 shows the efficiency of the silencing approach on the GO target gene. The marked downregulation of GO expression caused by VIGS (around 80%) resulted in substantial decrease of catharanthine, vindoline, and vindorosine accumulation but to an increased synthesis of ajmalicine and serpentine. This pattern of MIA accumulation tempted to demonstrate that GO is likely involved in the early steps of the synthesis of the precursor of *Aspidosperma* (tabersonine, vindoline, vindorosine) and *iboga* (catharanthine). As a consequence, the blocking of the *Aspidosperma* and iboga pathways led to a rerouting of the MIA precursors to other MIA biosynthetic branches such as the heteroyohimbine one ensuring ajmalicine and serpentine synthesis. Such a positioning in the pathway was finally confirmed by biochemical assays enabling to establish that GO was the second enzyme after strictosidine glucosidase in the pathway.

4 Notes

1. Spermidine integrity decreases even kept at −20 °C. To avoid compromising the VIGS procedure, a stock solution of 1 M should be kept at −20 °C followed by 0.1 M aliquots that should be renewed every 3–6 months.
2. Typically, for each round of experiment, 60 plantlets are prepared in order to guarantee the maximum homogeneity of the required developmental stages. VIGS assays should always be planted with at least ten plants per construct including empty vector, candidate gene, and reporter gene. Silencing of the reporter gene causes phenotype alterations (photobleaching for PDS) and is used as a positive control for the procedure and to determine the optimal period for sample harvesting. A second round of silencing is required to confirm results.
3. The ratio of absorbance values at 260 and 280 nm allows the prediction of DNA purity. A pure DNA has an A260/A280 ratio of 1.8–2.0; if the ratios are below this average, it indicates the presence of contaminants such as proteins.
4. To achieve optimal RT performance, take in consideration that a 20 μL RT reaction efficiently converts a maximum of 1 μg of total RNA to cDNA.
5. The target genes should be amplified with primers that generate fragments around 200–400 bp. To avoid cross-silencing of nontarget gene(s), only include sequences very specific to the target gene and that can generate any 21 mer sequence identical to any other gene. 5′ or 3′ untranslated region (UTR) of the target genes can also be included. Avoid homopolymeric regions (poly A or poly T). We use the example of the silencing

of geissoschizine oxidase (GO) from *C. roseus*, using the phytoene desaturase (PDS) reporter gene known to cause the leaf bleaching [28, 30]. The silencing of PDS in *R. tetraphylla* will be also described to illustrate the procedure, but other reporter genes may be also considered such as the one encoding the Protoporphyrin IX Magnesium chelatase subunit chlH (ChlH [31]) whose silencing results in very pale yellow-green leaf phenotypes.

6. For more options of restriction sites used to clone in this vector, *see* the multiple cloning site (MCS) of the pTRV2 vector online. The enzymes used are just suggestions.
7. Never clone the reporter gene (phytoene desaturase, PDS on this case) with your targeting gene. The reporter gene is known to interfere with secondary metabolite pathways.
8. Plasmid integrity should be confirmed by analytical electrophoresis gel performed before each VIGS experiments; plasmid degradation may compromise the experiment.
9. For ten bombardments (ten plant transformations), 6.25 mg of gold particles should be prepared.
10. Best results were achieved when vortex speed ranges 1000 rpm of intensity.
11. Once prepared, particles can be stored in −20 °C upon usage. The preparation of the golden particles does not depend on the species used. Prior to use the golden particles should be sonicated for 15 min with intercalated vortexing periods.
12. To achieve better results, short pulses of ultrasounds with sonic bath should be performed before the gold particles manipulation. During the coating step, vortexing while pipetting avoids precipitation of the beads.
13. Attention: do not exceed the volume of 2 μL of each plasmid—concentrate the plasmid solutions if required.
14. The best conditions allow performing bombardment without resulting in marked leaf damages. This method allows up to 90% of successful silencing of candidate genes in both *C. roseus* and *R. tetraphylla*.
15. On this step, the harvest is performed simultaneously for qPCR and MIA quantification. Thus, a previous identification of your tubes is required to avoid mixing of the samples.
16. Perform this step rapidly to avoid RNA degradation.
17. Be sure to always include duplicate "No-template Negative Controls" (NTC) in which Milli-Q water is used instead of cDNA.
18. The most important step to a success qPCR is high-quality DNA preparation. Integrity and purity of DNA template are

essential. Contaminants can interfere with fluorescence detection.

19. To know the best concentration of primers and matrix for a qPCR reaction, a calibration curve should be done in advance. To generate good calibration curves, different concentrations of DNA (we recommend a minimum of five) should be used. Each concentration should be run in triplicate.

20. At the metabolic level, consequences related to target gene silencing can be evaluated by measuring the qualitative and quantitative differences of MIA accumulation between plants silenced for a dedicated targeted gene and plants transformed with empty silencing vectors. Identifying up- or downregulated MIAs give remarkable clues regarding candidate gene function and positioning in the pathway. The method described for MIA extractions is fitted for a wide range of conditions, and species accumulating this type of compounds but specific adaptations may be required depending on the plant [32].

References

1. Le Quesne PW (1999) Alkaloids: biochemistry, ecology, and medicinal applications edited by Margaret F. Roberts (University of London) and Michael Wink (University of Heidelberg). Plenum Press, New York, NY. J Nat Prod 62 (4):664–664. https://doi.org/10.1021/np980259j
2. Hinse C, Stöckigt J (2000) The structure of the ring-opened N beta-propyl-ajmaline (Neo-Gilurytmal) at physiological pH is obviously responsible for its better absorption and bioavailability when compared with ajmaline (Gilurytmal). Pharmazie 55(7):531–532. PMID: 10944783
3. Ro D, Paradise EM, Ouellet M et al (2006) Production of the antimalarial drug precursor artemisinic acid in engineered yeast. Nature 440(7086):940–943. https://doi.org/10.1038/nature04640
4. Song MC, Kim EJ, Kim E et al (2014) Microbial biosynthesis of medicinally important plant secondary metabolites. Nat Prod Rep 31:1497–1509. https://doi.org/10.1039/c4np00057a
5. Galanie S, Thodey K, Trenchard IJ et al (2015) Complete biosynthesis of opioids in yeast. Science 349(6252):1095–1100. https://doi.org/10.1126/science.aac9373
6. Saleki R, Young PG, Lefebvre DD (1993) Mutants of *Arabidopsis thaliana* capable of germination under saline conditions. Plant Physiol 101:839–845. https://doi.org/10.1104/pp.101.3.839
7. Koiwa H, Bressan RA, Hasegawa PM (2006) Identification of plant stressresponsive determinants in Arabidopsis by large-scale forward genetic screens. J Exp Bot 57:1119–1128. https://doi.org/10.1093/jxb/erj093
8. Qu Y, Easson MEAM, Simionescu R et al (2018) Solution of the multistep pathway for assembly of corynanthean, strychnos, iboga, and aspidosperma monoterpenoid indole alkaloids from 19E-geissoschizine. Proc Natl Acad Sci U S A 115(12):3180–3185. https://doi.org/10.1073/pnas.1719979115
9. Williams D, Qu Y, Simionescu R et al (2019) The assembly of (+)-vincadifformine and (−)-tabersonine derived monoterpenoid indole alkaloids in *Catharanthus roseus* involve separate branch pathways. Plant J 99 (4):626–636. https://doi.org/10.1111/tpj.14346
10. Guo Y, Halfter U, Ishitani M et al (2001) Molecular characterization of functional domains in the protein kinase SOS2 that is required for plant salt tolerance. Plant Cell 13:1383–1399. https://doi.org/10.1105/tpc.13.6.1383
11. Burgos-Rivera B, Dawe RK (2012) An Arabidopsis tissue-specific RNAi method for studying genes essential to mitosis. PLoS One 7(12):

e51388. https://doi.org/10.1371/journal.pone.0051388
12. Ramegowda V, Mysore KS, Senthil-Kumar M (2014) Virus-induced gene silencing is a versatile tool for unraveling the functional relevance of multiple abiotic-stress-responsive genes in crop plants. Front Plant Sci 5:323. https://doi.org/10.3389/fpls.2014.00323
13. Unver T, Budak H (2009) Virus-induced gene silencing, a post transcriptional gene silencing method. Int J Plant Genomics 2009:198680. https://doi.org/10.1155/2009/198680
14. Voinnet O (2001) RNA silencing as a plant immune system against viruses. Trends Genet 17(8):449–459. https://doi.org/10.1016/S0168-9525(01)02367-8
15. Romano N, Macino G (1992) Quelling: transient inactivation of gene expression in Neurospora crassa by transformation with homologous sequences. Mol Microbiol 6 (22):3343–3353. https://doi.org/10.1111/j.1365-2958.1992.tb02202.x
16. Baulcombe DC (1999) Fast forward genetics based on virus-induced gene silencing. Curr Opin Plant Biol 2:109–113. https://doi.org/10.1016/S1369-5266(99)80022-3
17. Liu H, Reavy B, Swanson M et al (2002a) Functional replacement of the Tobacco rattle virus cysteine-rich protein by pathogenicity proteins from unrelated plant viruses. Virology 298:232–239. https://doi.org/10.1006/viro.2002.1421
18. Xu P, Zhang Y, Kang L et al (2006) Computational estimation and experimental verification of off-target silencing during posttranscriptional gene silencing in plants. Plant Physiol 142:429–440. https://doi.org/10.1104/pp.106.083295
19. Valentine T, Shaw J, Blok VC et al (2004) Efficient virus-induced gene silencing in roots using a modified Tobacco rattle virus vector. Plant Physiol 136:3999–4009. https://doi.org/10.1104/pp.104.051466
20. Ratcliff F, Martin-Hernandez AM, Baulcombe DC (2001) Technical Advance. Tobacco rattle virus as a vector for analysis of gene function by silencing. Plant J 25:237–245. https://doi.org/10.1046/j.0960-7412.2000.00942.x
21. Liu Y, Schiff M, Dinesh-Kumar SP (2002b) Virus-induced gene silencing in tomato. Plant J 31:777–786. https://doi.org/10.1046/j.1365-313X.2002.01394.x
22. Carqueijeiro I, Masini E, Foureau E et al (2015) Virus-induced gene silencing in *Catharanthus roseus* by biolistic inoculation of tobacco rattle virus vectors. Plant Biol 17:1242–1246. https://doi.org/10.1111/plb.12380
23. Corbin C, Lafontaine F, Sepúlveda LJ et al (2017) Virus-induced gene silencing in Rauwolfia species. Protoplasma 254:1813. https://doi.org/10.1007/s00709-017-1079-y
24. Dessens JT, Lomonossoff GP (1993) Cauliflower mosaic virus 35S promoter-controlled DNA copies of cowpea mosaic virus RNAs are infectious on plants. J Gen Virol 74:889–892. https://doi.org/10.1099/0022-1317-74-5-889
25. Jakab G, Droz E, Brigneti G et al (1997) Infectious in vivo and in vitro transcripts from a full-length cDNA clone of PVY-N605, a Swiss necrotic isolate of potato virus Y. J Gen Virol 78(12):3141–3145. https://doi.org/10.1099/0022-1317-78-12-3141
26. Foureau E, Carqueijeiro I, Dugé de Bernonville T et al (2016) Prequels to synthetic biology: from candidate gene identification and validation to enzyme subcellular localization in plant and yeast cells. In: O'Connor S (ed) Synthetic biology and metabolic engineering in plants and microbes. Part B: Metabolism in plants, Methods in enzymology, vol 576. Elsevier, Academic Press, Amsterdam, pp 167–206. https://doi.org/10.1016/bs.mie.2016.02.013
27. Lim HS, Vaira AM, Domier LL et al (2010) Efficiency of VIGS and gene expression in a novel bipartite potexvirus vector delivery system as a function of strength of TGB1 silencing suppression. Virology 402(1):149–163. https://doi.org/10.1016/j.virol.2010.03.022
28. Tatsis EC, Carqueijeiro I, Dugé de Bernonville T et al (2017) A three enzyme system to generate the Strychnos alkaloid scaffold from a central biosynthetic intermediate. Nat Commun 8 (1):316. https://doi.org/10.1038/s41467-017-00154-x
29. Caputi L, Franke J, Farrow SC et al (2018) Missing enzymes in the biosynthesis of the anticancer drug vinblastine in Madagascar periwinkle. Science 360(6394):1235–1239. https://doi.org/10.1126/science.aat4100

30. Salim V, Yu F, Altarejos J et al (2013) Virus-induced gene silencing identifies *Catharanthus roseus* 7-deoxyloganic acid-7-hydroxylase, a step in iridoid and monoterpene indole alkaloid biosynthesis. Plant J 76:754–765. https://doi.org/10.1111/tpj.12330
31. Liscombe DK, O'Connor SE (2011) A virus-induced gene silencing approach to understanding alkaloid metabolism in *Catharanthus roseus*. Phytochemistry 72(16):1969–1977. https://doi.org/10.1016/j.phytochem.2011.07.001
32. Besseau S, Kellner F, Lanoue A et al (2013) A pair of tabersonine 16-hydroxylases initiates the synthesis of vindoline in an organ-dependent manner in *Catharanthus roseus*. Plant Physiol 163(4):1792–1803. https://doi.org/10.1104/pp.113.222828

Chapter 9

Virus-Induced Gene Silencing in *Nepeta*

Lira Palmer and Sarah E. O'Connor

Abstract

Virus-induced gene silencing (VIGS) is a versatile tool for genetic studies that has been applied to a variety of plant species. With the advent of more accessible genomic and transcriptomic technology applied to an increasing range of plants, tools such as VIGS are being adapted to more non-model plants to explore genes relevant to agriculture and chemical discovery. In this protocol, we adapted VIGS technology to target genes in *Nepeta cataria* (catnip) and *Nepeta mussinii* (catmint). These plants carry biochemical and economical value for their production of nepetalactone, an iridoid which provokes a strong reaction in both house cats and aphids. We describe a method to target magnesium chelatase subunit H (CHlH), a gene often targeted as a visual marker for VIGS. Furthermore, we describe a method to simultaneously target two genes in a single plant, which aids in the study of genes found in key biochemical steps in the production of nepetalactone. This approach, which was successfully applied in two members of the Lamiaceae family (mint), could be adapted to other members of the mint family with economical and chemical value.

Key words VIGS, *Nepeta*, Catnip, Catmint, Mint, Lamiaceae, Nepetoideae, Iridoid, Nepetalactone

1 Introduction

Virus-induced gene silencing (VIGS) is a transient and versatile posttranscriptional genetic knockdown technique that uses a plant's innate RNA-mediated gene silencing defense system. Post-translational gene silencing (PTGS) was first discovered in early plant transformation efforts, as transgene and endogenous genes would often be silenced after transformation [1, 2]. Further research led to the development of silencing cassettes containing viral sequences from various plant viruses such as *Potato virus X*, *Tobacco rattle virus*, and CaLCuV, among others, that trigger the plant's posttranscriptional silencing machinery [1–8]. Along with the viral sequences, inclusion of short homologous sequences to endogenous host genes leads to the formation of siRNA transcripts that guide the silencing machinery to methylate, and thereby silence, these endogenous genes, albeit transiently [9].

Vincent Courdavault and Sébastien Besseau (eds.), *Virus-Induced Gene Silencing in Plants: Methods and Protocols*, Methods in Molecular Biology, vol. 2172, https://doi.org/10.1007/978-1-0716-0751-0_9,

The extent and effectiveness of silencing has been found to depend on the viral vector as well as the species of plant host. Often, it is necessary to use viral vectors taken from viruses that infect the preferred host. Such is the case of the use of the *Barley stripe mosaic virus* (BSMV) in monocots or the *Tobacco rattle virus* (TRV) and *Potato virus X* (PVX) in Solanaceae. (1) However, certain vectors have been shown to work across many plant families. TRV has proven to be particularly versatile and has been shown to induce endogenous gene silencing across the dicot families, as well as to be efficient at systematic silencing.

The TRV vector system is adapted from the plus-strand RNA *Tobacco rattle virus* with a bipartite genome. This type of virus can be introduced into *Agrobacterium tumefaciens* which is then used to infect the plant, allowing the viral sequences to trigger the host's posttranscriptional silencing machinery. The TRV vector system is comprised of two vectors (Fig. 1) which work in tandem: pTRV1, which contains viral replication machinery (RdRP) and movement protein (MP) sequences (Fig. 1a), and pTRV2, which contains the sequences for virion formation (CP) and a multiple cloning site to introduce the host sequence to be targeted for silencing (Fig. 1b). The elements of these vectors are under the 35S promoter from the *Cauliflower mosaic virus* for constitutive expression and are well suited for systematic transient silencing [3, 4].

Using the pTRV vector system, we were able to target genes for silencing in *Nepeta cataria* and *Nepeta mussinii*, plants in the Nepetoideae subfamily of the Lamiaceae. Members of the Lamiaceae family, colloquially known as the mint family, are of significant economic and cultural interest, as this family includes several species of well-known culinary herbs such as basil, mint, and lavender [10–12]. VIGS has been previously adapted for use in *Ocimum basilicum* (sweet basil) to study the specialized metabolome [13, 14]. *N. cataria* and *N. mussinii* are of particular interest due to the production of a novel type of iridoid, nepetalactone. Nepetalactone has potential industry importance due to its insect repellent activities and is also responsible for the euphoric response in cats [10]. Thus, we set out to transfer the VIGS system into these species for in vivo studies into specialized metabolism. We have also adapted this system to target multiple genes at once. The protocol below details the steps to take to target single genes and includes a guide for multigene targeting. Using transcriptomic data, we selected magnesium chelatase subunit H (CHlH) and geraniol synthase (GES) as genes to target for VIGS in *Nepeta cataria*. This protocol details the use of traditional molecular cloning techniques such as ligation and recombinase reactions to construct specific pTRV2 targeting vectors. We describe the use of *Agrobacterium tumefaciens* as a host for our pTRV2 constructs, as well as plant tissue management for optimal results.

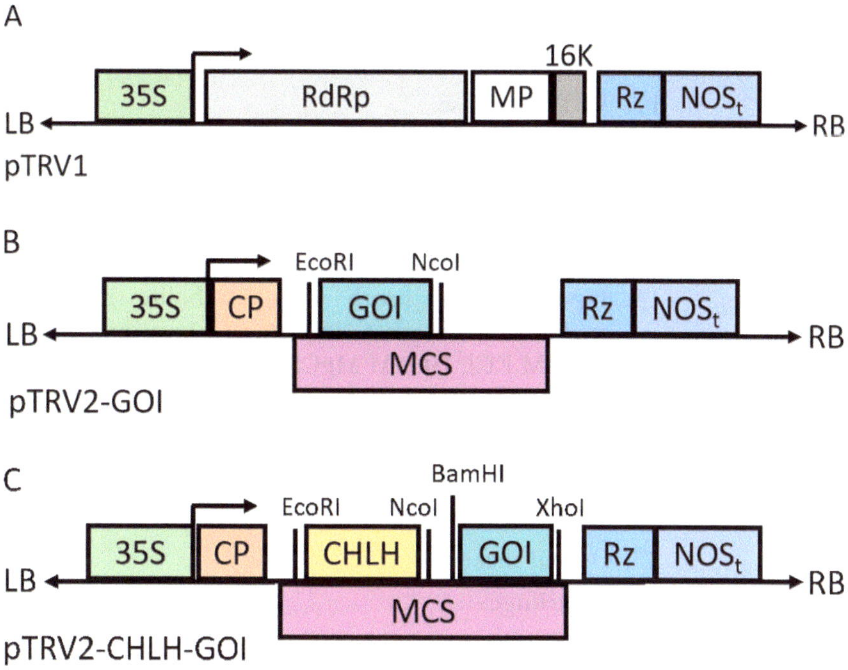

Fig. 1 Simplified maps of VIGS vectors used in this protocol. pTRV1 (**a**), pTRV2-GOI (**b**), and pTRV2-CHIH-GOI (**c**) are represented. To target one gene, a fragment of the GOI is cloned into the MCS of pTRV2 between BamHI and XhoI. To target two genes, one sequence (e.g., CHIH) is cloned into the MCS of pTRV2 between EcoRI and NcoI and the second (e.g., GOI) between BamHI and XhoI. *35S* CaMV promoter, *NOSt* nopaline synthase terminator, *RdRp* RNA-dependent RNA polymerase, *MP* movement protein, *16K* cysteine-rich protein, *CP* coat protein, *MCS* multiple cloning site, *Rz* self-cleaving ribosome

2 Materials

2.1 Vector Design

1. Mortar and pestle.
2. Plant RNA isolation kits.
3. Sterile microcentrifuge tubes and pipette tips, RNAse cleaner.
4. cDNA synthesis kits.
5. High-fidelity DNA polymerase and buffers.
6. CHlH primers (350 bp fragment):

   CHlH BamHI Forward 5′-CGAGGATCC-ACCAATGACAT-GAAGGCCAC-3′.

   CHlH XhoI reverse 5′-CGATCTCGAG-ACGCTGCTAA-CAACCCG-3′.

   (5′-pTRV2 overhang-CHlH fragment overhang-3′).
7. Thermal cyclers.
8. 1% agarose gel, 1×TAE, ethidium bromide, and electrophoresis power supply.

9. NanoDrop.
10. PCR cleanup and gel extraction kits.
11. T4 DNA ligase.
12. Recombinant enzyme and buffers: BamHI, XhoI, EcoRI, and/or NcoI.
13. Chemically competent *E. coli* strain (such as HST08 or DH10B).
14. Super optimal broth with catabolite repression (SOC) medium: 0.5% yeast extract, 2% tryptone, 10 nM NaCl, 2.5 mM KCl, 10 mM $MgCl_2$, 10 mM $MgSO_4$, 20 mM glucose added after autoclaving, pH 7.
15. LB (Miller) medium plates: 0.5% yeast extract, 1% tryptone, 10 nM NaCl, 1.1% Formedium agar, pH 7.
16. 50 g/L kanamycin stock solution.
17. Orbital shaker, 37 °C incubator.
18. Centrifuge.
19. High-fidelity Taq polymerase, dNTP.
20. pTRV2 empty vector and pTRV1 vector.
21. pTRV2 oligonucleotides:
    pTRV2 forward 5′-GATGGACATTGTTACTCAAGGAAGC-3′.
    pTRV2 reverse 5′-CAGTCGAGAATGTCAATCTCGTAGG-3′.
22. Plasmid minipreparation kit.

2.2 Agrobacterium Transformation

1. Electrocompetent *Agrobacterium tumefaciens* GV1303.
2. Electroporator equipment, 0.1 cm electroporation cuvettes.
3. Sterile microcentrifuge tubes and pipette tips.
4. Antibiotics: stock concentration at 50 g/L kanamycin, 50 g/L rifampicin, and 100 g/L gentamicin.
5. LB (Miller) agar plates (as in Subheading 2.1).
6. 30 °C incubator.
7. 50% glycerol.

2.3 Plant Infection

1. Plant material, *Nepeta cataria*: cuttings or seeds.
2. Scissors.
3. Pre-infection medium: LB medium with 50 mg/L kanamycin, 50 mg/L rifampicin, 100 mg/L gentamicin, 10 mM of MES, and 20 μM of acetosyringone.
4. Infiltration buffer: 10 mM NaCl, 1.75 mM of $CaCl_2$, and 100 μM of acetosyringone.

5. Pincer tweezers or syringe.
6. Falcon tubes, culture vials.
7. Soil, pots, and trays.
8. Growth chamber: 25 °C, 16 h:8 h of light on/off cycle.

3 Methods

3.1 Nepeta cataria Plant Material Preparation

1. Plant material can be from cuttings or from seedlings.
2. If making cuttings, cut the stem to get two or three nodes in a row (Fig. 2a). Cut the stem just below the node, as close as possible to the node. Clear the bottom node from any leaves, again as close as possible to the node. For the rest of the cutting, leave about one or two leaf pairs (*see* **Note 1**).
3. Place the cutting in water (Fig. 2b) (*see* **Note 2**). Wait for rooting, usually between 7 and 12 days of watering every 2 or 3 days or topping up the water to always keep the bottom naked nodes under water (Fig. 2b).
4. Once rooted, transfer to soil. Allow the plants to adapt to their new soil for 1 or 2 days before infection.
5. If growing from seedlings, place *Nepeta cataria* seeds in soil in 5 × 5 cm pots, with four seeds per pot to allow easier transfer once germinated. Lightly cover with soil and spray with water. Keep indoors with access to sunlight or in a growth chamber. Cover the soil with a plastic cover and keep the soil moist.
6. Allow the seedlings to grow until they have two or three aerial nodes before infection (*see* **Note 3**).

3.2 Vector Design and Transformation

1. Pre-cool pestle and mortar. Cut and flash-freeze less than 100 g of leaf tissue. Thoroughly grind it to a fine powder while keeping cool using liquid nitrogen.
2. Follow the protocol for plant RNA extraction from the available RNA extraction kit (*see* **Note 4**).
3. To check the quality of the RNA, run on a 1% agarose gel with MOPS buffer. For sharper bands, boil the RNA aliquot to be loaded on the gel in 5% formaldehyde for 2 min, cool on ice, and run for 45 min at 120 V and 400 A. Good quality RNA is assessed by the ratio of band intensity of 28/18S, which should be close to 2, and should not contain DNA in the gel wells.
4. Using as commercially available cDNA synthesis kit, carry out the reaction to obtain cDNA from 1.5 μg of RNA.
5. Using transcriptomic data, select a 200–350 bp region of the gene of interest (GOI), such as CHlH (*see* **Note 5**). This region can span any length of the cDNA and UTR but should not have

Fig. 2 Cutting rooting setup. *Nepeta cataria* can be easily rooted in water or coconut husk. Panel (**a**) shows the length and nodes needed for rooting. One or two nodes from the bottom should be completely removed of leaves and submerged in water (**b**). Cutting that has rooted after 13 days (**c**)

any matches to other transcripts of up to 20 bp. For the purpose of this protocol example, the sequence should not contain the restriction sites of BamHI and XhoI (*see* **Note 6**).

6. Design primers containing the cut sites, and a 15–25 bp overhang of the cDNA fragment of interest, such as those in Subheading 2.1, **item 6**.
7. Digest 2–5 μg of the pTRV2 plasmid using BamHI-HF and XhoI-HF overnight at 37C°. Gel purify (*see* **Note** 7).
8. Amplify via high-fidelity PCR the CHlH fragment from approximately 100 ng/μl of cDNA or a synthetic gene. If carrying out a recombinase reaction, purify the reaction via gel extraction. If carrying out a ligation reaction, digest the PCR product using BamHI-HF and XhiI-HF overnight at 37 °C. Gel purify.
9. Carry out the ligation reaction by adding 20–50 ng of linear pTRV2, 20–50 ng of the CHlH fragment, 1 μl of T4 DNA ligase, 2 μl of 10× T4 DNA ligase buffer, and nuclease-free water up to 20 μl. Incubate at 16 °C overnight. Alternatively, carry out a recombination reaction by adding 20–50 ng of linear pTRV2, 20–50 ng of the CHlH fragment, 1 μl of recombinase enzyme master mix, and nuclease-free water up to 5 μl; incubate at 50 °C for 15 min.
10. Thaw the chemically competent cells on ice, and then aliquot 30–50 μl per reaction. Add 2.5 μl of each ligation product to an aliquot and include a digested empty vector control. Keep on ice for 5–10 min.
11. Heat shock the cells at 42 °C for 45 s, and then cool on ice for 2 min.

12. Add 150 μl of SOC media and incubate at 37 °C for 30–60 min.
13. Plate 75 μl of each reaction on LB plates with 50 mg/L kanamycin. Incubate overnight at 37 °C.
14. Select colonies and check via colony PCR using the primers pTRV2 forward and pTRV2 reverse (Subheading 2.1, **item 21**) that the correct size insert has been introduced. Select the positive colonies and grow overnight in LB media with kanamycin at 37 °C.
15. Miniprep the cultures to obtain the plasmid. This can be done using a commercial kit. Confirm the plasmid via sequence verification using the primers pTRV2 forward and pTRV2 reverse.
16. Once confirmed, create an *E. coli* glycerol stock by adding 500 μl of LB culture to 500 μl of 50% glycerol to store at −80 °C, and keep a plasmid stock of pTRV1, pTRV2 empty, and pTRV2-CHlH.

3.3 Agrobacterium *Transformation*

1. Thaw 50 μl of *A. tumefaciens* electrocompetent cells on ice.
2. Add 100 ng of plasmid to the cells and mix gently by stirring the pipette tip or flicking the tube.
3. Leave cells on ice for 2 min and transfer in pre-cold electroporation cuvette.
4. Shock cells using an electroporator at 1–5 kV/cm.
5. Immediately add 150 μl of SOC media to the cells and incubate for 3 h at 30 °C under shaking.
6. Plate the mixture on LB media plates containing 50 mg/L kanamycin, 50 mg/L rifampicin, and 100 mg/L gentamicin. Leave plates for 2 days at 30 °C.
7. Do a colony PCR to check for the correct plasmid insert size as described in Subheading 3.2.
8. Inoculate the following strains into 1 ml of LB media containing 50 mg/L kanamycin, 50 mg/L rifampicin, and 50 mg/L gentamicin: pTRV1, pTRV2 empty vector, pTRV2-CHlH, and pTRV2-GOI. Allow to grow for 16–24 h under shaking to make a glycerol stock as detailed in Subheading 3.2, **step 16**.
9. Inoculate 100 ml of pre-infection medium with 1 ml of pre-culture of the *Agrobacterium* strains carrying pTRV2 empty vector, pTRV2-CHlH, or pTRV2-GOI and 300 ml of pre-infection medium with 3 ml of pre-culture of the *Agrobacterium* strain containing pTRV1. Allow to grow for 16–24 h at 30 °C under shaking.

3.4 Plant Infection

1. Centrifuge the culture at 3500 × *g* for 10 min to concentrate the agrobacteria.
2. Resuspend the agrobacteria into 10 ml of infiltration buffer.
3. Incubate at room temperature for 4–6 h.
4. Before infection, make 1:1 solution of *Agrobacterium* suspension pTRV1:pTRV2-EV, pTRV1-CHlH, and pTRV1:pTRV2-GOI, and transfer to a 50 ml Falcon tube. Each tube should contain about 15–20 ml of solution.
5. Plants should be about 10–15 cm tall, with two or three aerial nodes. Remove leaves to leave about two or three leaf pairs.
6. Briefly wash the plant from excess dirt in water.
7. Wound every node (including the crown) by pinching or inserting a sterile needle into both sides of the node, between the leaf and the stem (Fig. 3a).
8. Introduce the wounded plant into the Falcon tube containing *Agrobacterium* suspension, and gently rock for 30–60 s, making sure the solution covers the plant (Fig. 3b).
9. Remove the plant, blot dry, and replant in soil.
10. Make sure to water the plants consistently every 1 or 2 days while they recover from infection. Store in a long-day cycle CER for genetically modified organisms, with growing conditions as specified in Subheading 2.2, **item 8**.
11. After about 3–6 weeks, new leaves should emerge with the knockdown phenotype. If using a visual marker, such as magnesium chelatase subunit ChlH (CHlH) (Fig. 4), these new leaves should display the expected visual phenotype. Knockdown is not fully systemic. Some leaves will be affected more, less, or not at all. The plant eventually stops silencing the gene in new leaves.

3.5 Multigene Targeting

1. While Subheadings 3.2–3.4 detail the protocol to design vectors to target single genes, the pTRV vector system can be used to target multiple sites (Fig. 1c). This is especially useful when targeting genes that present no visual phenotype, such as those working in biochemical pathways. The following steps detail multigene targeting for the visual marker CHlH and geraniol synthase (GES). Knocking down both genes in the same plant allows for the targeted study of the metabolite landscape of the affected tissues.
2. If using restriction-ligase reactions, the vector can be digested at various entry points, including the BamHI or XhoI sites, as well as EcoRI and NcoI. This requires two digestions and ligation steps (*see* **Note 8**).

Fig. 3 Plant infection methodology. All plants to be infected should be rooted. Remove leaf pairs to leave only two or three young leaves. Puncture each node on two sites (opposite to each other) (**b**), using a syringe, or modified tweezers (**a**). Follow the black arrows (which has already been stripped to three leaf pairs). If the node still has a leaf, puncture the node between the leaf stem and the stem. Introduce the plant in a Falcon tube with the agrobacteria (**c**), and rock gently for 30 s

Fig. 4 CHLH as a visual marker for PTGS. Selection of *Nepeta cataria* leaves showing knocked-down CHLH phenotype 6 weeks post infection from the same plant. Although PTGS can spread through the plant, knockdown is patchy on affected areas. Some leaves will be completely unaffected, as those displayed in the bottom row

3. Using the same principles as detailed in Subheading 3.2, choose 200–350 bp of the genes of interest, for example, CHlH and GES. Design primers with overlapping regions for GES: GES EcoRI forward primer: 5′-CCGATTCT-AATC-CAACGGCTGGGAATCGG-3′, GES NcoI reverse primer: 5-′-CCCCATGGA-TCGCTACAAAGGCGAGGTGC-3′ (5′-pTRV2 overhang-gene fragment overhang-3′).
4. Follow **steps 7–16** of Subheading 3.2 to introduce CHlH.
5. Digest the pTRV2-CHLH vector with EcoRI and NcoI. Follow **steps 8–16** of Subheading 3.2 to introduce GES.
6. Send for sequencing to check if both regions have been introduced. Proceed with the rest of the protocol as normal.

4 Notes

1. When making cuttings, be aware that not all cuttings will root. Therefore, when accounting for sample sizes, collect a few more cuttings than needed.
2. Cuttings will also root in coconut husk. Make sure to keep the husk moist during rooting (7–12 days).
3. If growing from seedlings, note that the stem on these will be thinner and harder to infect, but still possible.
4. RNA can be easily degraded. To obtain high-quality RNA, make sure to thoroughly wipe clean any surfaces or equipment needed with ethanol and RNase spray cleaner. For a high volume of tissue, tissues can be flash-frozen in Eppendorf tubes, with two sterile 3 mm tungsten beads added and placed on a shaker to pulverize.
5. Visual markers of silencing should be used in any assay to test for silencing, for example, CHlH, crucial for chlorophyll biosynthesis, which upon silencing results in yellow leaf tissue. *N. cataria* is a tetraploid; therefore you should test if the region will overlap multiple transcripts of the same gene from different chromosomes, if both copies are active.
6. During the sequence selection, it is common to get 16 bp hits, which should not lead to off-site knockdown. Furthermore, if there is no 200–350 bp region which does not exactly match other transcripts above 20 bp and does not contain the BamHI or XhoI cut sites, then other restriction sites on pTRV2 can be used. UTR regions can be used for targeting.
7. Alkaline phosphatase can be used to avoid re-ligation during digestion.
8. The vector can be digested at various entry points, including the BamHI or XhoI sites, as well as EcoRI and NcoI.

References

1. Lange M, Yellina AL, Orashakova S, Becker A (2013) Virus-Induced Gene Silencing (VIGS) in plants: an overview of target species and the virus-derived vector systems. In: Becker A (ed) Virus-induced gene silencing: methods and protocols. Humana Press, Totowa, NJ, pp 1–14. https://doi.org/10.1007/978-1-62703-278-0_1
2. Robertson D (2004) VIGS vectors for gene silencing: many targets, many tools. Annu Rev Plant Biol 55(1):495–519. https://doi.org/10.1146/annurev.arplant.55.031903.141803
3. Liu Y, Schiff M, Dinesh-Kumar SP (2002) Virus-induced gene silencing in tomato. Plant J 31(6):777–786. https://doi.org/10.1046/j.1365-313X.2002.01394.x
4. Liu Y, Schiff M, Marathe R, Dinesh-Kumar SP (2002) Tobacco Rar1, EDS1 and NPR1/NIM1 like genes are required for N-mediated resistance to tobacco mosaic virus. Plant J 30 (4):415–429. https://doi.org/10.1046/j.1365-313X.2002.01297.x
5. Velásquez AC, Chakravarthy S, Martin GB (2009) Virus-induced gene silencing (VIGS) in Nicotiana benthamiana and tomato. J Vis Exp (28):1292. https://doi.org/10.3791/1292
6. Liu E, Page JE (2008) Optimized cDNA libraries for virus-induced gene silencing (VIGS) using tobacco rattle virus. Plant Methods 4(1):5. https://doi.org/10.1186/1746-4811-4-5
7. Golenberg EM, Sather DN, Hancock LC, Buckley KJ, Villafranco NM, Bisaro DM (2009) Development of a gene silencing DNA vector derived from a broad host range geminivirus. Plant Methods 5:9. https://doi.org/10.1186/1746-4811-5-9
8. Lu R, Peart JR, Malcuit I, Baulcombe DC (2003) Virus-induced gene silencing in plants. Methods 30:296–303. https://doi.org/10.1016/S1046-2023(03)00037-9
9. Geu-Flores F, Sherden NH, Courdavault V, Burlat V, Glenn WS, Wu C et al (2012) An alternative route to cyclic terpenes by reductive cyclization in iridoid biosynthesis. Nature 492:138. https://doi.org/10.1038/nature11692
10. Boachon B, Buell CR, Crisovan E, Dudareva N, Garcia N, Godden G et al (2018) Phylogenomic mining of the mints reveals multiple mechanisms contributing to the evolution of chemical diversity in Lamiaceae. Mol Plant 11(8):1084–1096. https://doi.org/10.1016/j.molp.2018.06.002
11. Weng J-K, Philippe RN, Noel JP (2012) The rise of chemodiversity in plants. Science 336 (6089):1667–1670. https://doi.org/10.1126/science.1217411
12. Lichman BR, Kamileen MO, Titchiner GR, Saalbach G, Stevenson CEM, Lawson DM, O'Connor SE (2018) Uncoupled activation and cyclization in catmint reductive terpenoid biosynthesis. Nat Chem Biol 15(1):71–79. https://doi.org/10.1101/391953
13. Misra RC, Sharma S, Sandeep, Garg A, Chanotiya CS, Ghosh S (2017) Two CYP716A subfamily cytochrome P450 monooxygenases of sweet basil play similar but nonredundant roles in ursane- and oleanane-type pentacyclic triterpene biosynthesis. New Phytol 214 (2):706–720. https://doi.org/10.1111/nph.14412
14. Chakraborty P (2018) Herbal genomics as tools for dissecting new metabolic pathways of unexplored medicinal plants and drug discovery. Biochim Open 6:9–16. https://doi.org/10.1016/j.biopen.2017.12.003

Chapter 10

Virus-Induced Gene Silencing in Sweet Basil (*Ocimum basilicum*)

Rajesh Chandra Misra, Shubha Sharma, Anchal Garg, and Sumit Ghosh

Abstract

Virus-induced gene silencing (VIGS) is a powerful reverse genetic tool for rapid functional analysis of plant genes. Over the last decade, VIGS has been widely used for conducting rapid gene knockdown experiment in plants and played a crucial role in advancing applied and basic research in plant science. VIGS was studied extensively in model plants *Arabidopsis* and tobacco. Moreover, several non-model plants such as *Papaver* (Hileman et al., Plant J 44:334–341, 2005), *Aquilegia* (Gould and Kramer, Plant Methods 3:6, 2007), *Catharanthus* (Liscombe and O'Connor, Phytochemistry 72:1969–1977, 2011), *Withania* (Singh et al., Plant Biol J 13:1287–1299, 2015), and *Ocimum* (Misra et al., New Phytol 214:706–720, 2017) were also successfully explored. We have recently developed a robust protocol for VIGS in sweet basil (*Ocimum basilicum*). Sweet basil, a popular medicinal/aromatic herb, is being studied for the diversity of specialized metabolites produced in it.

Key words *Tobacco rattle virus* (TRV), Virus-induced gene silencing (VIGS), Functional genomics, *Agrobacterium*, Transformation, Sweet basil

1 Introduction

Defining function of the genes has been of great interest to the researchers working in the area of functional genomics. For this, it is important to obtain sufficient experimental data that could provide direct evidence for the involvement of a gene or group of genes in a specific cellular pathway or process. In this context, gene knockout or knockdown methods are most useful for characterization of the genes and can give valuable insight into gene function. Virus-induced gene silencing (VIGS) method has been widely adapted and emerged as a powerful and cost-effective technique to accomplish rapid loss-of-function experiment in plants. VIGS takes place at the posttranscriptional level, empowering functional genomics in various plant species [1]. In plants, endogenous gene silencing pathway starts with the formation of double-stranded RNA (dsRNA) from single-stranded RNA (ssRNA) with the help

Vincent Courdavault and Sébastien Besseau (eds.), *Virus-Induced Gene Silencing in Plants: Methods and Protocols*, Methods in Molecular Biology, vol. 2172, https://doi.org/10.1007/978-1-0716-0751-0_10,

of viral RNA-dependent RNA polymerase (RdRP). These dsRNA molecules are cleaved by Dicer-like enzymes (DCL, type III endoribonucleases) to produce small double-stranded fragments (21–24 nt), referred to as small interfering RNAs (siRNAs). siRNAs assimilate into an RNA-induced silencing complex (RISC) with AGO family proteins, in which they act as guide molecules to degrade the target transcript and activate the antiviral RNA silencing pathway, resulting in target gene suppression (Fig. 1 [2–4]).

A. van Kammen first used the term VIGS in 1997, to explain how plant mount defense response against viral infection [5]. Silencing of gene by VIGS can be achieved by infecting plant with a recombinant VIGS vector that harbors the host-derived target gene sequence [6]. Several RNA and DNA viruses of plants (Table 1) were engineered to develop VIGS vectors for targeted silencing of plant genes by producing double-stranded RNA, leading to the activation of the antiviral RNA silencing pathway [1, 7, 8]. These are *Tobacco mosaic virus* (TMV) [9], *Potato virus X* (PVX) [6], *Tomato golden mosaic virus* (TGMV) [10], *Bean pod mottle virus* (BPMV) [11], *Turnip yellow mosaic virus* (TYMV) [12], *Apple latent spherical virus* (ALSV) [13–15], *Tobacco rattle virus* (TRV) [16], and *Barley stripe mosaic virus* (BSMV) [17, 18]. TMV is the first modified RNA virus used to develop VIGS method for plants. Phytoene desaturase (PDS) and protoporphyrin IX magnesium chelatase subunit H (ChlH), the enzymes of the carotenoid and chlorophyll biosynthetic pathways, respectively, are being generally used as VIGS marker genes because their silencing develops a visible photo-bleaching/yellowing phenotype in a relatively short period of time [9, 19].

Tobacco rattle virus (TRV)-based VIGS vectors have been extensively used to study silencing of plant genes and successfully tested in a variety of plant species including *Arabidopsis* [8, 20], *Nicotiana benthamiana* [7, 16, 21], *Petunia hybrida* [22, 23], *Solanum lycopersicum* [16, 24], *Capsicum annuum* [25], *Thalictrum* [26], *Fragaria ananassa* [27], *Eschscholzia californica* [28], *Aquilegia* [29], *Rosa hybrida* [30], *Mirabilis jalapa* [31], and *Jatropha curcas* [32]. TRV belongs to the *Tobravirus* genus [33], having a bipartite genome consisting of positive-sense single-stranded RNAs, named as TRV1 (or RNA1) and TRV2 (or RNA2). The RNA1 has an important role for viral movement. RNA1 genome encodes replicase proteins of 134 and 194 kDa, a movement protein (29 kDa), and a cysteine-rich protein (16 kDa). The RNA2 genome encodes the coat protein (CP) and nonstructural proteins with no role in plant infection. To modify and construct TRV-based VIGS vectors (i.e., pTRV1 and pTRV2), the viral sequences from TRV1 and TRV2 were inserted into T-DNA cassette in plant binary expression vector, carrying the duplicated *cauliflower mosaic virus* 35S promoter (2× CaMV 35S) and a *nopaline synthase* terminator (NOSt) [16, 34, 35]. *Agrobacterium*

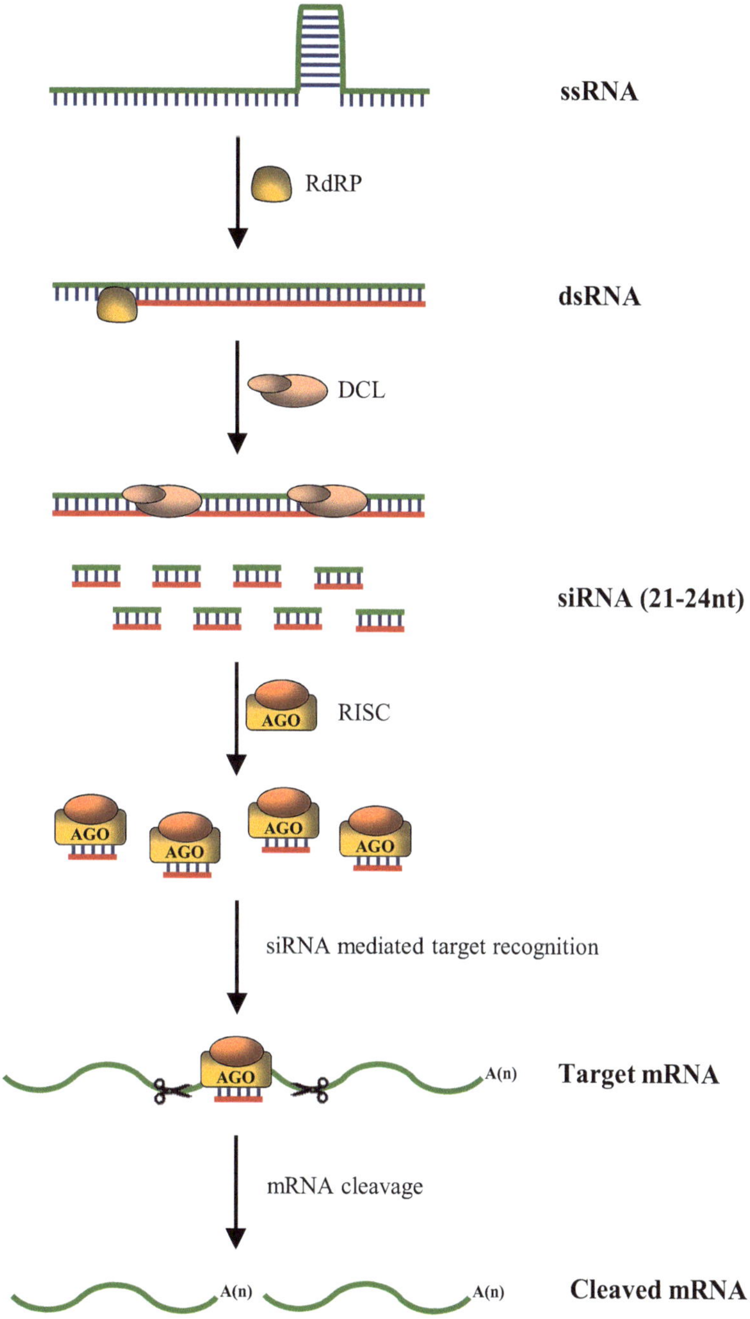

Fig. 1 Mechanism of virus-induced gene silencing. Viral RNA-dependent RNA polymerase (RdRP) catalyzes the replication of single-stranded RNA (ssRNA) to form double-stranded RNA (dsRNA). Dicer-like enzymes (DCL) recognize dsRNA and process dsRNA into small interfering RNAs (siRNAs) of about 21–24 nt. These siRNAs assemble into RNA-induced silencing complex (RISC) and unwound into ssRNA by AGO family proteins which act as guide molecules to target mRNA for degradation

Table 1
List of viruses used to develop VIGS vectors

Virus name	Target genes for silencing study	Host plant for gene silencing study	References
African cassava mosaic virus (ACMV)	PDS, SU, CYP79D2	*Nicotiana benthamiana*, *Manihot esculenta*	[48]
Tobacco curly shoot virus (TCSV)	GFP, SU, CHS, PCNA	*Solanum lycopersicon*, *Nicotiana tabacum*, *Petunia hybrida*, *Nicotiana benthamiana*	[49]
Tomato golden mosaic virus (TGMV)	SU, LUC	*Nicotiana benthamiana*	[10]
Cabbage leaf curl virus (CaLCuV)	GFP, CH42, PDS	*Arabidopsis thaliana*	[20]
Tomato yellow leaf curl china virus (TYLCCNV)	GFP, PDS, PCNA, SU	*Solanum lycopersicon*, *Nicotiana tabacum*, *Nicotiana glutinosa*, *Nicotiana benthamiana*	[50]
Cotton leaf crumple virus (CLCrV)	PDS, CHLI	*Gossypium hirsutum*	[51]
Rice tungro bacilliform virus (RTBV)	PDS, CHLH, XA21	*Oryza sativa*	[52]
Barley stripe mosaic virus (BSMV)	PDS, LR21, RAR1, SGT1, HSP90	*Hordeum vulgare*	[53]
Potato virus X (PVX)	PDS, GUS, DWARF, SSU, NFL, LFY	*Arabidopsis thaliana*, *Nicotiana benthamiana*	[6]
Bean pod mottle virus (BPMV)	PDS	*Glycine max*	[11]
Tobacco mosaic virus (TMV)	PDS, PSY	*Nicotiana benthamiana*, *Nicotiana tabacum*	[9]
Cucumber mosaic virus (CMV)	CHS, SF3′H1	*Glycine max*	[54]
Pea early browning virus (PEBV)	PDS, UNI, KOR	*Pisum sativum*, *Phaseolus vulgaris*	[36]
Tobacco rattle virus (TRV)	PDS, FLO/LFY (NFL), RBCS, CHLH	*Solanum lycopersicon*, *Arabidopsis thaliana*, *Nicotiana benthamiana*, *Ocimum basilicum*	[7, 16, 19]
Satellite tobacco mosaic virus (STMV)	PDS, RBCS, RBCL	*Nicotiana tabacum*	[55]
Turnip yellow mosaic virus (TYMV)	PDS, LFY	*Arabidopsis thaliana*	[12]

(continued)

Table 1
(continued)

Virus name	Target genes for silencing study	Host plant for gene silencing study	References
Poplar mosaic virus (PMV)	GFP	*Nicotiana benthamiana*	[56]
Apple latent spherical virus (ALSV)	PDS, SU, PCNA	*Solanum lycopersicon*, *Nicotiana tabacum*, *Nicotiana occidentalis*, *Nicotiana glutinosa*, *Nicotiana benthamiana*, *Arabidopsis thaliana*	[13]
Brome mosaic virus (BMV)	ACTIN 1, PDS, RUBISCO ACTIVASE	*Zea mays*, *Hordeum vulgare*, *Oryza sativa*	[57]
Pepper huasteco yellow veins virus (PHYVV)	COMT, PAMT, KAS	*Capsicum*	[58]
Grapevine Algerian latent virus (GALV)	GFP, CHLH	*Nicotiana benthamiana*, *Vitis vinifera*	[59, 60]

tumefaciens strains carrying pTRV1 and pTRV2 vectors are mixed in equal proportion and infiltrated into plant tissue to initiate VIGS. Different *Agrobacterium* strains (GV3101, LBA4404, EHA105) were successfully tested for VIGS in plants. Various *Agrobacterium* infection methods including spraying, vacuum infiltration, and syringe infiltration to the plants have been tested for VIGS assays in different plant species [36–39]. However, 100% silencing could not be achieved by following any of the above methods. Moreover, silencing efficiency varies experiment to experiment depending upon plant growth stage and method of inoculation [40–45].

We developed a protocol for VIGS in sweet basil (*Ocimum basilicum*) and evaluated VIGS efficiency by silencing *protoporphyrin IX magnesium chelatase subunit H* (*ObChlH*) [19]. The silencing of *ObChlH* resulted in leaf-yellowing symptoms, while empty vectors did not induce any symptoms in the agroinfiltrated leaves. In addition, by employing the VIGS protocol, we determined *in planta* role of cytochrome P450 monooxygenases (*CYP716A252* and *CYP716A253*) in the biosynthesis of ursolic acid and oleanolic acid, demonstrating the utility of TRV-based VIGS protocol to understand gene function in sweet basil [19]. This chapter describes a detailed protocol to perform VIGS assay in sweet basil using the TRV-based vectors.

2 Materials

2.1 Plant Materials

1. Sweet basil (cv. 'CIM-Saumya') seeds.

2.2 Amplification of cDNA Sequence of Target Genes

1. RNA extraction kit.
2. Reverse transcription kit.
3. Gene-specific oligonucleotides for PCR:

   ObChlH forward CCGAATTCATGACAGAGAAGAGG AAAATCT and reverse CATCTAGAAACTGTAGAGAT CTTGTCTTCA.

   CYP716A252 forward CCGAATTCATCCCCGCACAGCT-TAATTAA and reverse CATCTAGAGTACGAAAACGA TTTGTTTTT.

   CYP716A253 forward CCGAATTCTCCCTCACAAATTTT-GATCAC and reverse CATCTAGAAATCATCCTAA GAAAGATTCA.
4. High-fidelity DNA polymerase and buffer.
5. Thermal cycler.
6. Agarose gel 1.2%: 1.2 g of agarose in 100 mL 1× TAE buffer.
7. Gel electrophoresis apparatus.
8. 50× TAE buffer stock solution: 2 M Tris–HCl, pH 8.0, 1 M glacial acetic acid, 50 mM EDTA.
9. Spectrophotometer.

2.3 Cloning of cDNA Sequence into VIGS Vector

1. pTRV1 and pTRV2 vectors (Source: Arabidopsis Biological Resource Center).
2. Restriction enzymes: *EcoR*I and *Xba*I.
3. Gel extraction and PCR cleanup kit.
4. T4 DNA ligase.
5. *Escherichia coli* strain DH5α-competent cells.
6. Luria-Bertani broth: 10 g/L tryptone, 5 g/L yeast extract, 10 g/L NaCl. Combine reagents and dissolve in 800 mL deionized water, adjust pH to 7 using 1 M NaOH, and finally adjust volume to 1 L.
7. Luria-Bertani agar: 10 g/L tryptone, 5 g/L yeast extract, 10 g/L NaCl. Combine reagents and dissolve into 800 mL deionized water, and adjust pH to 7 using 1 M NaOH. Add 20 g of agar and finally adjust volume to 1 L.
8. 50 mg/mL kanamycin stock solution in sterilized Milli-Q water.
9. DNA sequencing kit.

10. Incubator shaker for 28 and 37 °C.
11. Refrigerated centrifuge.

2.4 Agroinfiltration

1. *Agrobacterium tumefaciens* strain GV3101-competent cells.
2. 25 mg/mL rifampicin stock solution in methanol.
3. YEP agar medium: 10 g/L yeast extract, 10 g/L Bacto peptone, 5 g/L NaCl. Adjust pH to 7. Add 20 g of agar and make up final volume to 1 L with deionized water.
4. 2-(N-Morpholino)ethanesulfonic acid (MES). Prepare 1 M stock solution in Milli-Q water.
5. 3′-5′-Dimethoxy 4′-hydroxyacetophenone (acetosyringone). Prepare 200 mM stock solution in dimethyl sulfoxide (DMSO).
6. Infiltration buffer: 10 mM $MgCl_2$, 10 mM MES, 200 μM acetosyringone, pH 5.6.
7. Vacuum desiccator (250 mm diameter).
8. Glass beakers (100 mL).
9. Vacuum pump.
10. Soil: perlite, peat moss, and vermiculite mix (1:1:1).
11. Growth chamber with 24 °C temperature and 16 h:8 h, light/dark cycle.

3 Methods

3.1 Plant Growth

1. Germinate seeds in pot with pre-sterilized soil [46], water gently, and keep in dark for 1 day (*see* **Note 1**).
2. Next day, transfer pot in growth room with 24 °C temperature and 16 h:8 h, light/dark cycle (*see* **Note 2**).
3. Seedlings at cotyledonary leaf stage (about 1-week-old seedlings) are used for VIGS assay using vacuum infiltration method (*see* **Note 3**).

3.2 Selection and Amplification of Target Genes

1. Isolate total RNA from 100 mg sweet basil leaf tissue using a RNA purification kit.
2. Quantify total RNA in spectrophotometer. Analyze RNA quality by determining A260:280 and A260:230 ratios, and confirm the integrity of RNA by resolving on 1.2% denaturing agarose gel [46].
3. Synthesize first-strand cDNAs from total RNA, using reverse transcriptase and oligo(dT), following manufacturer's instructions.

4. Identify sequences of the target genes for amplification. As examples, we considered a 3′-end fragment of sweet basil *protoporphyrin IX magnesium chelatase subunit H* (*ChlH*; Genbank Accession: KX679572) and cytochrome P450 monooxygenases (*CYP716A252* and *CYP716A253*; Genbank Accessions JQ958967 and JQ958968) [19] (*see* **Note 4**).
5. Amplify cDNA fragments using high-fidelity DNA polymerase and gene-specific oligonucleotides, incorporating *Eco*RI and *Xba*I restriction sites. (For more choice of restriction sites, *see* the multiple cloning site (MCS) of pTRV2.) PCR conditions were as follows: 1 min pre-incubation step at 98 °C; 35 cycles of 20 s at 98 °C (denaturation), 30 s at 54–57 °C (annealing, depending on oligonucleotide Tm), and 45 s at 72 °C (extension); and 10 min final extension step at 72 °C.
6. Analyze 5 μL of amplified PCR products by agarose gel electrophoresis to ensure the size (ObChlH 570 bp, CYP716A252 261 bp, CYP716A253 280 bp) and yield of specific DNA fragment.
7. Extract DNA from agarose gel using gel extraction kit. PCR product can also be directly purified using PCR cleanup kit.

3.3 pTRV2 Construct Preparation

1. Linearize 1 μg of pTRV2 vector with *Eco*RI and *Xba*I. Typically, a 50 μL reaction mixture contains 5 μL 10× buffer, 1 μg plasmid DNA, 1–5 U each of the restriction enzymes, and deionized water to make up total volume to 50 μL.
2. Incubate the reaction at 37 °C for 2 h.
3. Following the same protocol, digest amplified DNA fragments (500 ng) of target genes with *Eco*RI and *Xba*I.
4. Resolve the entire digested samples on 1% agarose gel, excise, and gel purify using gel extraction kit.
5. Ligate digested DNA fragments (50–100 ng) into *Eco*RI/*Xba*I sites of pTRV2 vector (50–100 ng) to create pTRV2-GOI (Fig. 2a). Add 1 μL of 10× ligase buffer, 50–100 ng of vector DNA, 50–100 ng of insert DNA, 1 unit T4 DNA ligase, and deionized water to a total reaction volume of 10 μL, and mix contents properly by pipetting up and down.
6. Incubate reactions at 16 °C for 12–16 h (overnight).
7. Transform 5 μL ligation product to chemically competent *E. coli* (strain DH5α) cells following standard heat shock method.
8. Select clones on Luria-Bertani agar media containing 50 μg/mL kanamycin.
9. Confirm positive constructs through colony PCR and restriction digestion followed by DNA sequencing.

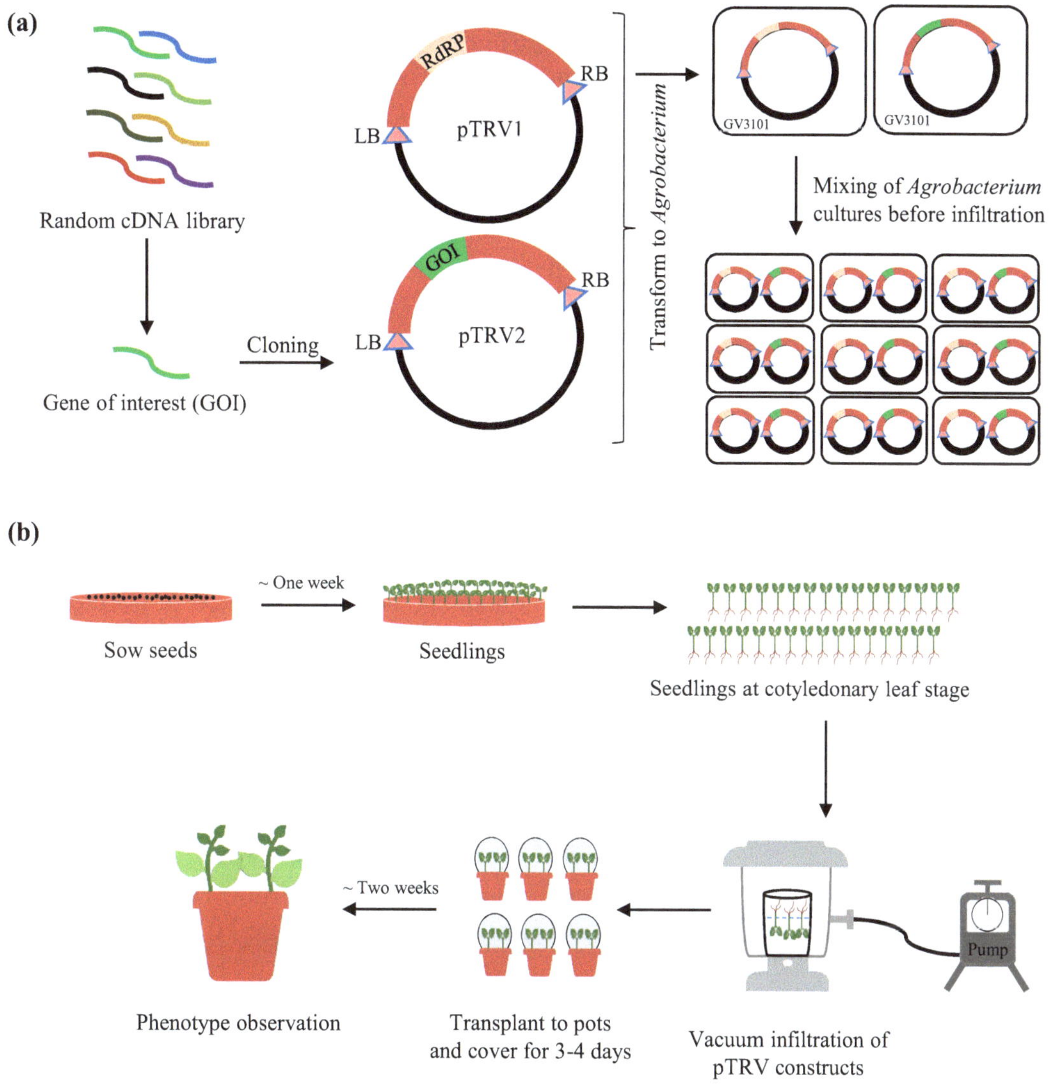

Fig. 2 Overview of *Tobacco rattle virus*-based virus-induced gene silencing in sweet basil. (**a**) VIGS is a multistep process. The gene sequence of interest (usually 250–500 base pairs) is first selected from random cDNA library and cloned into TRV-based vector (pTRV2) under the control of the 2× CaMV35S promoter. The plasmids (pTRV1 containing viral RNA-dependent RNA polymerase (RdRp) and a movement protein and pTRV2 containing the coat protein and the gene of interest) transformed separately in *Agrobacterium*, and strains are mixed in a 1:1 ratio before plant infiltration. (**b**) Sweet basil seedlings at cotyledonary leaf stage were considered for VIGS assay [19]. Seedlings were vacuum infiltrated with suspension of *Agrobacterium* strain harboring empty vectors (pTRV1/pTRV2) or pTRV1/pTRV2-GOI in a ratio of 1:1 (v/v) at an OD600 of 1.5. Vacuum infiltrated seedlings were transplanted into soil, covered with polyethylene bags, and maintained in the dark for 12 h. Seedlings were further grown at 22–24 °C temperature with a 16 h:8 h, light/dark cycle. Silencing phenotype usually develops at 10–14 days post-agroinfiltration

3.4 Agrobacterium *Transformation*

1. Transform each construct (pTRV2-GOI) individually into *Agrobacterium* by freeze-thaw method (Fig. 2a [47]). For use as negative control, transform pTRV2 empty vector in *Agrobacterium* (*see* **Note 5**).
2. Transform pTRV1 vector separately into *Agrobacterium* (Fig. 2a).
3. Select transformed *Agrobacterium* on YEP agar media containing 50 μg/mL kanamycin and 25 μg/mL rifampicin, and confirm positive colonies by colony PCR (*see* **Note 6**).

3.5 Preparation of Agrobacterium *Inoculum*

1. Separately inoculate *Agrobacterium* strains containing either pTRV1, pTRV2, and pTRV2-GOI plasmids into 5 mL Luria-Bertani broth medium containing 50 μg/mL kanamycin and 25 μg/mL rifampicin, and grow for overnight at 28 °C with continuous shaking (*see* **Note 7**).
2. Add 50 μL of overnight-grown culture of *Agrobacterium*, to secondary inoculum on next day, into 50 mL Luria-Bertani broth medium containing 10 mM MES, 20 μM acetosyringone, 50 μg/mL kanamycin, and 25 μg/mL rifampicin, and grow for overnight at 28 °C with continuous shaking (*see* **Note 8**).
3. Harvest *Agrobacterium* at 4000 × *g* by centrifugation for 15 min at 28 °C.
4. Resuspend the obtained pellet in infiltration buffer to achieve a final OD_{600} of 1.5 (*see* **Note 9**).
5. Incubate *Agrobacterium* suspension at room temperature without vigorous shaking for 3–4 h (*see* **Note 10**).
6. Mix *Agrobacterium* strains carrying pTRV1 and pTRV2-GOI together at the ratio of 1:1, just before agroinfiltration (Fig. 2a). For use as negative control, mix *Agrobacterium* strains carrying pTRV1 and pTRV2 empty vectors.

3.6 Infiltration of Agrobacterium *into Sweet Basil Seedlings*

1. Gently uproot seedlings having indication of emergence of first pair of true leaf [19] (Fig. 2b) (*see* **Note 11**).
2. Carefully wash roots under running tap water to avoid damage to roots. Wash whole seedling with RO water. To remove excess water, place them on tissue paper (*see* **Note 12**).
3. Submerge seedlings into *Agrobacterium* suspension (prepared by mixing at the ratio of 1:1 of *Agrobacterium* harboring pTRV1 and pTRV2-GOI) in a glass beaker (100 mL), and place the beaker inside the vacuum desiccator, connected with a vacuum pump (Figs. 2b and 3a) (*see* **Note 13**).
4. Turn ON the pump until maximum pressure reaches to 750 mm of mercury (Hg), and continue for an additional 90 s (*see* **Note 14**).

Fig. 3 Virus-induced gene silencing of sweet basil *ObChlH*. (**a**) Photograph of a vacuum pump connected with a desiccator used in VIGS study. (**b**) Silencing of *ObChlH*, a VIGS marker gene caused leaf-yellowing phenotype in sweet basil. Symptoms were clear on the first true leaf pair of agroinfiltrated sweet basil when compared with agroinfiltrated vector control plants. Photographs were taken about 2 weeks post-vacuum infiltration

5. Turn off the pump, wait for 3 min, and after that slowly release the pressure (*see* **Note 15**).
6. Transplant agroinfiltrated seedlings into pots with soil, cover with clear polyethylene bag, and keep in dark for 12 h (Fig. 2b) (*see* **Note 16**).
7. Transfer seedlings to a temperature (22–24 °C) and humidity (60%) controlled growth chamber with a 16 h: 8 h, light/dark cycle.

8. Remove cover after 3–4 days post-infiltration and ideally water them once a week (*see* **Note 17**).
9. Silencing phenotype could be clearly visible around 10–14 days post-infiltration (Figs. 2b and 3b) (*see* **Note 18**).

4 Notes

1. Bacterial and fungal diseases are the common problem of sweet basil. They mainly affect plant roots. To combat this problem, use pre-sterilized soil. Sweet basil seeds are myxospermic. When they come into contact with water, form a thick layer of mucilage and swell. Therefore, seeds do not need much water in order to germinate.
2. Sweet basil is hot climate herb. The germination efficiency will be higher above 24 °C. Insufficient light will lead to hypocotyl elongation, and seedlings will become skinny and fragile.
3. The selection of sweet basil seedlings' developmental stage is an important factor for successful gene silencing experiment. Select only those seedlings which have an indication of emergence of first pair of true leaf [19].
4. Select DNA fragments in the range of 250–500 bp for gene silencing. To avoid silencing of non-target gene(s), only include the sequences that are very specific to the target gene. 5′ or 3′ untranslated region (UTR) of the target gene can also be included. Avoid homopolymeric regions (poly A or poly T).
5. We tested *Agrobacterium* strain GV3101; however, other strains (LBA4404, EHA105) might also work. It is desirable to use both positive and negative controls to confirm VIGS effectiveness.
6. We confirm positive colonies of *Agrobacterium* through colony PCR analysis. However, other methods such as colony hybridization might also be followed.
7. Do not use older plates of agrobacteria for making an inoculum. Use freshly grown agrobacteria streaked from a glycerol stock frozen at −80 °C.
8. Try to avoid overgrowth of *Agrobacterium* culture. The optical density of secondary inoculum generally should not exceed an OD_{600} of 2. If the final OD_{600} is >2, the amount of dead bacterial cells will be higher which might lead to decreased VIGS efficiency.
9. Prepare fresh infiltration buffer on the day of infiltration and adjust pH 5.6 with KOH.
10. Acetosyringone induces the expression of *Vir* genes and enhances the transformation efficiency of *Agrobacterium*.

Therefore, incubation of *Agrobacterium* in infiltration medium is preferable.

11. Always select healthy and disease-free seedlings for infiltration. Carefully uproot seedlings to avoid damage to the roots.
12. It is necessary to wash the roots of uprooted seedlings to remove the soil adhering to the roots. The presence of soil will contaminate *Agrobacterium* suspension for infiltration. Do not leave uprooted seedlings on the tissue paper for a longer time, and immediately submerge into the *Agrobacterium* suspension.
13. Take maximum ten seedlings per *Agrobacterium* infiltration. Place them in inverted position (roots up and cotyledons down) in a glass beaker (do not exceed the *Agrobacterium* suspension volume more than 60 mL). Make sure that entire shoot is completely submerged into infiltration buffer containing *Agrobacterium* suspension (Figs. 2b and 3a). If handling with different pTRV2 constructs, change glass beaker to avoid cross contamination.
14. Pressure must be kept within the desirable limits. Very high or low pressure is not ideal for plant infiltration. Moreover, longer exposure time results in reduced survival of seedlings post-vacuum infiltration.
15. Quick release of pressure can contaminate desiccator with *Agrobacterium* suspension. Change *Agrobacterium* suspension if it turns greenish in color.
16. It is advisable to check the cotyledons after infiltration; slightly change in color is a mark of successful infiltration. After each agroinfiltration, change gloves or wipe with 70% ethanol.
17. Infiltrated seedlings typically recover quickly when covered with clear polyethylene bags for 3–4 days (12 h in dark and 3 days under light).
18. Silencing of marker gene (*protoporphyrin IX magnesium chelatase subunit H*) is visible as yellowish color in the first true leaf pair of infected plants (Fig. 3b).

Acknowledgments

This work was supported by the research grant to S.G. from the Department of Biotechnology (BT/08/IYBA/2014-13).

References

1. Burch-Smith TM, Anderson JC, Martin GB, Dinesh-Kumar SP (2004) Applications and advantages of virus-induced gene silencing for gene function studies in plants. Plant J 39:734–746
2. Tijsterman M, Ketting RF, Plasterk RHA (2002) The genetics of RNA silencing. Annu Rev Genet 36:489–519
3. Ziberman D, Cao X, Jacobsen SE (2003) ARGONAUTE4 control of locus specific siRNA accumulation and DNA and histone methylation. Science 299:716–719
4. Senthil-Kumar M, Ajith A, Uppalapati SR, Mysore KS (2008) Virus-induced gene silencing and its applications. CAB Rev Perspect Agri Vet Sci Nutri Nat Res 3:11–18
5. van Kammen A (1997) Virus-induced gene silencing in infected and transgenic plants. Trends Plant Sci 2:409–411
6. Ruiz MT, Voinnet O, Baulcombe DC (1998) Initiation and maintenance of virus induced gene silencing. Plant Cell 10:937–946
7. Ratcliff F, Martin-Hernandez AM, Baulcombe DC (2001) Technical advance. Tobacco rattle virus as a vector for analysis of gene function by silencing. Plant J 25:237–245
8. Burch-Smith TM, Schiff M, Liu Y, Dinesh-Kumar SP (2006) Efficient virus-induced gene silencing in Arabidopsis. Plant Physiol 142:21–27
9. Kumagai MH, Donson J, Della-Cioppa G, Harvey D, Hanley K, Grill LK (1995) Cytoplasmic inhibition of carotenoid biosynthesis with virus-derived RNA. Proc Natl Acad Sci U S A 92:1679–1683
10. Peele C, Jordan CV, Muangsan N, Turnage M, Egelkrout E, Eagle P, Hanley-Bowdoin L, Robertson D (2001) Silencing of a meristematic gene using geminivirus-derived vectors. Plant J 27:357–366
11. Zhang C, Ghabrial SA (2006) Development of Bean pod mottle virus-based vectors for stable protein expression and sequence-specific virus-induced gene silencing in soybean. Virology 344:401–411
12. Pflieger S, Blanchet S, Camborde L, Drugeon G, Rousseau A, Noizet M, Planchais S, Jupin I (2008) Efficient virus-induced gene silencing in Arabidopsis using a 'one-step' TYMV-derived vector. Plant J 56:678–690
13. Igarashi A, Yamagata K, Sugai T, Takahashi Y, Sugawara E, Tamura A, Yaegashi H, Yamagishi N, Takahashi T, Isogai M, Takahashi H, Yoshikawa N (2009) Apple latent spherical virus vectors for reliable and effective virus-induced gene silencing among a broad range of plants including tobacco, tomato, *Arabidopsis thaliana*, cucurbits, and legumes. Virology 386:407–416
14. Yamagishi N, Yoshikawa N (2009) Virus-induced gene silencing in soybean seeds and the emergence stage of soybean plants with Apple latent spherical virus vectors. Plant Mol Biol 71:15–24
15. Sasaki S, Yamagishi N, Yoshikawa N (2011) Efficient virus-induced gene silencing in apple, pear and Japanese pear using Apple latent spherical virus vectors. Plant Methods 7:15
16. Liu Y, Schiff M, Marathe R, Dinesh-Kumar SP (2002b) Tobacco *Rar1*, *EDS1* and *NPR1/NIM1* like genes are required for N-mediated resistance to tobacco mosaic virus. Plant J 30:415–429
17. Bruun-Rasmussen M, Madsen CT, Jessing S, Albrechtse M (2007) Stability of Barley stripe mosaic virus-induced gene silencing in barley. Mol Plant-Microbe Interact 20:1323–1333
18. Lee WS, Hammond-Kosack KE, Kanyuka K (2012) Barley stripe mosaic virus-mediated tools for investigating gene function in cereal plants and their pathogens: virus-induced gene silencing, host-mediated gene silencing, and virus-mediated overexpression of heterologous protein. Plant Physiol 160:582–590
19. Misra RC, Sharma S, Sandeep GA, Chanotiya CS, Ghosh S (2017) Two CYP716A subfamily cytochrome P450 monooxygenases of sweet basil play similar but nonredundant roles in ursane- and oleanane-type pentacyclic triterpene biosynthesis. New Phytol 214:706–720
20. Turnage MA, Muangsan N, Peele CG, Robertson D (2002) Geminivirus-based vectors for gene silencing in Arabidopsis. Plant J 30:107–114
21. Liu Y, Nakayama N, Schiff M, Litt A, Irish VF, Dinesh-Kumar SP (2004) Virus induced gene silencing of a DEFICIENS ortholog in *Nicotiana benthamiana*. Plant Mol Biol 54:701–711
22. Chen JC, Jiang CZ, Gookin TE, Hunter DA, Clark DG, Reid MS (2004) Chalcone synthase as a reporter in virus-induced gene silencing studies of flower senescence. Plant Mol Biol 55:521–530

23. Marton I, Zuker A, Shklarman E, Zeevi V, Tovkach A, Roffe S, Ovadis M, Tzfira T, Vainstein A (2010) Nontransgenic genome modification in plant cells. Plant Physiol 154:1079–1087
24. Fu DQ, Zhu BZ, Zhu HL, Jiang WB, Luo YB (2005) Virus-induced gene silencing in tomato fruit. Plant J 43:299–308
25. Chung E, Seong E, Kim YC, Chung EJ, Oh SK, Lee S, Park JM, Joung YH, Choi D (2004) A method of high frequency virus-induced gene silencing in chili pepper (*Capsicum annuum* L. cv. Bukang). Mol Cells 17:377–380
26. Di Stilio VS, Kumar RA, Oddone AM, Tolkin TR, Salles P, McCarty K (2010) Virus-induced gene silencing as a tool for comparative functional studies in Thalictrum. PLoS One 5: e12064
27. Jia HF, Chai YM, Li CL, Lu D, Luo JJ, Qin L, Shen YY (2011) Abscisic acid plays an important role in the regulation of strawberry fruit ripening. Plant Physiol 157:188–199
28. Wege S, Scholz A, Gleissberg S, Becker A (2007) Highly efficient virus-induced gene silencing (VIGS) in California poppy (*Eschscholzia californica*): an evaluation of VIGS as a strategy to obtain functional data from non-model plants. Ann Bot 100:641–649
29. Gould B, Kramer EM (2007) Virus-induced gene silencing as a tool for functional analyses in the emerging model plant *Aquilegia* (columbine, Ranunculaceae). Plant Methods 3:6
30. Ma N, Xue JQ, Li YH, Liu XJ, Dai FW, Jia WS, Luo YB, Gao JP (2008) *Rh-PIP2;1*, a rose aquaporin gene, is involved in ethylene-regulated petal expansion. Plant Physiol 148:894–907
31. Singh A, Liang YC, Kumar P, Reid M, Jiang CZ (2012) Co-silencing of the Mirabilis antiviral protein (MAP) permits virus-induced gene silencing (VIGS) of other genes in Four O'Clock plants (*Mirabilis jalapa*). J Hortic Sci Biotech 87:334–340
32. Ye J, Qu J, Bui HT, Chua NH (2009) Rapid analysis of *Jatropha curcas* gene functions by virus-induced gene silencing. Plant Biol J 7:964–976
33. MacFarlane SA (1999) Molecular biology of the tobraviruses. J Gen Virol 80:2799–2807
34. Liu Y, Schiff M, Dinesh-Kumar SP (2002) Virus-induced gene silencing in tomato. Plant J 6:777–786
35. Dinesh-Kumar SP, Anandalakshmi R, Marathe R, Schiff M, Liu Y (2003) Virus-induced gene silencing. Methods Mol Biol 236:287–294
36. Constantin GD, Krath BN, MacFarlane SA, Nicolaisen M, Johansen IE, Lund OS (2004) Virus-induced gene silencing as a tool for functional genomics in a legume species. Plant J 40:622–631
37. Padmanabhan M, Dinesh-Kumar SP (2009) Virus-induced gene silencing as a tool for delivery of dsRNA into plants. Cold Spring Harb Protoc 2:5139
38. Grønlund M, Olsen A, Johansen EI, Jakobsen I (2010) Protocol: using virus induced gene silencing to study the arbuscular mycorrhizal symbiosis in *Pisum sativum*. Plant Methods 6:28
39. Zhang CQ, Bradshaw JD, Whitham SA, Hill JH (2010) The development of an efficient multipurpose Bean pod mottle virus viral vector set for foreign gene expression and RNA silencing. Plant Physiol 153:52–65
40. Li C, Zhang ZC, Ghebremariam KM, Wang LH, Wu L, Liang Y (2014) A novel method for the evaluation of virus-induced gene silencing efficiency. Genet Mol Res 13:9443–9452
41. Kandoth PK, Heinz R, Yeckel G, Gross NW, Juvale PS, Hill J, Whitham SA, Baum TJ, Mitchum MG (2013) A virus-induced gene silencing method to study soybean cyst nematode parasitism in *Glycine max*. BMC Res Notes 6:255
42. Velásquez AC, Chakravarthy S, Martin GB (2009) Virus-induced gene silencing (VIGS) in *Nicotiana benthamiana* and tomato. J Vis Exp 28:1292
43. Broderick SR, Jones ML (2014) An optimized protocol to increase virus-induced gene silencing efficiency and minimize viral symptoms in Petunia. Plant Mol Biol Report 32:219–233
44. Senthil-Kumar M, Mysore KS (2011) New dimensions for VIGS in plant functional genomics. Trends Plant Sci 16:656–665
45. Kim KH, Lim S, Kang YJ, Yoon MY, Nam M, Jun TH, Seo MJ, Baek SB, Lee JH, Moon JK, Lee SH, Lee SH, Lim HS, Moon JS, Park CH (2016) Optimization of a virus-induced gene silencing system with Soybean yellow common mosaic virus for gene function studies in soybeans. Plant Pathol J 32:112–122
46. Misra RC, Maiti P, Chanotiya CS, Shanker K, Ghosh S (2014) Methyl jasmonate-elicited transcriptional responses and pentacyclic triterpene biosynthesis in sweet basil. Plant Physiol 164:1028–1044
47. Weigel D, Glazebrook J (2006) Transformation of *Agrobacterium* using the freeze–thaw method. CSH Protoc 7:4666
48. Fofana IB, Sangare A, Collier R, Taylor C, Fauquet CM (2004) A gemini virus induced gene

silencing system for gene function validation in cassava. Plant Mol Biol 56:613–624

49. Huang C, Xie Y, Zhou X (2009) Efficient virus-induced gene silencing in plants using a modified geminivirus DNA1 component. Plant Biotechnol J 7:254–265
50. Tao X, Zhou X (2004) A modified viral satellite DNA that suppresses gene expression in plants. Plant J 38:850–860
51. Tuttle JR, Idris AM, Brown JK, Haigler CH, Robertson D (2008) Geminivirus-mediated gene silencing from *Cotton leaf crumple virus* is enhanced by low temperature in cotton. Plant Physiol 148:41–50
52. Kant R, Dasgupta I (2017) Phenotyping of VIGS-mediated gene silencing in rice using a vector derived from a DNA virus. Plant Cell Rep 36:1159–1170
53. Holzberg S, Brosio P, Gross C, Pogue GP (2002) Barley stripe mosaic virus-induced gene silencing in a monocot plant. Plant J 30:315–327
54. Nagamatsu A, Masuta C, Senda M, Kasai HMA, Hong J, Kitamura K, Abe J, Kanazawa A (2007) Functional analysis of soybean genes involved in flavonoid biosynthesis by virus-induced gene silencing. Plant Biotechnol J 5:778–790
55. Gossele VV, Fache II, Meulewaeter F, Cornelissen M, Metzlaff M (2002) SVISS—a novel transient gene silencing system for gene function discovery and validation in tobacco. Plant J 32:859–866
56. Naylor M, Reeves J, Cooper JI, Edwards ML, Wang H (2005) Construction and properties of a gene-silencing vector based on *Poplar mosaic virus* (genus *Carlavirus*). J Virol Methods 124:27–36
57. Ding XS, Schneider WL, Chaluvadi SR, Rouf Mian RM, Nelson RS (2006) Characterization of a Brome mosaic virus strain and its use as a vector for gene silencing in monocotyledonous hosts. Mol Plant-Microbe Interact 19:1229–1239
58. Abraham-Juârez MR, Rocha-Granados MC, Lopez MG, Rivera-Bustamante RF, Ochoa-Alejo N (2008) Virus-induced silencing of *Comt*, *pAmt* and *Kas* genes results in a reduction of capsaicinoid accumulation in chili pepper fruits. Planta 227:681–695
59. Park SH, Choi H, Kim S, Cho WK, Kim KH (2016) Development of virus-induced gene expression and silencing vector derived from Grapevine Algerian latent virus. Plant Pathol J 32:371–376
60. Muruganantham M, Moskovitz Y, Haviv S, Horesh T, Fenigstein A, Preez J, Stephan D, Burger JT, Mawassi M (2009) Grapevine virusA-mediated gene silencing in *Nicotiana benthamiana* and *Vitis vinifera*. J Virol Methods 155:167–174
61. Hileman LC, Drea S, Martino G, Litt A, Irish VF (2005) Virus-induced gene silencing is an effective tool for assaying gene function in the basal eudicot species *Papaver somniferum* (opium poppy). Plant J 44:334–341
62. Liscombe DK, O'Connor SE (2011) A virus-induced gene silencing approach to understanding alkaloid metabolism in *Catharanthus roseus*. Phytochemistry 72:1969–1977
63. Singh AK, Dwivedi V, Rai A, Pal S, Reddy SG, Rao DK, Shasany AK, Nagegowda DA (2015) Virus-induced gene silencing of *Withania somnifera* squalene synthase negatively regulates sterol and defence-related genes resulting in reduced withanolides and biotic stress tolerance. Plant Biol J 13:1287–1299

Chapter 11

Virus-Induced Gene Silencing for Functional Genomics in *Withania somnifera*, an Important Indian Medicinal Plant

Dikki Pedenla Bomzan, H. B. Shilpashree, P. Anjali, Sarma Rajeev Kumar, and Dinesh A. Nagegowda

Abstract

Virus-induced gene silencing (VIGS) has emerged as a fast and efficient reverse and forward genetics tool to study gene function in model plants as well as in agriculturally important plants. In addition, VIGS approach has been successfully used to provide insights into the role of several genes and regulators involved in plant secondary metabolism. Ashwagandha (*Withania somnifera*) is an important Indian medicinal plant that accumulates pharmacologically important triterpenoid steroidal lactones, which are collectively termed as withanolides. *W. somnifera* being a highly recalcitrant plant for genetic transformation, *Tobacco rattle virus* (TRV)-mediated VIGS was established by our group to facilitate understanding of withanolides' pathway. Here, we describe a detailed procedure to carry out VIGS for gene function studies in *W. somnifera*.

Key words Ashwagandha, Functional genomics, pTRV vector, Secondary metabolism, Virus-induced gene silencing, *Withania somnifera*, Withanolides

1 Introduction

Withania somnifera (L.) Dunal, a member of Solanaceae family, commonly known as Ashwagandha, winter cherry, or Indian ginseng, is an important medicinal plant that has been used in Ayurvedic medicine for over 3000 years and also is relevant for modern medicine [1, 2]. The medicinal value of the plant is attributed to the presence of naturally occurring triterpenoid steroidal lactones which are collectively known as withanolides that exhibit diverse biological activities. A number of withanolides or extracts containing withanolides exhibit promising pharmacological activities like antitumor, anti-inflammation, cardioprotective, and neuroprotective properties. For instance, trials in cell culture and animal

Dikki Pedenla Bomzan and H. B. Shilpashree contributed equally.

Vincent Courdavault and Sébastien Besseau (eds.), *Virus-Induced Gene Silencing in Plants: Methods and Protocols*, Methods in Molecular Biology, vol. 2172, https://doi.org/10.1007/978-1-0716-0751-0_11,

research support the use of *W. somnifera* extracts for the treatment of Alzheimer's disease, and cognitive and neurological disorders [3, 4]. So far, more than 40 withanolides have been isolated from *W. somnifera* in which withaferin A and withanolide D have been shown to inhibit angiogenesis, Notch-1, and NFκB in cancer cells and induce apoptosis in breast cancer cells [5–8]. In addition, the diversity in structure and activities of withanolides makes them a promising source for novel compounds for nutraceutical use [9].

Despite the importance of withanolides, their application is still limited owing to their low *in planta* accumulation (0.001–0.5% dry weight) and the challenging nature of pure compound isolation from complex mixture. This can be overcome by metabolic engineering of plants, cell cultures, or heterologous microbial production platforms by synthetic biology approaches. However, the pathway for withanolides biosynthesis is unknown especially post-sterol formation, preventing their sustainable production via metabolic engineering. Moreover, *W. somnifera* is highly recalcitrant for genetic transformation making *in planta* gene function studies very difficult. Hence, an efficient *in planta* functional assay is very much necessary to characterize genes involved in withanolides biosynthesis. Lately, virus-induced gene silencing (VIGS) has emerged as a powerful reverse/forward genetic tool to study the function of genes and regulators. Using *Tobacco rattle virus* (TRV)-based approach, we have established a VIGS strategy in *W. somnifera* to facilitate *in planta* functional analysis of genes related to withanolides biosynthesis. VIGS in *W. somnifera* was optimized using phytoene desaturase (*PDS*) gene which served as a visual marker. Further, using the established protocol, the role of squalene synthase (*SQS*) and transcriptional regulator *WRKY1* in withanolides biosynthesis was studied [10, 11]. Subsequently, few other groups have also utilized our VIGS protocol to study function of genes involved in *W. somnifera* withanolides biosynthesis [9, 12–14] (Table 1). Here, we describe the detailed protocol of VIGS in *W. somnifera*. The established method can be used for the elucidation of biosynthetic steps involved in withanolides formation in *W. somnifera*. Figure 1 depicts the schematic workflow for achieving TRV-mediated VIGS in *W. somnifera*.

2 Materials

2.1 Bacterial Strains, Vectors, Plant, and Other Items

1. *Escherichia coli* strain XL1-Blue.
2. *Agrobacterium tumefaciens* strain GV3101.
3. Plasmids pTRV1 and pTRV2 (procured from ABRC, The Ohio State University, USA).
4. Seeds of *Withania somnifera* cv. Poshita or of any other Ashwagandha variety.

Table 1
List of genes studied in *W. somnifera* using VIGS approach

Gene(s)	In planta function	Effect of VIGS	Reference
Sterol Δ^{24}*-isomerase* (*24ISO*)	24ISO catalyzes the conversion of 24-methylenecholesterol to 24-methyldesmosterol	Silencing of *24ISO* resulted in the reduction of withaferin A	[9]
Squalene synthase (*SQS*)	SQS converts farnesyl diphosphate (FPP) to squalene, which is the precursor for phytosterols and withanolides	*SQS* silencing negatively affected sterol and defense-related genes leading to reduced squalene, phytosterols, withanolides, and biotic stress tolerance	[10]
WRKY1	*WRKY1* regulates genes of phytosterol and withanolide biosynthesis	VIGS of *WsWRKY1* negatively regulated phytosterol and withanolide biosynthesis and defense against biotic stress	[11]
Sterol glycosyltransferase (*SGT*) *SGTL1*, *SGTL2*, and *SGTL4*	SGTs catalyze transfer of glycon moiety to sterols. *SGTs* are involved in sterol modifications and participate in metabolic flexibility during stress	Silencing of *SGT* increased withanolide A, withaferin A, sitosterol, and stigmasterol accumulation and in turn reduced withanoside V content. In addition, *SGT* silencing also affected plant growth and development and resulted in higher susceptibility during heat stress	[12, 13]
1-Deoxy-D-xylulose-5-phosphate synthase (*DXS*) *1-Deoxy-D-xylulose-5-phosphate reductoisomerase* (*DXR*) *Hydroxymethylglutaryl-CoA reductase* (*HMGR*) *Farnesyl pyrophosphate synthase* (*FPPS*) *7-dehydrocholesterol reductase* (*DWF5*)	Genes of MEP and MVA pathway provide metabolic precursors for withanolides biosynthesis pathway	Silencing of these genes reduced total as well as specific withanolides in leaf tissues and the expression of other pathway genes. Development of unique phenotypes was observed in VIGS-silenced lines	[14]

5. Plastic trays and pots containing potting mixture: sterile soilrite, soil, and vermicompost in the ratio of 1:1:1.
6. Plastic cling wrap.
7. Cellulose acetate filter (0.2 μm).

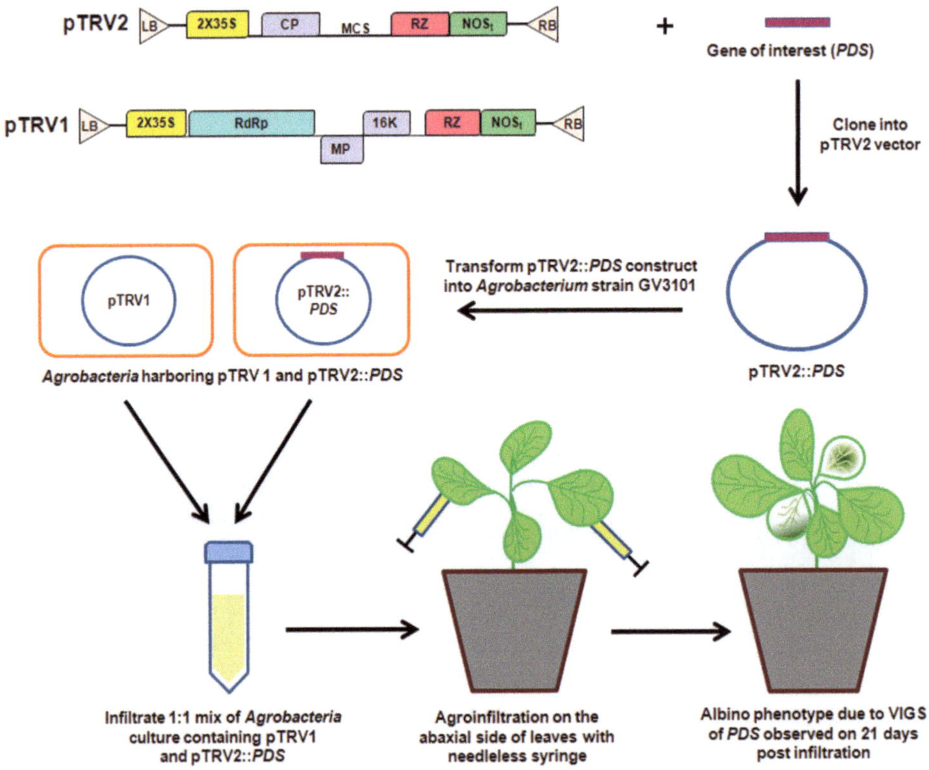

Fig. 1 Schematic representation of *Tobacco rattle virus* (TRV)-mediated silencing of *phytoene desaturase* (*PDS*) in *Withania somnifera*. Fragment of gene of interest (GOI, *PDS* in this case) is cloned into multiple cloning site of pTRV2 vector, and the resulting construct is transformed into *Agrobacterium tumefaciens* GV3101. *Agrobacteria* harboring pTRV1 and pTRV2::*PDS* are grown independently, and the cultures are mixed in 1:1 ratio for agroinfiltration [15]

2.2 Molecular Biology

1. cDNA of *W. somnifera*.
2. High-fidelity polymerase, buffer, and dNTP mix.
3. Blunt PCR cloning vector kit.
4. *PDS* primers: *WsPDS* F 5′-TCTAGAGAGATTGTTATTGCTGGTGCAGGT-3′ and *WsPDS* R 5′-CTCGAGAGGCACACCTTGCTTTCTCATCCA-3′.
5. T4 DNA ligase.
6. 50× TAE buffer stock.
7. DNA gel purification kit.
8. Plasmid purification kit.
9. RNA extraction reagent (e.g., TRIzol, Ambion, USA).
10. DNAse I (RNase-free).
11. High-capacity cDNA reverse transcription kit.

12. RT-qPCR primers: *WsPDS* RTF 5′-GCTTGGAGGGCAATCTTATG-3′ and *WsPDS* RTR 5′-GTTAAGCGCCTTTGACATGGC-3′.
13. SYBR Green qPCR Master Mix.

2.3 Equipment

1. Greenhouse.
2. Plant growth chamber at 20–22 °C, 70% relative humidity, and 16/8 h light-dark cycle.
3. Needleless syringe (5 mL).
4. Agarose gel electrophoresis unit.
5. Spectrophotometer.
6. Thermal cycler.
7. Refrigerated incubator shaker.
8. Refrigerated centrifuge.
9. Real-time PCR system.
10. Sonicating bath.
11. HPLC-DAD equipped with reverse phase C-18 (250 mm × 4.60 mm, 5 μm) column.
12. GC-FID equipped with HP-5 fused-silica capillary coated with 5% phenylmethyl siloxane column (30 m × 0.320 mm, 0.25 μm film thickness).

2.4 Buffers and Solutions

1. Infiltration buffer: weigh 0.952 g of $MgCl_2$ and 1.95 g of 2-(N-morpholino)ethanesulfonic acid (MES), and add both components into 950 mL sterile distilled water. To this, add 1 mL of 200 mM acetosyringone prepared in dimethyl sulfoxide (DMSO). Adjust the pH of the solution to 5.6 using 10 N NaOH. Finally make up the volume of the buffer to 1 L.
2. Yeast extract peptone (YEP) broth: for preparing 1 L of YEP medium, add 10 g of yeast extract, 10 g of peptone, and 5 g of NaCl into 1 L of distilled water. Sterilize the dissolved media by autoclaving at 121 °C with 15 psi for 15 min.
3. YEP agar media: for 1 L of medium, add 10 g yeast extract, 10 g of peptone, 5 g of NaCl, and 15 g of agar into 1 L of distilled water. Sterilize the dissolved media by autoclaving at 121 °C with 15 psi for 15 min.
4. Make stock solutions of 100 mg/mL ampicillin, 50 mg/mL kanamycin, and 50 mg/mL gentamicin in molecular biology grade water. Prepare stock solution of 60 mg/mL rifampicin in DMSO. Filter sterilize the stock solutions using cellulose acetate filter (0.2 μm) and store as aliquots in −20 °C.
5. 50 mM EDTA pH 8.0.

6. 50× TAE buffer stock: for 100 mL of 50× buffer, add 24.2 g of Tris base to 5.7 mL glacial acetic acid and 10 mL EDTA (0.5 M; pH 8.0). Make up the volume to 100 mL using distilled water.
7. Solvent A for HPLC: dissolve 1.02 mM of anhydrous potassium dihydrogen orthophosphate (KH_2PO_4) and 0.05% orthophosphoric acid (H_3PO_4) in HPLC grade water.
8. Solvent B for HPLC: gradient grade acetonitrile.
9. 20 mM calcium chloride solution.
10. HPLC grade water.
11. 6% KOH in methanol.
12. Chloroform.
13. 75% ethanol.
14. Isopropanol.
15. 100% and 70% methanol.
16. 7:3 (v/v) $CHCl_3$/methanol.
17. Hexane (GC grade).
18. Withanolide and phytosterol standards.
19. 40% glycerol.
20. Activated charcoal.
21. Sodium sulfate (Na_2SO_4).
22. Dichloromethane containing 120 ng of estrone (444.4 nM).

3 Methods

3.1 Seed Germination and Preparation of Plant Material for VIGS

1. Sow *W. somnifera* seeds by casting method in a seedbed of 360 × 280 × 75 (l × b × h mm) dimension containing potting mixture in a greenhouse (*see* **Note 1**).
2. The plant takes approximately 2–3 weeks to reach two- to three-leaf stage. At this stage, plants are transplanted from seedbed into individual pots filled with potting mixture. During transplantation gently uproot the plants using blunt end forceps without damaging roots, and carefully transfer to the pot and add topsoil. Moisture should be maintained by sprinkling water on a daily basis.
3. Two weeks after transplantation, the plants of four- to six-leaf stage can be used for agroinfiltration to silence specific target gene(s).

3.2 Construction of pTRV2 Vector and Transformation into *A. tumefaciens*

3.2.1 Cloning of the Gene of Interest (GOI) in pTRV2

1. The sequence of the GOI (*PDS* in this case) for VIGS is selected from the unique region (~500 bp) preferably from 5′ or 3′ untranslated region (UTR) to avoid the possibility of cross-silencing of genes having homologous sequences [15] (*see* **Note 2**). PCR amplify the selected sequence using *WsPDS* forward and reverse primers having *Xba*I and *Xho*I restriction sites, respectively. Perform PCR amplification with high-fidelity DNA polymerase using *W. somnifera* cDNA as template. PCR conditions are as follows: initial denaturation at 94 °C for 3 min, followed by 30 cycles of 94 °C for 30 s, 55 °C for 30 s, and 72 °C for 1 min along with a final extension of 72 °C for 10 min.
2. Clone the amplified *PDS* fragment into a PCR cloning vector (e.g., CloneJET PCR cloning kit, Thermo Scientific, USA) as per the manufacturer's guidelines. Transform the ligated product in to *E. coli* XL1-Blue competent cells.
3. Restriction digest the PCR cloning vector harboring the *PDS* fragment (GOI) and the pTRV2 vector using appropriate enzymes. Set up the digestion reaction in 20 μL with 2 μL of 10× tango buffer, 3 μL (500 ng) plasmid DNA, and restriction enzymes (1 U each of *Xho*I and *Xba*I in this case). Make up the volume with molecular biology grade water. Incubate the digestion mix at 37 °C for 2 h and resolve the products using 1% agarose gel.
4. Purify the digested *PDS* fragment and linearized pTRV2 vector from agarose gel using a dedicated kit. Set up ligation reaction in 10 μL using 1:3 ratio of purified restriction digested *PDS* fragment and linearized pTRV2 vector along with 1 μL of T4 DNA ligase (1 U/μL), 1× ligation buffer, and water. Incubate the ligation reaction at 16 °C for 12 h and then transform into *E. coli* XL1-Blue competent cells (Fig. 2). Plate the transformed cells on to a kanamycin selection media and incubate overnight at 37 °C.
5. Screen the colonies grown on selection media by colony PCR. Isolate plasmid DNA from PCR screened positive colony using plasmid purification kit. Confirm the presence of *PDS* fragment by resolving *Xho*I and *Xba*I restriction digested reaction products in 1% agarose gel.

3.2.2 Preparation of Agrobacterium *GV3101* Competent Cells

1. Streak one small loop of *A. tumefaciens* GV3101 on YEP agar plate with 60 mg/L rifampicin and 50 mg/L gentamicin antibiotic selection, and incubate at 28 °C.
2. Inoculate a single colony from **step 1** into 15 mL Falcon tube containing 5 mL of YEP broth with 60 mg/L rifampicin and 50 mg/L gentamicin, and allow the culture to grow overnight at 28 °C at 200 rpm (*see* **Note 3**).

Fig. 2 Agroinfiltration of *W. somnifera* leaves leading to VIGS of *PDS*. Agroinfiltration of four-leaf-staged *W. somnifera* plant on the abaxial side (**a**–**d**). Phenotypes of *W. somnifera* plant infiltrated with pTRV2::*PDS* construct at 30 and 50 dpi (**e**, **f**) [10]

3. Inoculate 5 μL of overnight grown culture into 250 mL Erlenmeyer flask containing 50 mL of YEP broth containing 60 mg/L rifampicin and 50 mg/L gentamicin, and allow the culture to grow overnight for 15–16 h at 28 °C at 100 rpm.
4. When the OD_{600} reaches 0.3–0.4, harvest the cells by centrifuging at 2800 × *g* for 5 min at 4 °C in a refrigerated centrifuge.
5. Wash the bacterial pellet twice with YEP and gently resuspend the pellet in 1 mL of sterile ice-cold 20 mM $CaCl_2$. Incubate the suspension on ice for 10 min.
6. Aliquot 100 μL of competent cells into pre-chilled microcentrifuge tubes and store at −80 °C (*see* **Note 4**).

3.2.3 Transformation of pTRV Vector into A. tumefaciens

1. Add 2.5 μg of plasmid DNA (pTRV1 or pTRV2 or pTRV2::*WsPDS*) (Fig. 2) to the microcentrifuge tube containing *A. tumefaciens* competent cells, and mix gently by tapping and incubate on ice for 20–30 min.
2. Freeze the DNA and competent cells mixture in liquid nitrogen and immediately thaw at 37 °C. Repeat this step three times.
3. Add 400 μL of YEP to the microcentrifuge tube and incubate at 28 °C at 100 rpm for 2–3 h (*see* **Note 5**). Plate 200 μL of the solution on YEP agar media containing 60 mg/L rifampicin, 50 mg/L gentamicin, and 50 mg/L kanamycin.

4. Screen the transformants by colony PCR using *PDS*-specific forward and reverse primers. Inoculate the PCR-positive colony in 5 mL YEP containing 60 mg/L rifampicin, 50 mg/L gentamicin, 50 mg/L kanamycin, and grow overnight at 28 °C in an orbital shaker maintained at 200 rpm.
5. Stocks were prepared with the overnight grown culture using sterile glycerol solution to get a final concentration of 20% glycerol. Flash freeze the stocks in liquid nitrogen and store in −80 °C freezer until use.

3.3 Preparation of Agrobacterium Suspension for VIGS and Agroinfiltration

1. Grow the primary culture by inoculating glycerol stocks of *A. tumefaciens* harboring pTRV1/pTRV2/pTRV2::*PDS* into 5 mL of YEP broth containing appropriate antibiotics as mentioned before. Grow the cultures overnight in an incubator shaker at 28 °C with 200 rpm speed.
2. Inoculate 1 mL of primary culture into 50 mL of YEP broth containing 10 mM MES, 200 μM acetosyringone, and antibiotics as mentioned in Subheading 3.2, **step 4**. Grow the cells overnight in an incubator shaker maintained at 28 °C and 200 rpm (*see* **Note 3**).
3. Harvest the overnight grown *Agrobacterium* (secondary culture) by centrifugation at 2800 × *g* in a refrigerated centrifuge maintained at 4 °C. Resuspend the bacterial pellet in infiltration buffer to obtain the OD_{600} of ~1.3. Incubate the bacterial suspension at 28 °C for 3–4 h with continuous shaking at 200 rpm (*see* **Notes 6** and 7).
4. Prior to infiltration, mix the *Agrobacterium* cultures of pTRV1and pTRV2 or its derivative in 1:1 ratio (Fig. 2).
5. Using a 5 mL needleless syringe, infiltrate the bacterial suspension into the abaxial side of top two to four fully expanded leaves of four to six-leaf-staged plant [11] (Fig. 3) (*see* **Notes 8**).

   pTRV1:pTRV2:: *WsPDS.*

   pTRV1:pTRV2 (empty vector control).

6. Post-infiltration, keep the plants in trays and cover with cling wrap to maintain humid condition, make small perforations in the wrap for proper aeration, and keep in dark for 48 h at 20–22 °C. The dark condition facilitates higher efficiency of agro-infection (*see* **Note 9**).
7. After 48 h post-infiltration, remove the cling wrap, and transfer the trays containing infiltrated plants to growth chamber maintained at 20–22 °C, 70% relative humidity, and 16/8 h light-dark cycle.
8. Characteristic curly phenotype due to viral infection is visible in the newly emerging first pair of leaves 21 days post-infiltration

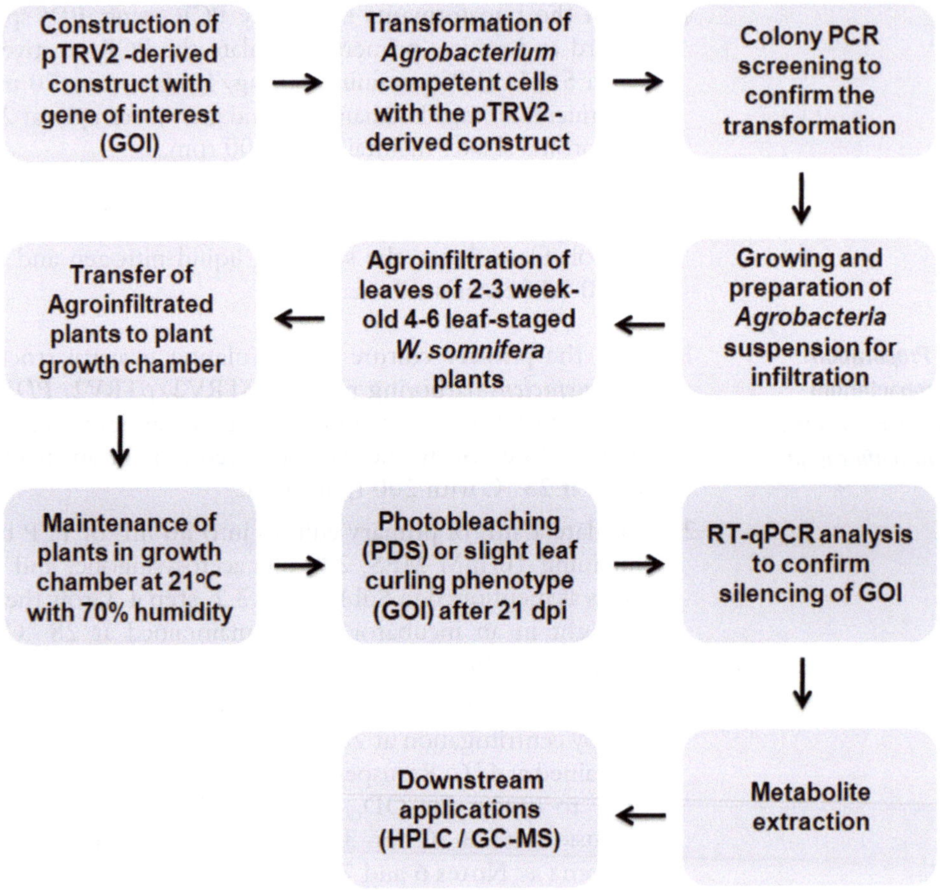

Fig. 3 Schematic workflow for TRV-mediated VIGS in *Withania somnifera*

(dpi). The leaves showing the typical viral infection phenotype can be harvested in liquid nitrogen and stored at −80 °C for further transcript and metabolite analyses. Silencing of *PDS* leads to a range of phenotype varying from mild photobleaching with green/white patches distributed on the leaf surface to strong uniform photobleaching in the entire leaf (Fig. 3) (*see* **Notes 10 and 11**).

3.4 Quantitative Reverse Transcriptase-PCR (RT-qPCR) Analysis

3.4.1 Isolation of RNA from Leaves

1. Take 50–100 mg of leaf samples (fresh or frozen in liquid nitrogen), and homogenize with 1 mL of RNA extraction reagent (e.g., TRIzol). Transfer the suspension to 1.5 mL microcentrifuge tube and incubate at room temperature for 10 min (*see* **Note 12**).
2. Add 0.2 mL of nuclease-free chloroform, and invert the tubes gently three to five times until it becomes a milky suspension.
3. Incubate the samples at room temperature for 10 min, and centrifuge the lysate at 13,000 × *g* for 10 min at 4 °C in a refrigerated centrifuge.

4. Transfer the clear supernatant (upper layer obtained after centrifugation) to a new tube, and add 0.5 mL of nuclease-free isopropanol to the aqueous phase. Incubate the mixture for 10 min at room temperature.
5. Centrifuge the samples at 13,000 × *g* for 10 min in a refrigerated centrifuge maintained at 4 °C. Carefully remove the supernatant using a micropipette. Total RNA will appear as a white gel-like pellet at the bottom of the microcentrifuge tube.
6. Wash the pellet with 1 mL of 75% ethanol, vortex the sample briefly, and then centrifuge at 5000 × *g* for 5 min at 4 °C.
7. Remove the supernatant carefully using a micropipette and air-dry the RNA pellet for 5–10 min. Resuspend the pellet in 26 μL of RNase-free water.
8. Add 3 μL of 10× DNase buffer and 1 U of RNase-free DNase (1 U/μL) to 26 μL RNA, and incubate for 30 min at room temperature.
9. After DNase treatment, add 1 μL of 50 mM EDTA and incubate at 72 °C for 1–2 min, and store the RNA at −80 °C until further use.

3.4.2 cDNA Synthesis and RT-qPCR Analysis

1. Thaw RNA samples on ice and quantify in spectrophotometer. Use 2 μg of total RNA for synthesis of first-strand cDNA with random hexamer primers using high-capacity cDNA reverse transcription kit as per manufacturer's instructions.
2. Perform RT-qPCR using cDNA prepared from RNA extracted from silenced and control leaves to check the expression level of the silenced gene. In RT-qPCR experiments, normalized cDNA concentration should be determined using *18S* rRNA (or any other housekeeping genes) as the endogenous control (*see* **Note 13**). The reaction mixture (5 μL) consists of 2× SYBR Green, 2 μM *WsPDS* RTF and *WsPDS* RTR primers (*see* **Note 14**), 0.5 μL of appropriately diluted cDNA, and molecular biology grade water. Determine the fold-change difference in gene expression by the comparative cycle threshold (*Ct*) method. RT-qPCR conditions were as follows: 95 °C for 5 min for 1 cycle, followed by 40 cycles of 95 °C for 15 s, 60 °C for 60 s [10, 11].

3.5 Metabolite Analysis

3.5.1 Extraction of Withanolides and HPLC Analysis

1. Extract 20 mg oven-dried *W. somnifera* leaf tissue three times with 1 mL methanol followed by sonicating 30 min for each extraction. After every sonication, centrifuge the extract at 2800 × *g* for 5 min and collect the supernatant. Pool the supernatant from each extraction and air-dry (*see* **Note 15**).
2. Resuspend the residue in 3 mL of 70% aqueous methanol, and then extract three times with equal volume of chloroform.

3. Pool the lower layers from each extraction, evaporate until the final volume is 1.5 mL, and decolorize the extract by adding 50 mg of activated charcoal (*see* **Note 16**).
4. Centrifuge at 2800 × *g* for 5 min to separate the charcoal and the supernatant [16].
5. Transfer the decolorized extract to a new tube, dry and dissolve in 200 μL chloroform:methanol (7:3, v/v), and use 40 μL for analysis by HPLC-DAD.
6. Gradient programming of the solvent system is carried out initially at 95% A and then changed to 55% A and 20% A at 18 and 25 min, respectively. The concentration of solvent A is maintained at 20% for next 10 min, increased to 55% at 35 min, followed by 95% at 40 min that is maintained till the run time reaches 45 min (*see* **Note 17**).
7. The flow rate is set at 1.5 mL/min and data is recorded at a wavelength of 227 nm.
8. The peak in each chromatogram can be compared to authentic withanolide standards and catharanthine as an internal standard.

3.5.2 Phytosterol Extraction and GC Analysis

1. Grind air-dried leaves (200 mg) to a fine powder and saponify in 5 mL of 6% methanolic KOH (w/v) and heat at 80 °C under reflux for 2 h (*see* **Note 18**).
2. Dilute the mixture with an equal volume of water and extract total sterols three times with three volumes of *n*-hexane.
3. Dry the pooled extract on 2–3 micro spatula of anhydrous Na_2SO_4 (*see* **Note 19**), evaporate completely, and dissolve in 30 μL of dichloromethane containing estrone (444.4 nM) as an internal standard [13, 17, 18] (*see* **Note 20**).
4. Inject 2 μL of phytosterol sample into a GC equipped with HP-5 fused-silica capillary coated with 5% phenylmethyl siloxane column (30 m × 0.320 mm, 0.25 μm film thickness) and flame ionization detection (FID) (*see* **Note 21**).
5. Adjust the oven temperature to 240 °C for 10 min and raised to 260 °C with an increment of 2 °C/min and maintained at 260 °C for 30 min with the injection port at 270 °C and detector at 300 °C.
6. Adjust the split ratio 15:1 with nitrogen as the carrier gas at a flow rate of 1 mL/min.
7. For quantification, plot a standard curve for individual standards separately, and determine the relative response factor (RRF) for each standard.
8. Subsequently, normalize the peak area corresponding to individual phytosterols with estrone.

9. The RRF and concentration of estrone are used to calculate the concentration of each phytosterol in the sample using the respective normalized peak area.

4 Notes

1. Care should be taken to prevent damping off of the seedlings and infestation by aphids and whiteflies.
2. The size of the target gene (to be silenced) should not exceed 500 bp. However, in other plants, we have also used gene fragment of approximately 250 bp which leads to efficient silencing.
3. Inoculation of bacterial culture (*E. coli* and *Agrobacterium)* into the media for culturing and infiltration should be done on a clean bench, and all the media, tubes, tips, etc. need to be sterilized by autoclaving.
4. Ideally, *Agrobacterium* competent cells should not be stored for more than a month or else the transformation efficiency reduces.
5. During *Agrobacterium* transformation, after freeze-thawing, culture should be incubated in shaker at 28 °C for 3–4 h with speed of not more than 100 rpm for optimal growth of the culture.
6. The derived *Agrobacterium* culture is resuspended in infiltration buffer to OD_{600} of 1.3. Lower OD_{600} of culture reduces the efficiency of silencing [10].
7. Incubate the infiltration buffer containing *Agrobacterium* culture for 3–4 h at 28 °C and 200 rpm before infiltration as it helps in the induction of *vir* genes.
8. It is advisable to use first and second pair of leaves of similar staged plants for infiltration to obtain efficient silencing. Infiltration is carried out on the abaxial side of the leaf, and much care should be taken not to give high pressure during injection of culture as this could damage the leaf [10, 11]. Approximately 5 mL of *Agrobacterium* culture is required to infiltrate three to four plants and if needed can scale up the culture based on the number of plants used.
9. While using different gene constructs for infiltration, it is always recommended to label the pots with plants properly, maintain them in separate trays, and use different syringes for each gene construct in order to avoid cross contamination.
10. Typical curly leaves can be harvested suitably 21–30 days post-infiltration or until the appearance of curly viral infection phenotype. In addition to agroinfiltration of GOI, a parallel set of

experiment should always be performed with agroinfiltration of pTRV1 and pTRV2 which serve as empty vector controls to know the specificity of GOI silencing and rule out the effects of viral propagation on gene expression and metabolite accumulation.

11. The plant biomass leftover after collecting the leaf tissues from silenced plants has to be decontaminated by autoclaving.
12. We use RNA extraction reagent (TRIzol) which gives high quality of RNA. However, RNA purification kit works equally well [10, 11].
13. The stability of different reference (housekeeping) genes should be determined before using in RT-qPCR analysis.
14. For determining silencing of the target gene in VIGS samples, primers for RT-qPCR should be designed on GOI outside of the region selected for cloning into pTRV2. In this case, we have used *PDS* as the GOI. PDS converts 15-*cis*-phytoene to lycopene that acts as an accessory pigment in photosynthesis. Other visual marker genes like *magnesium chelatase subunit H*, *cloroplastos alterados-1*, could also be used [19].
15. Extraction of withanolides should always be done with an equal amount of samples and a suitable internal standard from empty vector (EV) control and VIGS samples for proper comparison [16].
16. After extraction of withanolides, samples should be completely decolorized as colored samples can choke the column and subsequently reduce column's life.
17. The aqueous mobile phase should always be filtered and sonicated prior to use. Any suspended minor particles or air bubbles could block the column and lead to increased column pressure. For preparation of the aqueous mobile phase, only HPLC grade water should be used. Use other grades of water drift negative baseline in the chromatogram.
18. There are possibilities of methanolic KOH evaporation during reflux in a water bath. Hence, the level of methanolic KOH should be monitored at frequent intervals and should not be allowed to evaporate completely. In any case, an equal quantity of methanolic KOH can be added in all samples during reflux.
19. Phytosterol samples for GC should be dried over Na_2SO_4 to absorb moisture. The presence of moisture can interfere with GC analysis [13, 17, 18].
20. Dichloromethane is highly volatile, and GC vials containing phytosterols in dichloromethane should be closed tightly.
21. Cholesterol, campesterol, stigmasterol, β-sitosterol, and cycloartenol can be used as phytosterol standards.

Acknowledgments

This work was supported by the GAP376 project (SERB funded project EMR/2016/002746) of CSIR-Central Institute of Medicinal and Aromatic Plants. D.P.B., H.B.S., and S.R.K. are the recipients of the Research Fellowships from the University Grants Commission (UGC), Indian Council of Medical Research (ICMR), and Council of Scientific and Industrial Research (CSIR), respectively. The institutional communication number for this article is CIMAP/PUB/2019/May/40.

References

1. Mishra LC, Singh BB, Dagenais S (2000) Scientific basis for the therapeutic use of *Withania somnifera* (Ashwagandha). Altern Med Rev 5:334–346
2. Mirjalili MH, Moyano E, Bonfill M, Cusido RM, Palazon J (2009) Steroidal lactones from *Withania somnifera*, an ancient plant for novel medicine. Molecules 14:2373–2393
3. Sehgal N, Gupta A, Valli RK, Joshi SD, Mills JT, Hamel E, Khanna P, Jain SC, Thakur SS, Ravindranath V (2012) *Withania somnifera* reverses Alzheimer's disease pathology by enhancing low-density lipoprotein receptor-related protein in liver. Proc Natl Acad Sci U S A 109:3510–3515
4. Kaur G, Kaur T, Gupta M, Manchanda S (2017) Neuromodulatory role of *Withania somnifera*. In: Kaul S, Wadhwa R (eds) Science of Ashwagandha: preventive and therapeutic potentials. Springer, Cham
5. Kaileh M, Berghe WV, Heyerick A, Horion J, Piette J, Libert C, Keukeleire DD, Essawi T, Haegeman G (2007) Withaferin A strongly elicits IκB kinase b hyperphosphorylation concomitant with potent inhibition of its kinase activity. J Biol Chem 282:4253–4264
6. Koduru S, Kumar R, Srinivasan S, Evers MB, Damodaran C (2010) Notch-1 inhibition by Withaferin-A: a therapeutic target against colon carcinogenesis. Mol Cancer Ther 9:202–210
7. Hahm ER, Moura MB, Kelley EE, Houten BV, Singh SV (2011) Withaferin A-induced apoptosis in human breast cancer cells is mediated by reactive oxygen species. PLoS One 6:23354
8. Sehrawat A, Samanta SK, Hahm ER, St Croix C, Watkins SSV (2019) Withaferin A-mediated apoptosis in breast cancer cells is associated with alterations in mitochondrial dynamics. Mitochondrion 47:282–293.
9. Knoch E, Sugawara S, Mori T, Poulsen C, Fukushima A, Harholt J, Fujimoto Y, Umemoto N, Saito K (2018) Third DWF1 paralog in Solanaceae, sterol Δ^{24}-isomerase, branches withanolide biosynthesis from the general phytosterol pathway. Proc Natl Acad Sci U S A 115:8096–8103.
10. Singh AK, Dwivedi V, Rai A, Pal S, Reddy SG, Rao DK, Shasany AK, Nagegowda DA (2015) Virus-induced gene silencing of *Withania somnifera* squalene synthase negatively regulates sterol and defense-related genes resulting in reduced withanolides and biotic stress tolerance. Plant Biotechnol J 13:1287–1299
11. Singh AK, Kumar SR, Dwivedi V, Rai A, Pal S, Shasany AK, Nagegowda DA (2017) A WRKY transcription factor from *Withania somnifera* regulates triterpenoid withanolides accumulation and biotic stress tolerance through modulation of phytosterol and defense pathways. New Phytol 215:1115–1131
12. Singh G, Tiwari M, Singh SP, Singh S, Trivedi PK, Misra P (2016) Silencing of sterol glycosyltransferases modulates the withanolide biosynthesis and leads to compromised basal immunity of *Withania somnifera*. Sci Rep 6:25562
13. Singh G, Tiwari M, Singh SP, Singh R, Singh S, Shirke PA, Trivedi PK, Misra P (2017) Sterol glycosyltransferases required for adaptation of *Withania somnifera* at high temperature. Physiol Plant 160:297–311
14. Agarwal AV, Singh D, Dhar YV, Michael R, Gupta P, Chandra D, Trivedi PK (2018) Virus-induced silencing of key genes leads to a differential impact on withanolide biosynthesis in the medicinal plant *Withania somnifera*. Plant Cell Physiol 59:262–274
15. Dinesh-Kumar SP, Anandalakshmi R, Marathe R, Schiff M, Liu Y (2003) Virus-induced gene silencing. In: Grotewold E

(ed) Plant functional genomics. Humana Press, Totowa

16. Qin B, Eagles J, Mellon FA, Mylona P, Rodriguez L, Osbourn AE (2010) High throughput screening of mutants of oat that are defective in triterpene synthesis. Phytochemistry 71:1245–1252
17. Darnet S, Rahier A (2004) Plant sterol biosynthesis: identification of two distinct families of sterol 4alpha-methyl oxidases. Biochem J 378:889–898
18. Singh S, Pal S, Shanker K, Chanotiya CS, Gupta MM, Dwivedi UN, Shasany AK (2014) Sterol partitioning by HMGR and DXR for routing intermediates toward withanolide biosynthesis. Physiol Plant 152:617–633
19. Liu N, Xie K, Jia Q, Zhao J, Chen T, Li H, Wei X, Diao X, Hong Y, Liu Y (2016) Foxtail Mosaic Virus-induced gene silencing in monocot plants. Physiol Plant 171:1801–1807

Chapter 12

A *Prunus necrotic ringspot virus* (PNRSV)-Based Viral Vector for Characterization of Gene Functions in *Prunus* Fruit Trees

Hongguang Cui, Yinzi Li, and Aiming Wang

Abstract

Virus-induced gene silencing (VIGS) is a gene silencing mechanism by which an invading virus targets and silences the endogenous genes that have significant sequence similarity with the virus. It opens the door for us to develop viruses as powerful viral vectors and modify them for molecular characterization of gene functions in plants. In the past two decades, VIGS has been studied extensively in plants, and various VIGS vectors have been developed. Despite the fact that VIGS is in particular practical for functional genomic study of perennial woody vines and trees with a long life cycle and recalcitrant to genetic transformation, not many studies have been reported in this area. Here, we describe a protocol for the use of a *Prunus necrotic ringspot virus* (PNRSV)-based VIGS vector we have recently developed for functional genomic studies in *Prunus* fruit trees.

Key words Virus-induced gene silencing, RNA silencing, *Prunus necrotic ringspot virus*, Viral vector, Peach, Fruit trees

1 Introduction

RNA silencing or RNA interference (RNAi) is a well-studied fundamental, evolutionarily conserved mechanism in eukaryotes that is triggered by double-stranded RNA (dsRNA) to regulate gene expression [1–3]. This mechanism has been implemented by researchers to molecularly characterize gene functions in plants. A common approach used to achieve this goal is to generate transgenic plants through stable genetic transformation to express an inverted repeat (IR) double-stranded RNA (dsRNA) or artificial microRNA (amiRNA) that shares high sequence homology to the targeted gene [4, 5]. Unfortunately, there are several common disadvantages that are associated with this transgene-induced silencing approach, i.e., complexity of operation procedures, labor-intensive screening, and low transformation efficiency

Vincent Courdavault and Sébastien Besseau (eds.), *Virus-Induced Gene Silencing in Plants: Methods and Protocols*, Methods in Molecular Biology, vol. 2172, https://doi.org/10.1007/978-1-0716-0751-0_12,

[6, 7]. Besides, it is scarcely possible to obtain transformants if the lethal phenotype in plant growth and development is induced via mutagenesis or silencing of the target plant gene [7]. In addition, this approach is inefficient or not applicable to some plant species such as perennial woody fruit trees that are recalcitrant to genetic transformation. Such unavailability and limitations severely hinder genetic and genomic studies in these species and largely affect breeding efforts for fruit improvement and sustainable production [1]. To cope with these restrictions, a promising and potent alternative is to develop a modified viral vector that harbors small gene fragment to direct virus-induced gene silencing (VIGS) to knock down genes of interest [1].

VIGS functions via a posttranscriptional gene silencing (PTGS) mechanism to silence the targeted gene in a sequence-specific manner [8–10]. Typically, a virus-derived infectious clone suitable for VIGS in defined plant species, termed viral vector, is modified to contain a short sequence of a targeted plant gene. The modified VIGS vector is introduced into plants via either *Agrobacterium*-mediated transient expression or biolistic bombardment. Infection by the recombinant virus leads to the generation of dsRNA, which occurs during virus replication (for RNA viruses), bidirectional transcription of the viral genome (for DNA viruses), intramolecular pairing of viral RNA, and de novo biosynthesis of dsRNA by endogenous RNA-dependent RNA polymerases (RDRs) [11]. As the dsRNA contains the integrated gene-specific sequence, the induced RNA silencing not only degrades the viral transcripts but also the targeted endogenous gene mRNA. In the past two decades, a number of VIGS vectors have been developed using either RNA viruses such as *Potato virus X* (PVX), *Tobacco rattle virus* (TRV), and *Brome mosaic virus* (BMV) or DNA viruses such as *Cotton leaf crumple virus* (CLCrV) and *Cabbage leaf curl virus* (CaLCuV) [12–17]. Most of the reported VIGS vectors have a host range limited to herbaceous species, and very few can be used for functional genomic studies in allogamous woody perennials and vines [1].

Prunus necrotic ringspot virus (PNRSV) is a species of the genus *Ilarvirus* in the family *Bromoviridae* of plant RNA virus [18] and has a viral genome consisting of three RNA segments. RNA1 and RNA2 are monocistronic and encode replicase proteins P1 and P2, respectively (Fig. 1a). RNA3 is bicistronic and has two open reading frames (ORF), of which the 5′-proximal and 3′-proximal ones encode movement protein (MP) and coat protein (CP), respectively (Fig. 1a). PNRSV infects a variety of *Prunus* species worldwide and induces mild/latent symptoms in certain natural *Prunus* hosts [19–22]. These unique properties warrant PNRSV to be a suitable candidate to be developed as a VIGS vector for characterization of gene functions in *Prunus* spp. [1].

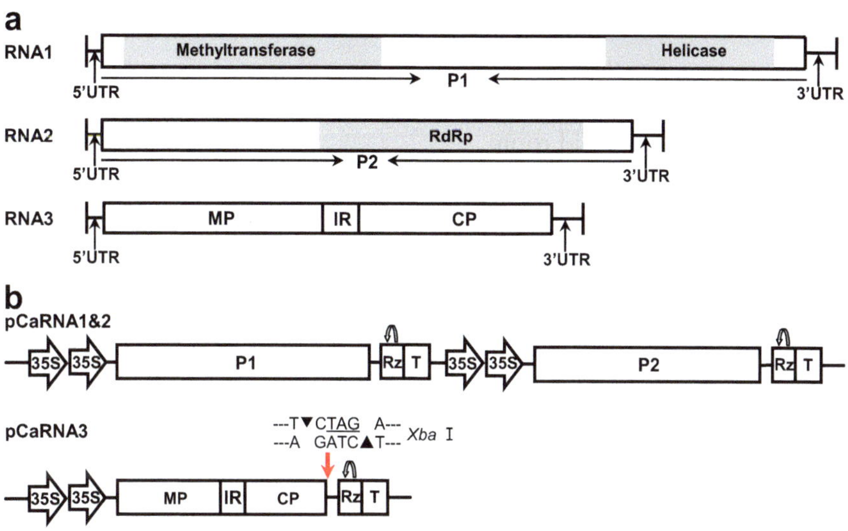

Fig. 1 Schematic diagram of *Prunus necrotic ringspot virus* (PNRSV)-based VIGS vector. (**a**) Schematic representation of the genomic organization of PNRSV. Genomic segments RNA1 and RNA2 are monocistronic and encode replicase proteins P1 and P2, respectively. P1 contains a methyltransferase domain and a helicase domain at its 5′- and 3′-proximal parts, respectively, whereas P2 contains an RNA-dependent RNA polymerase (RdRp) domain. RNA3 encodes the 5′-proximal movement protein (MP) and the 3′-proximal coat protein (CP), which are separated by an intergenic region (IR). (**b**) Schematic diagram of the PNRSV-based VIGS vectors. Both expression cassettes of genomic RNA1 and RNA2 were integrated into the same T-DNA backbone (pCass4-Rz) to generate the vector pCaRNA1&2. The expression cassette of genomic RNA3 was independently integrated into pCass4-Rz to generate pCaRNA3. The natural *Xba* I restriction sequence "TCTAGA" (TAG, stop codon of CP gene), located at the junction of CP and 3′ UTR in genomic RNA3, was employed to integrate foreign insert for construction PNRSV-based VIGS vectors. The self-cleavage of ribozyme (Rz) is indicated by a bent arrow; T, 35S terminator

The construction of viral infectious cDNA clone is a critical step for the development of a VIGS vector. Initially, the full-length cDNAs of RNA1, RNA2, and RNA3 of PNRSV isolate Pch12 were integrated into T-DNA vector (pCass4Rz), respectively [23]. However, the infection efficiency of the resulting PNRSV clones was only ~30% in peach [21]. To overcome the above deficiency, both RNA1 and RNA2 expression cassettes were integrated into the same binary vector, which, together with pCaRNA3, were biolistically transferred into plant cells (Fig. 1b). As expected, the infectivity rate of the improved PNRSV vectors (Fig. 1b) reached up to ~75% [1]. We thus further modified the RNA3 plasmid and developed the PNRSV-based VIGS system [1]. This chapter describes in detail the protocol to perform VIGS assay in peach, one *Prunus* species, using the PNRSV-based viral vector.

2 Materials

2.1 Plant Material

1. Peach seeds.
2. Cold room.
3. Nutcracker.
4. 20% sodium hypochlorite (commercial bleach).
5. Sterilized double-distilled water.
6. Germination medium: $^{1}/_{2}$ Quoirin and Lepoivre (QL) medium, 10 g/L fructose, pH 5.7, 7 g/L agar.
7. Pots and soil.
8. Growth cabinet in a regime of 16 h of light at 20 °C and 8 h of darkness at 16 °C.

2.2 Construction of PNRSV-Based VIGS Vectors

1. pCaRNA1&2 and pCaRNA3 vectors [1].
2. *Escherichia coli* (DH10B strain) competent cells.
3. Luria broth (LB) medium liquid and agar.
4. 100 mg/mL kanamycin stock solution.
5. Sterilized flasks (100 mL).
6. Sterilized centrifuge tubes (50 mL).
7. Plasmid purification kit.
8. PCR purification kit.
9. Spectrophotometer (e.g., Nanodrop).
10. Restriction endonuclease and buffer: *Xba*I.
11. Calf intestinal alkaline phosphatase (CIP) and buffer.
12. High-fidelity DNA polymerase, buffer, and dNTP.
13. Primer set: RNA3-CP-F 5′-AGGTCTTGGTTAGGGATTTG-3′ and RNA3-3′UTR-R 5′-GCTTCCCTAACGGGGCATCC-3′.
14. Gene-specific primers Spec-F/Spec-R.
15. PCR cycler equipment.

2.3 Biolistic Inoculation

1. Helios Gene Gun System (Bio-Rad).
2. Ultrasonic cleaner.
3. Purified plasmid DNA.
4. Gold microcarriers (1.0 μm).
5. 0.05 M spermidine.
6. 0.5 M PVP (360,000 MW) in absolute ethanol.
7. 1 M $CaCl_2$.
8. Fresh 100% ethanol.

9. Microfuge.
10. Syringe (5 mL).
11. Gold-coated tubing.

3 Methods

3.1 Preparation of Peach Seedlings

1. Store peach seeds in cold room (at 4 °C in darkness) for at least 1 month (*see* **Note 1**).
2. Remove seed shell with a nutcracker.
3. Sterilize shell-free seeds with 70% ethanol for 5 min, followed by 20% bleach for 20 min (*see* **Note 2**). Rinse them with sterilized double-distilled water three times (*see* **Note 2**).
4. Soak the surface-sterilized seeds in sterilized double-distilled water for 2 days, and remove seed coat carefully using forceps (*see* **Note 2**).
5. Culture the seeds without the testa on the germination medium (*see* **Note 2**).
6. Transplant the rooted peach seedlings from the medium into soil pots.
7. Maintain all peach seedlings in a growth cabinet in a regime of 16 h of light at 20 °C and 8 h of darkness at 16 °C (*see* **Note 3**).

3.2 Modification of PNRSV-Based VIGS Vectors

1. Transform pCaRNA1&2 and pCaRNA3 into *E. coli* (DH10B strain) competent cells. Streak the cultures onto LB plates containing 100 μg/mL kanamycin (*see* **Note 4**).
2. Inoculate single colonies containing either pCaRNA1&2 or pCaRNA3 into 10 mL LB liquid medium containing 100 μg/mL kanamycin, and then propagate them overnight at 37 °C, 220 rpm (*see* **Note 5**).
3. Divide each culture into three portions for the extraction of plasmids with a plasmid purification kit. Determine the concentration of the obtained plasmids (pCaRNA1&2 and pCaRNA3) using spectrophotometer (*see* **Note 6**).
4. Digest approximately 1 μg of pCaRNA3 with *Xba*I in 37 °C for 3 h, followed by the treatment with CIP (*see* **Note 7**).
5. Amplify a 100–200 bp fragment of the gene that is targeted to be silenced using gene-specific primer set Spec-F/Spec-R (*see* **Note 8**). The PCR product is purified with PCR purification kit. Digest the PCR fragment with *Spe*I, purify, and ligate it into *Xba*I/CIP-treated pCaRNA3 with the molar ratio of plasmid/insert as 1–3. Transform the pCaRNA3 derivatives into DH10B competent cells.

6. Screen positive transformants harboring a single-copy fragment insert in the sense orientation by DNA sequencing (*see* **Note 9**).
7. Propagate and extract the plasmid pCaRNA3 derivatives harboring the correct fragment insert, essentially as described in **steps 2** and **3**.

3.3 Biolistic Inoculation of Peach Seedlings

1. Weigh out 8.5 mg gold microcarriers in a 1.5 mL microfuge tube.
2. Add 100 μL of 0.05 M spermidine.
3. Vortex the gold and spermidine mixture for 30 s, and then sonicate it for 15 s (*see* **Note 10**).
4. Adjust the concentrations of pCaRNA1&2 and the modified pCaRNA3 derivative plasmids, respectively, to 1 μg/μL. Add 8.5 μL of each of pCaRNA1&2 and pCaRNA3 derivative plasmids into the gold and spermidine mixture. Mix DNA, spermidine, and gold by vortexing for 5 s.
5. Add 100 μL of 1 M $CaCl_2$ dropwise to the above mixture while vortexing at a moderate rate on a variable speed vortexer (*see* **Note 11**). Allow the mixture to be precipitated at room temperature for 10 min.
6. Spin the microcarrier solution in a microfuge for 15 s to pellet the coated gold, and then remove and discard the supernatant.
7. Resuspend the pellet in the remaining supernatant by vortexing briefly (*see* **Note 12**). Wash the pellet three times with 1 mL of fresh 100% ethanol each time.
8. Resuspend the pellet in 1 mL of PVP in absolute ethanol at its working concentration as 0.05 mg/mL (*see* **Note 13**).
9. Set up the Tubing Prep Station and connect it to a nitrogen tank.
10. Cut off a piece of tubing (approximately 30″) with one end well-connected to one syringe (5 mL).
11. Vortex the DNA/microcarrier solution for 10 s, promptly transfer it into the middle region of the tubing by syringe, and carefully bring and slide the loaded tubing into the Tubing Prep Station equipment at a horizontal position.
12. Let the tubing stand for 5 min, and then slowly remove the ethanol from the tubing using syringe without disturbing the microcarrier.
13. Detach the syringe from the tubing and dry the tubing via allowing nitrogen flow through the gold-coated tubing (0.35 LPM) for 5 min.
14. Cut the gold-coated tubing into approximately 17 cartridges (0.5″ each) and load into the cartridge holder.

15. Shot two cartridges into different young leaves for each peach seedling (3–4 weeks old and approximately 10–15 cm in height) using the Helios Gene Gun (*see* **Note 14**).
16. Monitor the silencing efficiency of target host gene by phenotypic observation, as well as molecular methods such as real-time quantitative RT-PCR and Northern blot hybridization (*see* **Note 15**).

4 Notes

1. Peach seeds should be collected in recent two growing seasons. Cold treatment of seeds at 4 °C in darkness for at least 1 month facilitates to break dormancy and is an indispensable step to increase the germination efficiency of peach seeds.
2. **Steps 3–5** (*see* Subheading 3.1) should be carefully operated in a flow clean bench. As long as the contamination occurs in any step, the germination of seeds might be seriously affected. Usually, it will take 2–3 weeks for the germination and rooting of seeds in germination medium.
3. The efficiency of PNRSV-mediated gene silencing is enhanced for peach seedlings under low temperature. Therefore, the growth condition of peach seedlings is suggested to be 16 h of light at 20 °C and 8 h of darkness at 16 °C for PNRSV-based VIGS experiments.
4. Both pCaRNA1&2 and pCaRNA3 plasmids contain kanamycin-resistant gene; therefore, kanamycin and its concentration should be correctly used for the screening of positive clones.
5. 100 mL sterilized flasks or 50 mL centrifuge tubes are suggested to use for the rapid and large-quantity propagation of *E. coli* cells. Thus, sufficient plasmids could be yielded for biolistic bombardment.
6. The plasmids should be kept in −20 °C freezer for the use in 6 months or −80 °C for long-time storage.
7. CIP treatment is an indispensable step, which could avoid the self-ligation of vectors during the ligation reaction.
8. The 5′ terminus of either primer Spec-F or Spec-R must contain *Xba*I sequence "TCTAGA" and several protective bases such as "ATATA." PNRSV-based vectors are able to tolerate a 100–200 bp fragment. A longer insert will result in the inactivation of virus and failure to silence the target host gene.
9. Primers RNA3-CP-F and RNA3-3′UTR-R are correspondingly homologous and complementary to nt 1683–1702 and 1932–1951 of PNRSV genomic RNA3 (GenBank Accession

#JN416776). Besides, they are located upstream and downstream of *Xba*I site of pCaRNA3, respectively (*see* Fig. 1a). Therefore, RNA3-CP-F/RNA3-3′UTR-R, combined with gene-specific primers Spec-F/Spec-R, could be used to identify the copy number(s) and orientation of foreign insert. In brief, copy number(s) of foreign insert could be determined by PCR with primer set RNA3-CP-F/RNA3-3′UTR-R. PCR with primer set Spec-F/RNA3-3′UTR-R or RNA3-CP-F/Spec-R is used to detect the orientation of foreign insert. It is worth to mention that only vectors harboring a single-copy foreign insert in the sense orientation are able to induce gene silencing in the PNRSV-based VIGS system in peach.

10. Sonication facilitates to break up gold clumps, contributing to highly efficient binding of plasmid DNAs with gold microcarriers.
11. The volume of 1 M $CaCl_2$ should be equivalent to that of the spermidine (*see* Subheading 3.3, **step 2**). The 1 M $CaCl_2$ should be dropwise added into the microfuge tube. Otherwise, the gold microcarriers might form clumps and could not be completely exposed to bind plasmid DNA.
12. The remaining supernatant should be at least 20 μL left in the microfuge tube, so the gold microcarriers could be sufficiently suspended and well washed.
13. It is suggested the freshly prepared DNA/microcarrier solution is used for tube preparation immediately.
14. The gun nozzle should get close to but not touch leaf surface during biolistic bombardment. The helium pressure should be adjusted to 160–180 psi when shooting.
15. It might take a varied length of time to silence different host target genes using PNRSV-based VIGS vectors. For example, eIF(*iso*)4E in peach could be efficiently silenced at 16 days post inoculation (dpi), whereas silencing of PDS gene will take more than 3 months [1].

Acknowledgments

This work was supported in part by grants from the National Natural Science Foundation of China (grant #31860487) to H.C. and from Agriculture and Agri-Food Canada to A.W.

References

1. Cui H, Wang A (2017) An efficient viral vector for functional genomic studies of Prunus fruit trees and its induced resistance to Plum pox virus via silencing of a host factor gene. Plant Biotechnol J 15:344–356
2. Baulcombe D (2015) VIGS, HIGS and FIGS: small RNA silencing in the interactions of viruses or filamentous organisms with their plant hosts. Curr Opin Plant Biol 26:141–146
3. Li F, Wang A (2019) RNA-targeted antiviral immunity: more than just RNA silencing. Trends Microbiol 27(9):792–805. https://doi.org/10.1016/j.tim.2019.05.007
4. Vaucheret H, Beclin C, Elmayan T, Feuerbach F, Godon C, Morel JB, Mourrain P, Palauqui JC, Vernhettes S (1998) Transgene-induced gene silencing in plants. Plant J 16:651–659
5. Jin Y, Guo HS (2015) Transgene-induced gene silencing in plants. Methods Mol Biol 1287:105–117
6. Benedito VA, Visser PB, Angenent GC, Krens FA (2004) The potential of virus-induced gene silencing for speeding up functional characterization of plant genes. Genet Mol Res 3:323–341
7. Burch-Smith TM, Anderson JC, Martin GB, Dinesh-Kumar SP (2004) Applications and advantages of virus-induced gene silencing for gene function studies in plants. Plant J 39:734–746
8. Becker A, Lange M (2010) VIGS—genomics goes functional. Trends Plant Sci 15:1–4
9. Dolja VV, Koonin EV (2014) The closterovirus-derived gene expression and RNA interference vectors as tools for research and plant biotechnology. Front Microbiol 4:83
10. Senthil-Kumar M, Mysore KS (2011) New dimensions for VIGS in plant functional genomics. Trends Plant Sci 16:656
11. Li F, Wang A (2018) RNA decay is an antiviral defense in plants that is counteracted by viral RNA silencing suppressors. PLoS Pathog 14 (8):e1007228
12. Angell SM, Baulcombe DC (1999) Potato virus X amplicon-mediated silencing of nuclear genes. Plant J 20:357–362
13. Liu Y, Schiff M, Dinesh-Kumar SP (2002) Virus-induced gene silencing in tomato. Plant J 31:777–786
14. Ratcliff F, Martin-Hernandez AM, Baulcombe DC (2001) Tobacco rattle virus as a vector for analysis of gene function by silencing. Plant J 25:237–245
15. Ding XS, Schneider WL, Chaluvadi SR, Mian MR, Nelson RS (2006) Characterization of a Brome mosaic virus strain and its use as a vector for gene silencing in monocotyledonous hosts. Mol Plant-Microbe Interact 19:1229–1239
16. Tuttle JR, Idris AM, Brown JK, Haigler CH, Robertson D (2008) Geminivirus-mediated gene silencing from Cotton leaf crumple virus is enhanced by low temperature in cotton. Plant Physiol 148:41–50
17. Turnage MA, Muangsan N, Peele CG, Robertson D (2002) Geminivirus-based vectors for gene silencing in Arabidopsis. Plant J 30:107–114
18. Bujarski JJ, Gallitelli D, García-Arenal F, Pallás VB, Palukaitis P, Reddy MK, Wang A, ICTV Report Consortium (2019) ICTV virus taxonomy profile: *Bromoviridae*. J Gen Virol 100:1206–1207. https://doi.org/10.1099/jgv.0.001282
19. Pallas V, Aparicio F, Herranz MC, Amari K, Sanchez-Pina MA, Myrta A et al (2012) Ilarviruses of *Prunus* spp.: a continued concern for fruit trees. Phytopathology 102:1108–1120
20. Cui H, Hong N, Wang G, Wang A (2012) Detection and genetic diversity of prunus necrotic ringspot virus in the Niagara Fruit Belt, Canada. Can J Plant Pathol 34:104–113
21. Cui H, Hong N, Wang G, Wang A (2013) Genomic segments RNA1 and RNA2 of prunus necrotic ringspot virus codetermine viral pathogenicity to adapt to alternating natural *Prunus* hosts. Mol Plant-Microbe Interact 26:515–527
22. Cui H, Liu H, Chen J, Zhou J, Qu L, Su J et al (2015) Genetic diversity of Prunus necrotic ringspot virus infecting stone fruit trees grown at seven regions in China and differentiation of three phylogroups by multiplex RT-PCR. Crop Prot 74:30–36
23. Annamalai P, Rao ALN (2005) Replication-independent expression of genome components and capsid protein of *Brome mosaic virus* in planta: a functional role for viral replicase in RNA packaging. Virology 338:96–111

Chapter 13

Virus-Induced Gene Silencing in Olive Tree (Oleaceae)

Konstantinos Koudounas, Margarita Thomopoulou, Elisavet Angeli, Dikran Tsitsekian, Stamatis Rigas, and Polydefkis Hatzopoulos

Abstract

Research on gene functions in non-model tree species is hampered by a number of difficulties such as time-consuming genetic transformation protocols and extended period for the production of healthy transformed offspring, among others. Virus-induced gene silencing (VIGS) is an alternative approach to transiently knock out an endogenous gene of interest (GOI) by the introduction of viral sequences encompassing a fragment of the GOI and to exploit the posttranscriptional gene silencing (PTGS) mechanism of the plant, thus triggering silencing of the GOI. Here we describe the successful application of *Tobacco rattle virus* (TRV)-mediated VIGS through agroinoculation of olive plantlets. This methodology is expected to serve as a fast tracking and powerful tool enabling researchers from diversified fields to perform functional genomic analyses in the olive tree.

Key words Virus-induced gene silencing (VIGS), Posttranscriptional gene silencing (PTGS), RNA interference (RNAi), Functional genomics, Olive, *Olea europaea*, *Agrobacterium tumefaciens*, *Tobacco rattle virus* (TRV)

1 Introduction

Olive tree (*Olea europaea* L., Oleaceae) is an emblematic crop and its cultivation in the Mediterranean basin goes back millennia. Recent advances in sequencing technologies have led to the availability of an enormous amount of genomic and transcriptomic datasets [1, 2] giving researchers the opportunity to focus on various aspects of olive tree biology. Functional characterization of a gene should ideally be initiated with a knockout mutation and complemented with observation of the phenotype. However, this approach is rather complicated when somebody is working with a non-model plant such as olive, member of a plant family that comprises perennial woody species [3]. Thus, precise characterization of biological/physiological mechanisms governing economically important traits (e.g., juvenile period [4], alternate bearing [5]) or Oleaceae-specific traits (e.g., biosynthesis of

Vincent Courdavault and Sébastien Besseau (eds.), *Virus-Induced Gene Silencing in Plants: Methods and Protocols*, Methods in Molecular Biology, vol. 2172, https://doi.org/10.1007/978-1-0716-0751-0_13,

oleosidic secoiridoids [6–9]) or even olive-specific traits (e.g., genes involved in olive oil quality [10], responses to (a)biotic stresses [11, 12], and morphogenesis of uncommon type of trichomes [13, 14]) is hampered. Few examples of successful genetic transformation of olives and development up to the stage of plantlet have been reported [15–17], even though these approaches are heavily time-consuming—a common problem for all the perennial trees recalcitrant to transformation.

A very powerful tool to study the functional attitudes of a gene/protein as well as the signaling process is the approach of reverse genetics using virus-induced gene silencing (VIGS). This approach is an alternative fast-track route in model plants and in plant species where stable genetic transformation is difficult or even impossible. This strategy has been successfully recruited in numerous species from several plant families [18, 19]. VIGS takes advantage of the posttranscriptional gene silencing (PTGS) mechanism that plants enable as a defense mechanism against viruses. In order to silence an endogenous plant gene of interest (GOI), a small fragment of this gene is cloned in a plasmid which encompasses viral sequences, and this sequence is delivered in the plant either through bombardment or via *Agrobacterium*. During viral replication, dsRNA is produced triggering PTGS and therefore systemic silencing of the mRNAs of the gene of interest. The major advantages of this approach are: there is no need for stable transformation; the transgene is not incorporated into the plant genome—except in agroinfiltrated tissues; and the silencing of the gene of interest is transient, and in many cases the virus is not transmitted to the offsprings [18, 20].

Due to the aforementioned advantages of this approach, VIGS has already gained attention as a tool to functionally characterize gene/protein function and signaling processes in numerous plant species. Although few examples were initially reported in which VIGS was recruited as a tool in woody tree species, recent advances provide additional approaches and strategies to study economically important crops. Up to date, VIGS has been successfully applied in different species of the *Prunus*, *Pyrus*, and *Malus* genera (Rosaceae) [21, 22], species of the *Jatropha* and *Vernicia* genera (Euphorbiaceae) [23, 24], species of the *Xanthoceras* and *Litchi* genera (Sapindaceae) [25, 26], and species of the *Populus* (Salicaceae) [27], *Citrus* (Rutaceae) [28], *Actinidia* (Actinidiaceae) [29], *Camptotheca* (Nyssaceae) [30], and *Morus* (Moraceae) [31] genera. This number is expected to drastically increase in the following years bridging the research in woody tree species and the field of functional genomics.

As a proof of concept, and in order to estimate when and where VIGS is taking place in a plant, silencing of a gene that results in distinguishable phenotype is performed in parallel with silencing of a GOI. Typically, constructs designed to trigger silencing of either the phytoene desaturase (*PDS*) or the H subunit of Mg-protoporphyrin chelatase (*ChlH*) gene serve as a positive control of successful VIGS process [32]. PDS is involved in the biosynthesis of carotenoids, and silencing of this gene reduces the photo-protective carotenoids, thus causing a photo-bleaching phenotype due to chlorophyll photooxidation [33]. ChlH is involved in the biosynthesis of chlorophyll, and successful silencing causes a yellowish leaf phenotype due to chlorophyll reduction [34]. Although silencing of *PDS* is often used as a visual marker in VIGS studies, it is accepted that silencing of *ChlH* serves as a higher sensitivity visual marker when compared to *PDS* [35–37].

We have successfully applied VIGS in olive plantlets with *Agrobacterium* harboring plasmids encoding the bipartite RNA genome of *Tobacco rattle virus* (TRV). Although it is unknown whether olive is naturally susceptible to the nematode-transmitted TRV, we chose to perform VIGS studies with TRV-based constructs for two reasons. First, TRV is known to have a wide host range and can infect about 400 species in more than 50 plant families [38]. Second, reports of TRV-infected plants from other genera of the Oleaceae family can be found in the literature [39–41]. From a taxonomic point of view, it is worth noting that olive is the fifth plant example that VIGS has been successfully applied in Lamiales, an order with almost 24,000 species distributed in more than 1000 genera and 24 families [42]. The other four reported cases are the sweet basil (*Ocimum basilicum*; Lamiaceae) [43], the snapdragon (*Antirrhinum majus*; Plantaginaceae) [44], the yellow monkeyflower (*Mimulus guttatus*; Phrymaceae) [45], and the purple witchweed (*Striga hermonthica*; Orobanchaceae) [19, 46].

This detailed protocol describes the pipeline to successfully silence endogenous genes in olives by cloning a small fragment of the GOI in a binary vector encompassing the TRV2 genome. Co-inoculation of *Agrobacterium* cells harboring the TRV1 genome with cells harboring the TRV2:GOI genome results in reconstitution of the bipartite genome of TRV in plant cells eliciting the viral infection. This triggers the PTGS mechanism of the plant and thus systemic silencing of the GOI. This approach provides researchers an additional tool to precisely and easily study gene/protein function, contributes to the understanding of signaling processes, implements strategies to tailor-change biochemical pathways, and is expected to boost functional genomic analysis in olive, a perennial woody tree species recalcitrant to genetic transformation. Taking into consideration that VIGS is known to persist *in planta* for several years [47] and can be induced after grafting [48], this approach could potentially be used as a tool to manipulate agronomic traits or even fortify olives against emerging (a)biotic challenges in already established orchards, thus giving new dimensions in sustainable agriculture of tomorrow.

2 Materials

2.1 Vector Construction

1. Sterilized tips, 0.2 and 1.5 mL tubes.
2. Agarose gel apparatus (gel tank, gel tray, well combs, and casting tray) and relative equipment (power supply, microwave oven, UV transilluminator, and gel documentation system).
3. 6× gel loading buffer: 0.25% bromophenol blue, 0.25% xylene cyanol FF, 40% sucrose.
4. 50× Tris-acetate buffer (TAE): mix 242 g Tris-base, 57.1 mL glacial acetic acid, and 100 mL of 0.5 M EDTA (pH 8.0) in 1 L of ddH_2O.
5. Powdered agarose.
6. Ethidium bromide (10 mg/mL, stored in room temperature and protected from light). Working solution is 0.5 μg/mL.
7. Restriction enzymes (*Kpn*I, *Xba*I, *Sma*I, and any other enzyme needed for cloning of the GOI) and the corresponding buffers.
8. Heat blocks or water baths.
9. pTRV1 and pTRV2-MCS vectors: VIGS plasmids encoding the bipartite RNA genome of *Tobacco rattle virus* (TRV) can be obtained from the Arabidopsis Biological Resource Center (ABRC, www.arabidopsis.org) under the stock numbers CD3-1039 (pTRV1) and CD3-1040 (pTRV2-MCS).
10. Liquid nitrogen, mortar, and pestle.
11. RNA extraction buffer: 100 mM Tris–HCl pH 9.5, 0.5% SDS.
12. Phenol, equilibrated to pH 8.
13. 24:1 (v/v) chloroform/isoamyl alcohol.
14. 3 M sodium acetate, pH 4.8.
15. Ethanol.
16. Spectrophotometer.
17. DNase I treatment kit and RNase inhibitor.
18. cDNA synthesis kit including oligo$(dT)_{17}$.
19. Proofreading DNA polymerase and buffer.
20. 10 mM dNTPs set.
21. GOI-specific primers including desired restriction site(s) (*see* **Note 1**):

    OePDS (GenBank Acc. no.: GABQ01079853.1) (232 bp fragment) primers OePDS-F 5′-GATGGCAATCCACCAGAAAGAC-3′ and OePDS-R 5′-ACTGTATCTCCTTCCAGTCCTC-3′.

    OeChlH (GenBank Acc. no.: GABQ01080755.1) (297 bp fragment) primers OeChlH-F 5′-GCAGCTTATTATTCCTATGTGG-3′ and OeChlH-R 5′-GGACGAGATCAACTCACTCAAC-3′.

22. PCR machine.
23. T4 DNA ligase.
24. PCR purification and miniprep kits.
25. Chemically competent *Escherichia coli* (DH5α) cells.
26. Luria-Bertani broth medium: 1% peptone, 0.5% yeast extract, and 1% NaCl. In case of solid medium, agar is added in a final concentration of 1.4% before autoclaving.
27. Kanamycin, 1000× stock solution: 50 mg/mL prepared in ddH_2O and stored at −20 °C.
28. Orbital shakers.
29. Centrifuges.
30. Incubators at 37 °C.

2.2 Handling of Agrobacterium Cells

1. 10% (v/v) sterile glycerol.
2. *Agrobacterium tumefaciens* strain C58C1 Rif^R (GV3101) containing the T-DNA-deficient Ti plasmid pMP90.
3. LB medium.
4. Antibiotic stock solutions: 5 g/L rifampicin resuspended in methanol (100×), 50 g/L gentamicin resuspended in ddH_2O (1000×), 50 g/L kanamycin resuspended in ddH_2O (1000×). Store all antibiotic stocks at −20 °C. Working solution for each antibiotic is 50 mg/L.
5. Electroporator and electroporation cuvettes of a 0.1 cm gap.
6. Refrigerated centrifuge, centrifugation bottles and tubes.
7. Erlenmeyer flasks, shakers, and incubators at 28 °C.
8. Dilution buffer: prepare stock solutions of 100 mM 2-N-morpholino-ethanesulfonic acid (MES) pH 5.6 and 100 mM $MgCl_2$. Prepare fresh dilution buffer (10 mM MES, pH 5.6, 10 mM $MgCl_2$) prior agroinoculations.
9. 100 mM acetosyringone stock solution in DMSO.
10. Syringes of 1 mL with needle.
11. Sterile glycerol (100%).

2.3 Preparation of Olive Plantlets

1. Growth chamber maintained at 22 °C with a 16/8 h light/dark cycle equipped with artificial lighting.
2. Olive seeds collected from *O. europaea* L. cv. 'Koroneiki' trees.
3. Solution of 10% NaOH.
4. A bench vice or a pipe cutter.
5. Solution of 20% NaClO containing 0.01% Triton-X.
6. Soil, pots, and trays.

2.4 Real-Time PCR

1. Reagents for RNA extraction (*see* Subheading 2.1, **items 10–15**).
2. DNAse I treatment kit.
3. Single-strand cDNA synthesis kit including oligo(dT)$_{17}$.
4. GOI-specific primers for qPCR analysis (*see* **Note 2**):
   OePDS: OePDSrt-F 5′-AAACTCCAAGGTCCGTCTATAA-3′ and OePDSrt-R 5′-GCTTTGTGTAATCACCAGCTAAA-3′ primers.
   OeChlH: OeChlHrt-F 5′-GTACACTTTCGGAGACGGTAAG-3′ and OeChlHrt-R 5′-CTTGTCCTGAAGTTGCACTCCA-3′ primers.
   OeActin (GenBank Acc. no.: GABQ01079399.1): OeActin-F 5′-GTATGTTGCTATCCAGGCTGTT-3′ and OeActin-R 5′-AAATGGGTACTGTGTGACTCAC-3′ primers.
5. 2× SYBR Select Master Mix.
6. 96-Well reaction plates with optically transparent lids.
7. Real-time PCR cycler.

3 Methods

3.1 Cloning Procedures and Vector Construction

1. Grind young leaves (and/or other desired plant material) from *Olea europaea* L. cv. 'Koroneiki' plants into powder using a mortar and pestle in the presence of liquid nitrogen.
2. Transfer 200 mg of powdered tissue in a sterile and pre-chilled tube, and add 200 μL of RNA extraction buffer and 200 μL of phenol (pH 8.0). Mix thoroughly by vortexing and centrifuge the sample for 5 min at 13,000 × *g*.
3. Transfer the aqueous (upper) phase in a new tube and add an equal volume of phenol. Mix thoroughly and centrifuge the sample for 5 min at 13,000 × *g*.
4. Transfer the aqueous phase in a new tube, and add an equal volume of phenol/chloroform/isoamyl alcohol (25:24:1). Mix thoroughly and centrifuge the sample for 5 min at 13,000 × *g*.
5. Transfer the aqueous phase in a new tube, and add an equal volume of chloroform/isoamyl alcohol (24:1). Mix thoroughly and centrifuge the sample for 5 min at 13,000 × *g*.
6. Transfer the aqueous phase in a new tube, and add 1/10 volume of 3 M sodium acetate (pH 4.8) and 2.5 volumes of ice-cold ethanol (100%). Mix gently and precipitate the nucleic acids by incubating the sample at −20 °C for 16 h.
7. Centrifuge the sample at 13,000 × *g* for 20 min at 4 °C. Discard the supernatant, air-dry the pellet, and resuspend the nucleic acids at 30 μL of RNase-free ddH$_2$O.

8. Determine the quantity and the quality of the extracted nucleic acids using a spectrophotometer and agarose gel electrophoresis, respectively. The sample can be stored at −80 °C.
9. Proceed with RNase-free DNase I treatment to remove the DNA using 5–15 μg of the extracted nucleic acids and following the manufacturer's instructions. The reaction is incubated at 37 °C for 1 h. When the reaction is finished, add ddH_2O in a total volume of 400 μL, and proceed with phenol/chloroform/isoamyl alcohol (25:24:1) extraction and ethanol/sodium acetate precipitation (**steps 4–7**) to recover the DNA-free RNA sample(s).
10. Determine the quantity and the quality of the extracted DNA-free RNA(s) using a spectrophotometer and agarose gel electrophoresis, respectively.
11. Reverse transcribe 1.5 μg of RNA sample(s) in the presence of the oligo$(dT)_{17}$ primer using a standard reverse transcription enzyme or commercial kits following the manufacturer's instructions. The synthesized cDNA can be stored at −20 °C until use.
12. Dilute a fraction of the synthesized single-strand cDNA(s) five times, and use this as a template for PCR amplifications with a proofreading polymerase following the manufacturer's instructions. Typically a PCR reaction is set up in a final volume of 25 μL containing 5 μL of 5× polymerase buffer, 500 nM of each gene-specific primer, 200 μM of dNTPs (each), 0.5 U of proofreading polymerase, and 10–50 ng of cDNA.
13. PCR amplify a 200–400 bp fragment of the gene of interest (GOI) with sequence-specific primers that introduce the desired restriction site(s) at the ends of the fragment. The fragment can be a partial part of either the coding sequence (CDS) or the untranslated regions (UTRs) of the transcribed GOI. Select the fragment carefully to avoid triggering any off-target silencing (*see* **Note 3**). Amplify also the *OeChlH* and/or the *OePDS* fragments using the respective primers OeChlH-F/OeChlH-R and OePDS-F/OePDS-R. Set up the PCR machine to perform up to 35 cycles for the amplification of the fragments to avoid introducing point mutations.
14. Digest and ligate the fragments in the corresponding restriction site(s) of the multiple cloning site (MCS) of pTRV2 in the antisense orientation relative to the TRV coat protein using common molecular cloning protocols, and verify the constructs by sequencing. The restriction sites available in the multiple cloning site region of pTRV2 are *EcoR*I, *Xba*I, *Stu*I, *Nco*I, *Bam*HI, *Kpn*I, *Sac*I, *Mlu*I, *Xho*I, *Srf*I, and *Sma*I.

3.2 Handling of *Agrobacterium* Cells

3.2.1 Preparation of Electro-competent *Agrobacterium tumefaciens*

1. Streak the *Agrobacterium tumefaciens* strain C58C1 Rif^R (GV3101) containing the T-DNA-deficient Ti plasmid pMP90 Gen^R on LB plate supplemented with 50 mg/L rifampicin and 50 mg/L gentamicin, and grow for 36–48 h at 28 °C.
2. Inoculate a single colony in 5 mL liquid LB medium supplemented with 50 mg/L rifampicin and 50 mg/L gentamicin, and grow at 28 °C under continuous shaking (200–250 rpm) until the stationary phase (at least 16 h).
3. Use 1 mL of the culture to inoculate a 1 L flask containing 100 mL of LB medium supplemented with 50 mg/L rifampicin and 50 mg/L gentamicin, and grow at 28 °C under continuous shaking (200–250 rpm) until the $OD_{600} = 0.5$ (typically 5–8 h).
4. Place the flask containing the culture on ice for 15 min, transfer the culture in ice-cold centrifuge bottles, and centrifuge at 1500 × *g* for 5 min at 4 °C.
5. Remove the supernatant carefully, and gently resuspend the pellet in 100 mL of ice-cold 10% glycerol solution. Centrifuge at 1500 × *g* for 5 min at 4 °C.
6. Remove the supernatant carefully, and gently resuspend the pellet in 50 mL of ice-cold 10% glycerol solution. Centrifuge at 1500 × *g* for 5 min at 4 °C.
7. Remove the supernatant carefully, and gently resuspend the pellet in 2 mL of ice-cold 10% glycerol solution. Divide the cell suspension in aliquots of 40 μL in ice-cold Eppendorf tubes, freeze immediately the cells in liquid nitrogen, and store at −80 °C.

3.2.2 Transformation of *Agrobacterium tumefaciens* Through Electroporation

1. Defreeze the appropriate number of *Agrobacterium* electrocompetent cells to be transformed by letting the Eppendorfs on ice for 5–10 min (i.e., do not warm the cells above 4 °C).
2. Add 10–50 ng of plasmid DNA in 40 μL of cells, incubate on ice for 1 min, and transfer the mix in an ice-cold electroporation cuvette.
3. Adjust the settings of the electroporation apparatus to deliver a pulse of 1.8 kV, 400 Ω resistance, and 25 μF capacitance for electroporation cuvettes of 0.1 cm gap size of electrodes.
4. Dry well the outer part of the cuvette with a paper, place it in the cuvette holder of the electroporation apparatus, and deliver the electric pulse. A time constant of approximately 5 ms with an electric field strength of 18 kV/cm should register on the apparatus (in case of arcing, *see* **Note 4**).
5. Immediately add 1 mL of LB (without antibiotics) to the cuvette, transfer the cell suspension to an Eppendorf, and incubate for 3 h at 28 °C under slow agitation.

6. Centrifuge the cell suspension at 4000 × *g* for 5 min; resuspend gently the pellet in 200 μL of LB; streak the cell suspension on LB plates supplemented with 50 mg/L rifampicin, 50 mg/L gentamicin and 50 mg/L kanamycin; and incubate at 28 °C. Transformed colonies are visible after 2–3 days.
7. Inoculate at least one colony per transformation in 5 mL liquid LB supplemented with 50 mg/L rifampicin, 50 mg/L gentamicin, and 50 mg/L kanamycin. Grow at 28 °C under continuous shaking (200–250 rpm) for 36–48 h, and prepare glycerol stocks in sterilized Eppendorfs by mixing 650 μL of each culture with 350 μL of glycerol (100%) for archiving the strains at −80 °C.

3.3 Agroinoculation of Olive Plantlets

3.3.1 Preparation of Olive Plantlets

1. Harvest ripe fruits from *Olea europaea* L. cv. 'Koroneiki' naturally growing trees (*see* **Note 5**).
2. Remove the mesocarps, and subject the woody endocarps in 10% NaOH for 10 min.
3. Wash thoroughly the woody endocarps with water to remove any fleshy remnants.
4. Dry naturally the woody endocarps (Fig. 1a), and store them in cool and ventilated area until use. Stones can be stored for several months without loss of germination rate.
5. Carefully crack the woody endocarps with a bench vice (Fig. 1b) or a pipe cutter to remove seeds (Fig. 1c). Apply pressure gently to avoid damaging seeds.
6. Surface-sterilize the seeds with 20% NaClO containing 0.01% Triton-X for 5 min, and wash three to four times with sterile ddH_2O (Fig. 1d, e).
7. After 7 days of stratification at 4 °C in sterile ddH_2O, transfer the seeds in soil (Fig. 1f), and grow at 22 °C in a growth chamber with a 16/8 h light/dark cycle and 100 μmol/m^2/s light intensity. Typically, olive plantlets develop the first pair of true leaves after 1 month.

3.3.2 Preparation of *Agrobacterium* Strains Harboring the VIGS Vectors for Agroinoculation

1. Streak each of the *Agrobacterium* strains in LB plates supplemented with 50 mg/L rifampicin, 50 mg/L gentamicin, and 50 mg/L kanamycin, and grow for 2 days at 28 °C. The strains to be used are transformed with the following plasmids:

   pTRV1.

   pTRV2-EV.

   pTRV2-OePDS or *pTRV2-OeChlH*.

   pTRV2-GOI.
2. Inoculate single colonies (*see* **Note 6**) from each strain in 5 mL of LB supplemented with 50 mg/L rifampicin, 50 mg/L

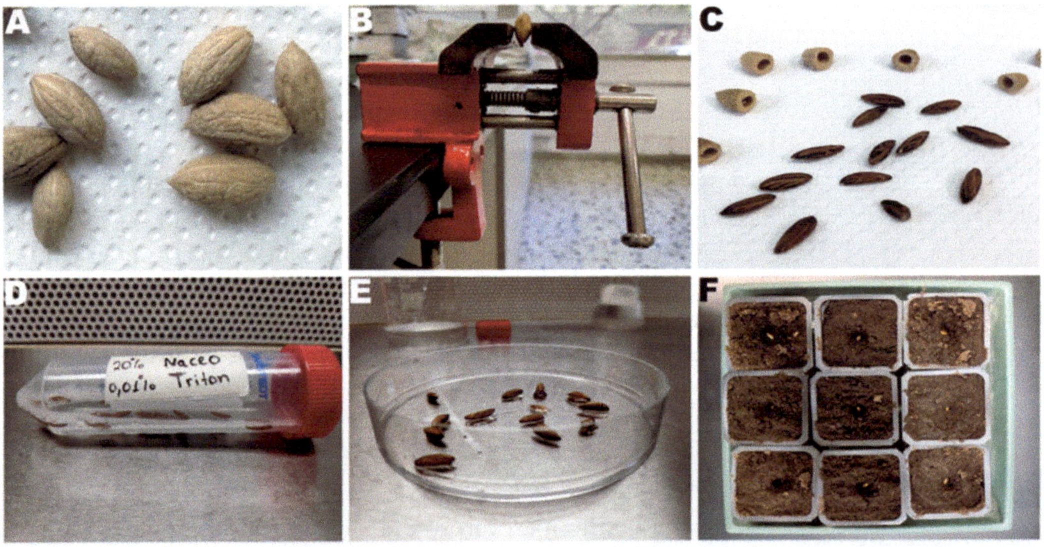

Fig. 1 Preparation of olive plantlets. Dried olive stones (**a**) were carefully cracked with a bench vice (**b**) in order to remove seeds (**c**). After surface sterilization (**d**, **e**), the seeds were potted in soil (**f**) for germination

gentamicin, and 50 mg/L kanamycin, and grow at 28 °C for ~36–48 h under continuous shaking (200–250 rpm).

3. Transfer the *Agrobacterium* cultures in Falcon tubes, and centrifuge for 10 min at 1000 × *g* in room temperature.
4. Discard the supernatant and carefully wash the cells with 2.5 mL dilution buffer (1/2 of the initial volume of the culture) without disrupting the pellet in order to remove any remaining LB medium.
5. Gently resuspend the pellet in 1 mL of dilution buffer (1/5 of the initial culture), and measure the OD_{600} of a diluted fraction for each cell suspension.
6. Mix each of the cell suspensions harboring a variant of pTRV2 in a 1:1 ratio with the cell suspension harboring the pTRV1 plasmid so that each culture has a final OD_{600} of 3 in dilution buffer supplemented with acetosyringone to a final concentration of 150 μM.
7. Incubate the cell suspensions at 28 °C for 3 h under slow agitation.

3.3.3 Agroinoculation

1. Use sterile 1 mL syringes (Fig. 2a) to inoculate the *Agrobacterium* cell suspensions prepared. Use a different syringe for each combination of strains to be used.
2. Olive plantlets developing the first true leaf pair should be used for agroinoculation (Fig. 2b).

Fig. 2 Agroinoculation of TRV constructs in olive plantlets. *Agrobacterium* cell suspensions (**a**) were inoculated on olive plantlets that have developed the first true leaf pair (**b**) by gently traumatizing the abaxial side of a leaf (**c**). Agrobacteria were also inoculated at the crown region (**d**) and drenched onto the shoot apical meristem of the plant (**e**). Panels (**c**) and (**d**) show older plantlets on the fifth round of booster inoculations

3. Attach the needle and inoculate the abaxial side of a leaf by gently traumatizing the leaf (Fig. 2c) or by pricking. Avoid harming the veins or puncturing across the leaf.
4. Agroinoculate also the crown region (Fig. 2d), and drench the remaining cell suspension (i.e., without traumatizing the plantlets) onto the shoot apical meristem of the plant (Fig. 2e).
5. In order to boost the TRV infection, repeat the agroinoculation process on the same plants every 2 weeks until the desired phenotype is observed.

3.4 Evaluation of GOI Silencing by Real-Time PCR

1. Typically, after ~3–4 months the plants will exhibit the expected phenotype (Fig. 3). Agroinoculating plants with the silencing markers (either *pTRV2-OePDS* or *pTRV2-OeChlH*) in parallel with *pTRV2-GOI* helps to roughly estimate when and in which leaf nodes the GOI is expected to be silenced (*see* **Note 7**).
2. Collect leaves from the two pairs of leaves that emerged above the agroinoculated region, freeze immediately in liquid nitrogen, and store at −80 °C for RNA extraction.
3. Grind each leaf into powder using a mortar and pestle in the presence of liquid nitrogen, and extract RNA using either a phenol/chloroform protocol (Subheading 3.1) or a commercial kit following the manufacturer's instructions. Proceed with RNase-free DNase I treatment to remove the DNA and determine the quantity and the quality of the extracted DNA-free RNAs using a spectrophotometer and agarose gel electrophoresis, respectively.
4. Reverse transcribe equal quantities of each RNA sample in the presence of the oligo(dT)$_{17}$ primer using a standard reverse transcription kit following the manufacturer's instructions.
5. Set up the real-time PCR reactions according to the manufacturer's instructions. Typically, each reaction is set up in a final

Fig. 3 Phenotypes of representative olive plants agroinoculated with a combination of pTRV1 and either pTRV2-OePDS or pTRV2-EV or pTRV2-OeChlH constructs

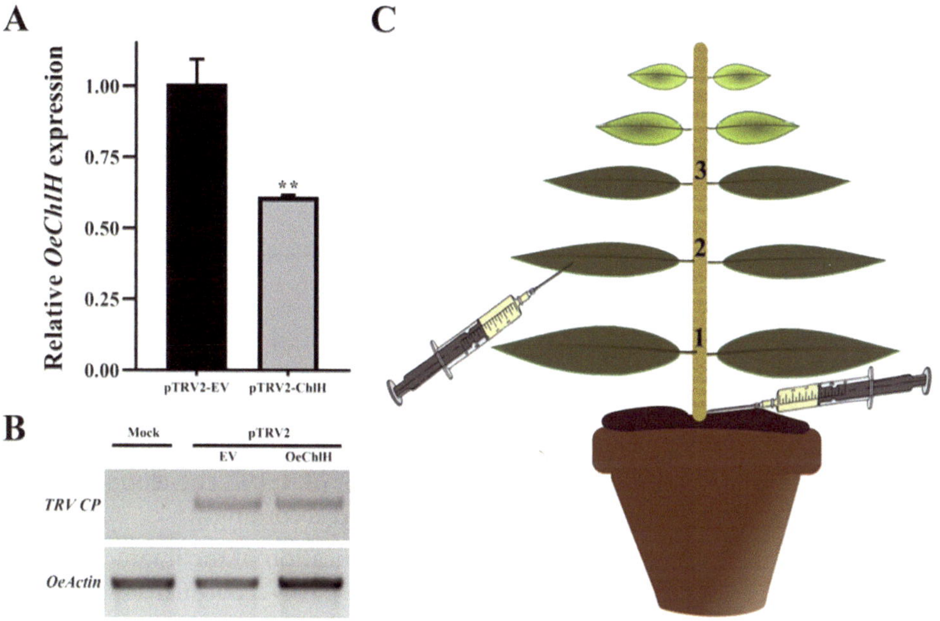

Fig. 4 Evaluation of GOI silencing. (**a**) Real-time PCR analysis of OeChlH expression level between plants agroinoculated with pTRV2-EV or pTRV2-OeChlH constructs. Three technical replicates of three biological replicates per treatment were analyzed. Mean $\pm$ SE and asterisks denote statistical significance ($P \leq 0.01$, Student's *t*-test). (**b**) RT-PCR of the TRV coat protein (CP) in the newly emerged (e.g., non-treated) leaves per treatment, as indicated. (**c**) Graphical representation of agroinoculation of an olive plantlet. Agrobacteria harboring either the pTRV1 or pTRV2-ChlH plasmid were co-inoculated, by using a syringe, onto the abaxial side of the leaf and on the crown region. Agroinoculation starts when olive plantlets have developed the first true leaf pair at one leaf at a time. The procedure is repeated after 2 weeks onto the other leaf of each pair and on the crown region. Typically silencing is observed after 3 months when all six leaves have been inoculated. Dark-green leaves represent inoculated non-silenced leaves, and light-green leaves represent non-agroinoculated (non-treated) gene-silenced leaves. Numbers 1, 2, and 3 represent the pairs of leaves used to perform initial and booster inoculations

volume of 20 μL containing 10 μL of 2× SYBR Select Master Mix, 300 nM for each gene-specific primer, and 5 ng of cDNA. The transcript of *OeActin* [13] can be used as an endogenous control for all the samples using the respective primers OeActin-F and OeActin-R (*see* **Note 2**).

6. The cycling conditions typically used in the real-time PCR machine are initial denaturation at 95 °C for 7 min followed by 40 cycles of denaturation at 95 °C for 5 s, annealing at 58 °C for 30 s, and extension at 72 °C for 30 s.
7. Perform the quantitative gene expression analysis (Fig. 4a) in an appropriate real-time PCR cycler, and analyze the data by the ΔΔCt method [49] (Fig. 4a).
8. If needed, perform semiquantitative RT-PCR to screen for the presence of the TRV coat protein (Fig. 4b) (*see* **Note 8**) [32].

9. The same plants can be used for further boosting the TRV infection by agroinoculating again with the corresponding strains until the desired silencing is observed (Fig. 4c). Use booster inoculations (i.e., starting from the pair of leaves of the first node, then the second and then the third) every 2 weeks (*see* **Note 9**).

4 Notes

1. Both sets of primers do not introduce any restriction sites; thus the PCR products need to be first blunt-end ligated in a *Sma*I digested pUC19 plasmid, and then the fragments can be directly subcloned in pTRV2-MCS (EV) utilizing the *Kpn*I and *Xba*I restriction sites of pUC19. The cloning strategy for any GOI should be adjusted accordingly.
2. The primers used for amplification of the fragment of the GOI to be silenced (i.e., for the pTRV2 constructs) should not be used for qPCR analysis. At least one primer should hybridize upstream or downstream the region used to trigger silencing of the GOI.
3. Several bioinformatic approaches have been developed in order to assist in the prediction of specific regions of a GOI that is expected to produce efficient and specific siRNAs [50]. An easy-to-use software that is very helpful when working with non-model plants is si-Fi (http://labtools.ipk-gatersleben.de/) since researchers can create local databases with in-house and/or publicly available datasets of unigenes.
4. Arcing is usually caused due to the presence of ions resulting in increased conductivity of the solution. The transformation efficiency is dramatically reduced due to uneven transfer of the electrical charge through the cuvette. Use a commercial kit to purify the plasmids that will be electroporated, and resuspend plasmids in salt-free water. If needed, increase the washing steps during the preparation of the electro-competent cells, and prepare the buffers in ddH_20 of low electrical conductance. Another reason for arcing could emerge by the presence of moisture in the outer (side) of the electroporation cuvette. Ensure that the cuvette is well dried before the delivery of the pulse.
5. In order to ensure that the seeds are mature and able to germinate, collect a number of fruits at their full ripening stage when the pulp and the peel have a similar purple-black color.
6. The number of colonies and/or the volume of LB to be inoculated is adjustable and depends on the number of plants

to be agroinoculated and the desired optical density of the cultures to be mixed in the final solution. Note that the strain harboring the pTRV1 plasmid will be mixed in 1:1 ratio with each one of the strains harboring the pTRV2 constructs, so ensure that sufficient number of cultures are initiated.

7. In our experience, silencing of *OeChlH* results in a more intense and easily to identify phenotype than silencing of *OePDS*.
8. A convenient way to narrow down the number of the samples expected to exhibit the higher degree of silencing of the GOI is to screen the plants for the presence of transcripts of the TRV coat protein. Detection of the TRV coat protein can be performed by semiquantitative RT-PCR using the TRV2cp-F 5-′-CTGGGTTACTAGCGGCACTGAATA-3′ and TRV2cp-R 5′-TCCACCAAACTTAATCCCGAATAC-3′ set of primers (GenBank Acc. no.: AF406991.1).
9. When the experiment is finished, ensure that the plants are disposed only after autoclaving.

Acknowledgments

P.H. acknowledges funding from General Secretariat of Research and Technology, Greece, National Emblematic Action "Olive Routes." M.T. acknowledges funding from Greece and the European Union (European Social Fund, ESF) through the Operational Programme "Human Resources Development, Education and Lifelong Learning" in the context of the project "Strengthening Human Resources Research Potential via Doctorate Research" (MIS-5000432), implemented by the State Scholarships Foundation (IKY). We also acknowledge the editors of this volume for the helpful suggestions.

References

1. Cruz F, Julca I, Gómez-Garrido J, Loska D, Marcet-Houben M, Cano E, Galán B, Frias L, Ribeca P, Derdak S, Gut M, Sánchez-Fernández M, García JL, Gut IG, Vargas P, Alioto TS, Gabaldón T (2016) Genome sequence of the olive tree, *Olea europaea*. GigaScience 5(1):1–12. https://doi.org/10.1186/s13742-016-0134-5
2. Unver T, Wu Z, Sterck L, Turktas M, Lohaus R, Li Z, Yang M, He L, Deng T, Escalante FJ, Llorens C, Roig FJ, Parmaksiz I, Dundar E, Xie F, Zhang B, Ipek A, Uranbey S, Erayman M, Ilhan E, Badad O, Ghazal H, Lightfoot DA, Kasarla P, Colantonio V, Tombuloglu H, Hernandez P, Mete N, Cetin O, Van Montagu M, Yang H, Gao Q, Dorado G, Van de Peer Y (2017) Genome of wild olive and the evolution of oil biosynthesis. Proc Natl Acad Sci U S A 114 (44):E9413–E9422. https://doi.org/10.1073/pnas.1708621114
3. Green PS (2004) Oleaceae. In: Kadereit JW (ed) Flowering Plants · Dicotyledons: Lamiales (except Acanthaceae including Avicenniaceae). Springer Berlin Heidelberg, Berlin, Heidelberg, pp 296–306. https://doi.org/10.1007/978-3-642-18617-2_16

4. Fernández-Ocaña A, Carmen García-López M, Jiménez-Ruiz J, Saniger L, Macías D, Navarro F, Oya R, Belaj A, de la Rosa R, Corpas FJ, Bautista Barroso J, Luque F (2010) Identification of a gene involved in the juvenile-to-adult transition (JAT) in cultivated olive trees. Tree Genet Genomes 6 (6):891–903. https://doi.org/10.1007/s11295-010-0299-5
5. Yanik H, Turktas M, Dundar E, Hernandez P, Dorado G, Unver T (2013) Genome-wide identification of alternate bearing-associated microRNAs (miRNAs) in olive (*Olea europaea* L.). BMC Plant Biol 13(1):10. https://doi.org/10.1186/1471-2229-13-10
6. Koudounas K, Banilas G, Michaelidis C, Demoliou C, Rigas S, Hatzopoulos P (2015) A defence-related *Olea europaea* β-glucosidase hydrolyses and activates oleuropein into a potent protein cross-linking agent. J Exp Bot 66(7):2093–2106. https://doi.org/10.1093/jxb/erv002
7. Koudounas K, Thomopoulou M, Michaelidis C, Zevgiti E, Papakostas G, Tserou P, Daras G, Hatzopoulos P (2017) The C-domain of oleuropein β-glucosidase assists in protein folding and sequesters the enzyme in nucleus. Plant Physiol 174 (3):1371–1383. https://doi.org/10.1104/pp.17.00512
8. Alagna F, Geu-Flores F, Kries H, Panara F, Baldoni L, O'Connor SE, Osbourn A (2016) Identification and characterization of the iridoid synthase involved in oleuropein biosynthesis in olive (*Olea europaea*) fruits. J Biol Chem 291(11):5542–5554. https://doi.org/10.1074/jbc.M115.701276
9. Volk J, Sarafeddinov A, Unver T, Marx S, Tretzel J, Zotzel J, Warzecha H (2019) Two novel methylesterases from *Olea europaea* con tribute to the catabolism of oleoside-type secoiridoid esters. Planta 250(6):2083–2097. https://doi.org/10.1007/s00425-019-03286-0
10. Banilas G, Hatzopoulos P (2013) Genetics and molecular biology of olives. In: Aparicio R, Harwood J (eds) Handbook of olive oil: analysis and properties. Springer US, Boston, MA, pp 129–161. https://doi.org/10.1007/978-1-4614-7777-8_5
11. Corrado G, Garonna A, Cabanás CG-L, Gregoriou M, Martelli GP, Mathiopoulos KD, Mercado-Blanco J, Saponari M, Tsoumani KT, Rao R (2016) Host response to biotic stresses. In: Rugini E, Baldoni L, Muleo R, Sebastiani L (eds) The olive tree genome. Springer International Publishing, Cham, pp 75–98. https://doi.org/10.1007/978-3-319-48887-5_6
12. Sebastiani L, Gucci R, Kerem Z, Fernández JE (2016) Physiological responses to abiotic stresses. In: Rugini E, Baldoni L, Muleo R, Sebastiani L (eds) The olive tree genome. Springer International Publishing, Cham, pp 99–122. https://doi.org/10.1007/978-3-319-48887-5_7
13. Koudounas K, Manioudaki ME, Kourti A, Banilas G, Hatzopoulos P (2015) Transcriptional profiling unravels potential metabolic activities of the olive leaf non-glandular trichome. Front Plant Sci 6:10. https://doi.org/10.3389/fpls.2015.00633
14. Roka L, Koudounas K, Daras G, Zoidakis J, Vlahou A, Kalaitzis P, Hatzopoulos P (2018) Proteome of olive non-glandular trichomes reveals protective protein network against (a) biotic challenge. J Plant Physiol 231:210–218. https://doi.org/10.1016/j.jplph.2018.09.016
15. Torreblanca R, Cerezo S, Palomo-Ríos E, Mercado JA, Pliego-Alfaro F (2010) Development of a high throughput system for genetic transformation of olive (*Olea europaea* L.) plants. Plant Cell Tissue Organ Cult 103(1):61–69. https://doi.org/10.1007/s11240-010-9754-0
16. Rugini E, Cristofori V, Silvestri C (2016) Genetic improvement of olive (*Olea europaea* L.) by conventional and in vitro biotechnology methods. Biotechnol Adv 34(5):687–696. https://doi.org/10.1016/j.biotechadv.2016.03.004
17. Haberman A, Bakhshian O, Cerezo-Medina S, Paltiel J, Adler C, Ben-Ari G, Mercado JA, Pliego-Alfaro F, Lavee S, Samach A (2017) A possible role for flowering locus T-encoding genes in interpreting environmental and internal cues affecting olive (*Olea europaea* L.) flower induction. Plant Cell Environ 40 (8):1263–1280. https://doi.org/10.1111/pce.12922
18. Becker A, Lange M (2010) VIGS – genomics goes functional. Trends Plant Sci 15(1):1–4. https://doi.org/10.1016/j.tplants.2009.09.002
19. Dommes AB, Gross T, Herbert DB, Kivivirta KI, Becker A (2018) Virus-induced gene silencing: empowering genetics in non-model organisms. J Exp Bot 70(3):757–770. https://doi.org/10.1093/jxb/ery411
20. Senthil-Kumar M, Mysore KS (2011) New dimensions for VIGS in plant functional genomics. Trends Plant Sci 16(12):656–665. https://doi.org/10.1016/j.tplants.2011.08.006

21. Sasaki S, Yamagishi N, Yoshikawa N (2011) Efficient virus-induced gene silencing in apple, pear and Japanese pear using Apple latent spherical virus vectors. Plant Methods 7 (1):15. https://doi.org/10.1186/1746-4811-7-15
22. Kawai T, Gonoi A, Nitta M, Yamagishi N, Yoshikawa N, Tao R (2016) Virus-induced gene silencing in various Prunus species with the Apple latent spherical virus vector. Sci Hortic 199:103–113. https://doi.org/10.1016/j.scienta.2015.12.031
23. Ye J, Qu J, Bui HTN, Chua NH (2009) Rapid analysis of *Jatropha curcas* gene functions by virus-induced gene silencing. Plant Biotechnol J 7(9):964–976. https://doi.org/10.1111/j.1467-7652.2009.00457.x
24. Jiang Y, Ye S, Wang L, Duan Y, Lu W, Liu H, Fan D, Zhang F, Luo K (2014) Heterologous gene silencing induced by tobacco rattle virus (TRV) is efficient for pursuing functional genomics studies in woody plants. Plant Cell Tissue Organ Cult 116(2):163–174. https://doi.org/10.1007/s11240-013-0393-0
25. Zhao N, Zhang Y, Li Q, Li R, Xia X, Qin X, Guo H (2015) Identification and expression of a stearoyl-ACP desaturase gene responsible for oleic acid accumulation in *Xanthoceras sorbifolia* seeds. Plant Physiol Biochem 87:9–16. https://doi.org/10.1016/j.plaphy.2014.12.009
26. Li X-J, Zhang J-Q, Wu Z-C, Lai B, Huang X-M, Qin Y-H, Wang H-C, Hu G-B (2016) Functional characterization of a glucosyltransferase gene, LcUFGT1, involved in the formation of cyanidin glucoside in the pericarp of *Litchi chinensis*. Physiol Plant 156 (2):139–149. https://doi.org/10.1111/ppl.12391
27. Shen Z, Sun J, Yao J, Wang S, Ding M, Zhang H, Qian Z, Zhao N, Sa G, Zhao R, Shen X, Polle A, Chen S (2015) High rates of virus-induced gene silencing by tobacco rattle virus in Populus. Tree Physiol 35 (9):1016–1029. https://doi.org/10.1093/treephys/tpv064
28. Agüero J, Vives MC, Velázquez K, Pina JA, Navarro L, Moreno P, Guerri J (2014) Effectiveness of gene silencing induced by viral vectors based on Citrus leaf blotch virus is different in Nicotiana benthamiana and citrus plants. Virology 460-461:154–164. https://doi.org/10.1016/j.virol.2014.04.017
29. Liu Y, Zhou B, Qi Y, Liu C, Liu Z, Ren X (2018) Biochemical and functional characterization of AcUFGT3a, a galactosyltransferase involved in anthocyanin biosynthesis in the red-fleshed kiwifruit (*Actinidia chinensis*). Physiol Plant 162(4):409–426. https://doi.org/10.1111/ppl.12655
30. Jin Z, Yan T, Chang C, Liu Z, Wang Y, Tang Z, Yu F (2016) Application of virus-induced gene silencing approach in *Camptotheca acuminata*. Plant Cell Tissue Organ Cult 126(3):533–540. https://doi.org/10.1007/s11240-016-1022-5
31. Li R, Liu L, Dominic K, Wang T, Fan T, Hu F, Wang Y, Zhang L, Li L, Zhao W (2018) Mulberry (*Morus alba*) MmSK gene enhances tolerance to drought stress in transgenic mulberry. Plant Physiol Biochem 132:603–611. https://doi.org/10.1016/j.plaphy.2018.10.007
32. Senthil-Kumar M, Mysore KS (2014) Tobacco rattle virus–based virus-induced gene silencing in Nicotiana benthamiana. Nat Protoc 9 (7):1549–1562. https://doi.org/10.1038/nprot.2014.092
33. Kumagai MH, Donson J, della-Cioppa G, Harvey D, Hanley K, Grill LK (1995) Cytoplasmic inhibition of carotenoid biosynthesis with virus-derived RNA. Proc Natl Acad Sci U S A 92(5):1679. https://doi.org/10.1073/pnas.92.5.1679
34. Hiriart J-B, Lehto K, Tyystjärvi E, Junttila T, Aro E-M (2002) Suppression of a key gene involved in chlorophyll biosynthesis by means of virus-inducing gene silencing. Plant Mol Biol 50(2):213–224. https://doi.org/10.1023/A:1016000627231
35. Ryu CM, Anand A, Kang L, Mysore Kirankumar S (2004) Agrodrench: a novel and effective agroinoculation method for virus-induced gene silencing in roots and diverse Solanaceous species. Plant J 40(2):322–331. https://doi.org/10.1111/j.1365-313X.2004.02211.x
36. Liu H, Fu D, Zhu B, Yan H, Shen X, Zuo J, Zhu Y, Luo Y (2012) Virus-induced gene silencing in eggplant (*Solanum melongena*). J Integr Plant Biol 54(6):422–429. https://doi.org/10.1111/j.1744-7909.2012.01102.x
37. Liu Y, Sun W, Zeng S, Huang W, Liu D, Hu W, Shen X, Wang Y (2014) Virus-induced gene silencing in two novel functional plants, *Lycium barbarum* L. and *Lycium ruthenicum* Murr. Sci Hortic 170:267–274. https://doi.org/10.1016/j.scienta.2014.03.023
38. Harrison BD, Robinson DJ (1978) The tobraviruses. In: Lauffer MA, Bang FB, Maramorosch K, Smith KM (eds) Advances in virus research, vol 23. Academic Press, New York, pp 25–77. https://doi.org/10.1016/S0065-3527(08)60097-4

39. Cooper JI (1979) Virus diseases of trees and shrubs. Institute of Terrestrial Ecology, Cambridge
40. Kitajima EW, Rodrigues JCV, Freitas-Astua J (2010) An annotated list of ornamentals naturally found infected by Brevipalpus mite-transmitted viruses. Sci Agric 67:348–371. https://doi.org/10.1590/S0103-90162010000300014
41. Büttner C, von Bargen S, Bandte M, Mühlbach H-P (2013) Forest diseases caused by viruses. In: Gonthier P, Nicolotti G (eds) Infectious forest diseases. CAB International, Wallingford, Oxfordshire, pp 50–75. https://doi.org/10.1079/9781780640402.0050
42. Group TAP, Chase MW, Christenhusz MJM, Fay MF, Byng JW, Judd WS, Soltis DE, Mabberley DJ, Sennikov AN, Soltis PS, Stevens PF (2016) An update of the Angiosperm Phylogeny Group classification for the orders and families of flowering plants: APG IV. Bot J Linn Soc 181(1):1–20. https://doi.org/10.1111/boj.12385
43. Misra RC, Sharma S, Sandeep GA, Chanotiya CS, Ghosh S (2017) Two CYP716A subfamily cytochrome P450 monooxygenases of sweet basil play similar but nonredundant roles in ursane- and oleanane-type pentacyclic triterpene biosynthesis. New Phytol 214(2):706–720. https://doi.org/10.1111/nph.14412
44. Kim BM, Inaba J-i, Masuta C (2011) Virus induced gene silencing in *Antirrhinum majus* using the Cucumber mosaic virus vector: Functional analysis of the AINTEGUMENTA (Am-ANT) gene of *A. majus*. Hortic Environ Biotechnol 52(2):176. https://doi.org/10.1007/s13580-011-0172-y
45. Preston JC, Barnett LL, Kost MA, Oborny NJ, Hileman LC (2014) Optimization of virus-induced gene silencing to facilitate evo-devo studies in the emerging model species *Mimulus guttatus* (Phrymaceae). Ann MO Bot Gard 99(3):301–312. https://doi.org/10.3417/2010120
46. Kirigia D, Runo S, Alakonya A (2014) A virus-induced gene silencing (VIGS) system for functional genomics in the parasitic plant *Striga hermonthica*. Plant Methods 10(1):16. https://doi.org/10.1186/1746-4811-10-16
47. Senthil-Kumar M, Mysore KS (2011) Virus-induced gene silencing can persist for more than 2 years and also be transmitted to progeny seedlings in Nicotiana benthamiana and tomato. Plant Biotechnol J 9(7):797–806. https://doi.org/10.1111/j.1467-7652.2011.00589.x
48. Yan H, Shi S, Ma N, Cao X, Zhang H, Qiu X, Wang Q, Jian H, Zhou N, Zhang Z, Tang K (2018) Graft-accelerated virus-induced gene silencing facilitates functional genomics in rose flowers. J Integr Plant Biol 60(1):34–44. https://doi.org/10.1111/jipb.12599
49. Schmittgen TD, Livak KJ (2008) Analyzing real-time PCR data by the comparative C_T method. Nat Protoc 3:1101. https://doi.org/10.1038/nprot.2008.73
50. Ahmed F, Dai X, Zhao PX (2015) Bioinformatics tools for achieving better gene silencing in plants. In: Mysore KS, Senthil-Kumar M (eds) Plant gene silencing: methods and protocols. Springer, New York, pp 43–60. https://doi.org/10.1007/978-1-4939-2453-0_3

Chapter 14

ALSV-Based Virus-Induced Gene Silencing in Apple Tree (*Malus* × *domestica* L.)

Carolina Werner Ribeiro, Thomas Dugé de Bernonville, Gaëlle Glévarec, Arnaud Lanoue, Audrey Oudin, Olivier Pichon, Benoit St-Pierre, Vincent Courdavault, and Sébastien Besseau

Abstract

Virus-induced gene silencing (VIGS) is a fast and efficient tool to investigate gene function in plant as an alternative to knock down/out transgenic lines, especially in plant species difficult to transform and challenging to regenerate such as perennial woody plants. In apple tree, a VIGS vector has been previously developed based on the *Apple latent spherical virus* (ALSV) and an efficient inoculation method has been optimized using biolistics. This report described detailed step-by-step procedure to design and silence a gene of interest (GOI) in apple tree tissues using the ALSV-based vector.

Key words ALSV, VIGS, Apple tree (*Malus* × *domestica* L.), Quinoa (*Chenopodium quinoa*), Rub-inoculation, Biolistics

1 Introduction

Virus-induced gene silencing (VIGS) is a genetic tool dedicated to functional genomics [1]. VIGS takes advantage of cellular defense mechanisms against viruses [2], which derivate from posttranscriptional gene silencing [3], to transiently downregulate the expression of host genes using the virus as a vector. To this aim, a fragment of the targeted gene of interest (GOI) is included in virus genome, and the RNA-mediated gene silencing is promoted by double-stranded RNAs produced during recombinant virus replication [4]. The virus ability to spread and colonize plant tissues allows gene silencing establishment almost in most parts of the plant, outside inoculated organs. Therefore, VIGS method offers a fast and efficient alternative to stably knock down transgenic lines especially in plant species difficult to transform and challenging to regenerate such as perennial woody plants.

Vincent Courdavault and Sébastien Besseau (eds.), *Virus-Induced Gene Silencing in Plants: Methods and Protocols*, Methods in Molecular Biology, vol. 2172, https://doi.org/10.1007/978-1-0716-0751-0_14,

Tobacco rattle virus (TRV) is the most widely used virus to perform VIGS approaches due to its large host spectrum and asymptomatic replication [4]. However, VIGS vectors were developed along the years with additional viruses to better fit with host plant diversities and their specific interactions [5, 6]. In this context, the *Apple latent spherical virus* (ALSV) was employed to initiate VIGS in the apple and pear trees, constituting one of the first examples of VIGS in a woody plant [7]. The ALSV-based vectors were further demonstrated as efficient for VIGS in several dicotyledonous hosts (e.g., leguminous plants and Rosaceae) [8, 9] and can also be used to overexpress foreign genes [10]. The ALSV is asymptomatic in most of infected species with few exceptions such as in *C. quinoa* and is able to uniformly and systematically infect plant tissues including meristems as observed for TRV [7, 8].

The ALSV is a bipartite positive ssRNA virus. RNA1 and RNA2 are 6815 and 3384 nucleotides long, respectively, bear 3′ poly-A tail, and both harbor a single ORF encoding polyprotein which is further maturated by proteases [11]. The virus particles are spherical (25 nm in diameter), formed by three isomeric coat proteins. The ALSV-based vectors pEALSR1 and pEALSR2L5R5 include the complete RNA1 and RNA2 genomic sequences, respectively, under the control of an enhanced CaMV 35S promoter and the nopaline synthase terminator (Fig. 1). RNA2 genomic sequence was slightly modified by the insertion of a cloning site and an additional protease cleavage site between the MP and Vp25 sequences.

The method described in this report is restricted to the use of ALSV to perform VIGS in apple tree, targeting the small subunit of ribulose-1,5-bisphosphate carboxylase (rbcS, XM_008370438) as a reporter gene because its silencing generates a visual phenotype of

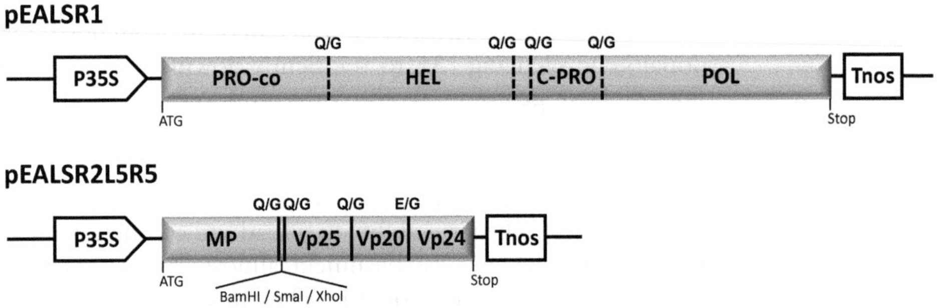

Fig. 1 VIGS vectors based on the *Apple latent spherical virus* (ALSV). pEALSR1 and pEALSR2L5R5, respectively, contain the RNA1 and RNA2 genomic sequences of ALSV, cloned under the control of an enhanced CaMV 35S promoter (P35S) and the nopaline synthase terminator (Tnos). Multiple cloning sites are located between the MP and Vp25 sequences in pEALSR2L5R5. Confirmed and putative protease cleavage sites of the translated polyproteins are shown with full and dash lines, respectively. *PRO-co* protease cofactor, *HEL* NTP-binding helicase, *C-PRO* cysteine protease, *POL* RNA-dependent RNA polymerase, *MP* movement protein, *Vp25* Vp20, *Vp24* capsid proteins

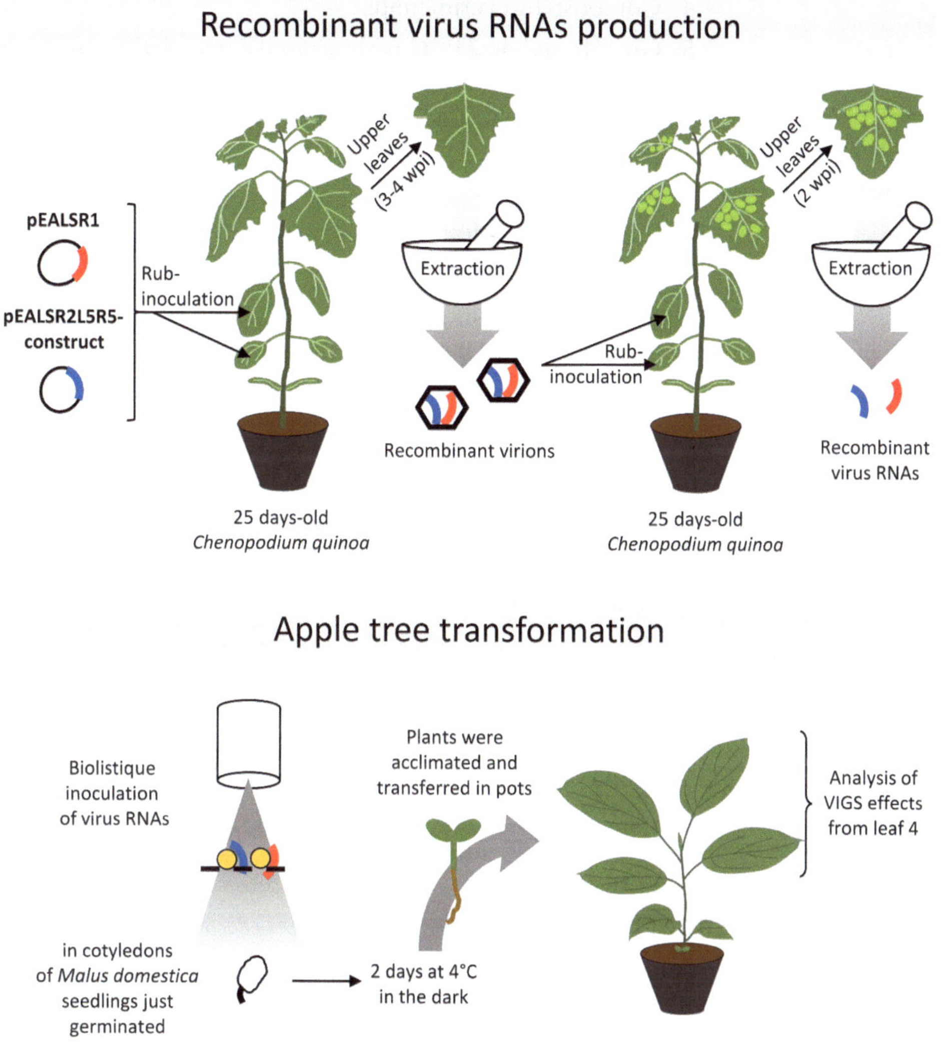

Fig. 2 Overview of the VIGS methods in apple tree using ALSV

leaf chlorosis [12]. To this aim, a two-step rub-inoculation of VIGS vector constructs is conducted in *C. quinoa* in order to amplify recombinant virus (Fig. 2). The delivery of virus constructs in apple tree is further performed on germinated seeds by biolistics using virus RNA.

2 Materials

2.1 Plant Materials and Culture

1. *Malus* × *domestica* L. seeds (e.g., cultivar 'Golden Delicious').
2. *Chenopodium quinoa* seeds.
3. Pots, trays, and clear covers.

4. Compost and vermiculite.
5. Greenhouse at 21 °C with around 50% relative humidity and light/dark cycle of 14/10 h.

2.2 Apple Tree Seed Germination

1. Laminar hood.
2. 50 ml plastic tubes.
3. 0.2% (w/v) sodium hypochlorite.
4. Sterile water.
5. Petri dishes (130 mm diameter) with Murashige and Skoog (MS) medium 10 g/L agar (without sucrose).
6. Sterile forceps.

2.3 Plasmid Construction

1. Vectors pEALSRl and pEALSR2L5R5 are available under request to Dr. Yoshikawa (Iwate University, Japan).
2. Reverse transcription kit, high-fidelity DNA polymerase kit, restriction enzymes and buffers (XhoI, BamHI, and/or SmaI), T4 DNA ligase kit.
3. Primers designed to amplify the fragment of the target gene of interest (GOI) used to induce silencing, including restriction sites (e.g., rbcS forward primer 5′-TACATCTCGAGGGT ACCGTGGCTACAGTTTCCGC-3′ and reverse primer 5′-T ACATGGATCCAGGAAGGTAAGAGAGGGTCTCG-3′).
4. pEASL2R5L5 forward primer 5′-TCTGACCCAAATCTGCT AGAAGG-3′ and reverse primer 5’-AAGTTTCGTTCCAC GACCG-3′.
5. PCR product purification kit.
6. *Escherichia coli* chimiocompetent cells (e.g., DH5α).
7. Spectrophotometer.
8. Agarose gel electrophoresis equipment, agarose, 1× TAE buffer, and ethidium bromide.
9. Luria-Bertani (LB) medium (liquid and solid).
10. 50 mg/ml ampicillin stock solution in water.
11. Incubator for bacterial cultivation at 37 °C.

2.4 RNA Extractions

1. Liquid nitrogen.
2. Sterile mortars, pestles, and spatula.
3. RNA extraction buffer: 100 mM Tris–HCl, pH 8, 1.4 M NaCl, 20 mM EDTA, 2% CTAB.
4. β-Mercaptoethanol.
5. Cold 90% acetone.
6. 50 ml plastic tubes.

7. Büchner vacuum filtration system with cellulose filter (e.g., Whatman n°1).
8. 25:24:1 (v/v/v) phenol/chloroform/isoamyl alcohol.
9. 10 M lithium chloride.
10. 3 M sodium acetate pH 5.2.
11. 100% and 70% ethanol.
12. Lysis buffer: 4 M guanidine isothiocyanate, pH 5, 0.2 M sodium acetate, 25 mM EDTA, 2.5% PVP-40.
13. 20% sarkosyl.
14. RNA purification kit.
15. Absolute ethanol.
16. Refrigerated centrifuge.
17. Vacuum rotary evaporator (SpeedVac).
18. Water bath at 65 °C.

2.5 Virus RNA Amplification

1. Incubator for bacterial cultivation at 37 °C with shaker.
2. High-volume plasmid purification kit.
3. 10 mM Tris–HCl pH 7.5.
4. Latex gloves.
5. Carborundum 400 mesh.
6. Virus inoculation buffer: 100 mM Tris–HCl, 100 mM NaCl, 50 mM $MgCl_2$. Adjust pH to 7.5.
7. Liquid nitrogen.
8. Sterile mortars and pestles.
9. Reverse transcription kit, Taq polymerase kit, SYBR Green Master Mix Kit.
10. Thermocycler and qPCR thermocycler.
11. CqEF1α forward primer 5′-GTACGCATGGGTGCTTGACAAACTC-3′ and reverse primer 5′-ATCAGCCTGGGAGGTACCAGTAAT-3′.
12. Purified pEASL2R5L5-GOI and purified PCR product of CqEF1α (obtain with EF1α ORF forward primer 5′-ATGGGTAAGGAAAAGATTCATATCAG-3′ and reverse primer 5′-TCACTTCTTCTTCAAAGCAGC-3′ on quinoa cDNA).

2.6 Biolistic Inoculation of Virus

1. Bio-Rad PDS1000/He delivery system.
2. 0.6 μm gold particles.
3. 100% and 70% ethanol.
4. RNAse-free water.
5. 5 M ammonium acetate.

6. Isopropanol.
7. Biolistic macrocarriers, stopping screens, and rupture disks 1100 psi.
8. Centrifuge (16,000 × *g*).
9. Water sonicator bath.
10. Freezer (−80 °C).
11. Sterile forceps.
12. Small metallic grid.
13. Small petri dishes (60 mm diameter) and large petri dishes (130 mm diameter) containing wet cotton.

3 Methods

3.1 Preparation of pEALSR2L5R5 Constructs

3.1.1 Design of the Target Sequence (See Note 1)

The ALSV genome RNA1 and RNA2 are translated into polyproteins which are further processed in mature viral proteins by post-translational cleavages. The cloning site of the target sequence in the ALSV vectors is located within the ALSV RNA2 genomic sequence, between the MP protein and the Vp proteins (Fig. 1). Consequently, the cloning design of the target sequences must respect the following rules:

1. The use of a specific sequence of around 201 bp of the target gene is recommended. A shorter fragment reduces silencing efficiency, whereas a longer one leads to virus instability.
2. The target sequence must be cloned in frame with the upstream MP sequence and the downstream Vp25 sequence.
3. Avoid stop codon in frame in the sequence used.

3.1.2 Efficient RNA Extraction from Apple Tree Tissues (See Note 2)

1. Harvest up to 1 g of plant tissues (e.g., young or old leaves, young stems, etc.) and freeze into liquid nitrogen. Store at −80 °C or proceed extraction directly.
2. Grind the frozen tissues to powder with liquid nitrogen using sterilized mortar and pestle.
3. Transfer the frozen powder in a pre-frozen 50 ml Falcon tube.
4. Add immediately 30 ml of cold 90% acetone and vortex during 20 min (*see* **Note 3**).
5. Acetone is removed from the powder by vacuum filtration on Whatman n°1.
6. Powder is washed once again in 90% acetone and then resuspended in 10 ml RNA extraction buffer (add 200 μl β-mercaptoethanol just before buffer use).

7. After 1-h incubation at 65 °C, one volume of 25:24:1 (v/v/v) phenol/chloroform/isoamyl alcohol is added. Vortex 1 min and centrifuge 10 min at 14,500 × *g*.
8. Collect 8 ml of aqueous phase, and precipitate nucleic acids using 3.3 ml of 10 M lithium chloride, overnight on ice.
9. Centrifuge 20 min at 14,500 × *g* and 4 °C. The pellet is resuspended in 500 μl of DEPC water, and extract again with one volume of 25:24:1 (v/v/v) phenol/chloroform/isoamyl alcohol.
10. After centrifugation, perform a second nucleic acid precipitation on 550 μl of aqueous phase using 200 μl of 3 M sodium acetate pH 5.2 and 1300 μl of 100% ethanol.
11. Centrifuge 20 min at 14,500 × *g* and 4 °C, wash the pellet with 500 μl of 70% ethanol, dry it using a SpeedVac, and dissolve the pellet in 50 μl of DEPC water.
12. Control the RNA quality and quantity by spectrophotometry.

3.1.3 Target Sequence Cloning into pEALSR2L5R5

1. cDNAs are synthesized using 1 μg of RNA, oligo dT18, and a reverse transcriptase, following manufacturer's instructions.
2. PCR amplify the target sequence of GOI from cDNA using gene-specific primers containing the appropriate restriction site and a high-fidelity DNA polymerase (*see* **Note 4**).
3. PCR product is purified using a PCR cleanup kit and eluted in 50 μl of water.
4. The digestion of the purified PCR product and 5 μg of pEALSR2L5R5 are conducted separately with the appropriate restriction enzymes. The target sequence can be cloned in pEALSR2L5R5 using the cohesive-end restriction endonucleases *XhoI* and/or *BamHI*, depending on the target sequence cloned. Alternatively, it could be cloned with blunt ends using *SmaI* restriction site of the vector.
5. Both vector and insert are purified using a PCR cleanup kit and eluted in 30 μl of water. DNA amounts are estimated by spectrophotometry and can be analyzed on 1% agarose gel electrophoresis.
6. Perform a ligation reaction of 20 μl using 100 ng of the purified vector and inserts in a 1:3 ratio. Leave the reaction 1 h on the bench, transform 5 μl of the ligation mixture in 50 μl of *E. coli* chimiocompetent cells, and spread on LB agar plate containing 50 μg/mL ampicillin. Grow overnight at 37 °C.
7. Clones are screened for the insertion by colony PCR using taq polymerase, pEASL2R5L5 forward and reverse primers (*see* **Note 5**).

3.2 Recombinant Virus RNA Amplification (See Note 6)

3.2.1 Culture of *Chenopodium quinoa*

1. Quinoa seeds are incubated overnight in water and sown on sol surface to germinate.
2. Seven days after sowing, seedlings are transferred in individual pots and grown in greenhouse.

3.2.2 Rub-Inoculation of Plasmids (See Note 7)

1. A 100 ml LB medium containing 50 μg/mL ampicillin is used to culture *E. coli* cells harboring pEALSR1 vector or pEALSR2L5R5-GOI construct. The whole cultures are used to purify plasmid using maxiprep kit. At the end of the purification, plasmid pellets are recovered in 100 μl of 10 mM Tris–HCl pH 7.5.
2. Plasmid concentrations are estimated using spectrophotometry and adjusted to 2 μg/μl in 10 mM Tris–HCl pH 7.5.
3. pEALSR1 and pEALSR2L5R5 vectors are equally mixed (1:1).
4. Wear latex gloves and dip a finger in carborundum dust. A small drop (5–8 μl) of plasmid mixture is added on the carborundum.
5. Slightly rub-inoculate one leaf of 25-day-old quinoa and repeat **step 4** to inoculate further leaves (*see* **Note 8**).
6. Once leaf surfaces are dried, wash them with water flux to remove carborundum.
7. Weak mosaic and chlorosis symptoms appeared 3–4 weeks post-inoculation (Fig. 3) in the upper non-inoculated leaves (*see* **Note 9**).

3.2.3 Virus Re-inoculation

1. Leaves with the most pronounced symptoms are collected and immediately ground with mortar and pestle in 50% w/v of virus inoculation buffer (add 3 μl β-mercaptoethanol for 1 ml buffer just before use) (*see* **Note 10**).
2. The virus extracts are collected and vortexed prior rub-inoculated in quinoa leaves as described previously for plasmids (Subheading 3.2.1, **steps 4–6**) (*see* **Note 11**).
3. Severe symptoms appear 2 weeks post-inoculation (Fig. 3) in the upper non-inoculated leaves (*see* **Note 12**).
4. Leaves with the most pronounced symptoms are collected, snap frozen in liquid nitrogen, and stored at −80 °C.

3.2.4 RNA Extraction in Infected Quinoa

1. Infected quinoa leaves are ground under nitrogen to fine powder with mortar and pestle.
2. Using a frozen spatula, transfer up to 50 mg of powder in a pre-frozen 2 ml microcentrifuge tube (*see* **Note 13**).

Fig. 3 Amplification of recombinant ALSV in *C. quinoa*. The rub-inoculation of ALSV vectors (first inoculation) or virus extracts (re-inoculation) is performed on the four upper leaves of 25-day-old quinoa as indicated with black arrows (**a**). The upper non-inoculated leaves (indicated with red arrows) accumulate virus particles after 2–4 weeks post-inoculation, depending on the inoculum used (**b**). Chlorosis and mosaic symptoms (**e**) emerge from the petiole side of the leaves (**d**) compared to uninfected leaf (**c**)

3. Immediately add 600 μl of lysis buffer (add 6 μl β-mercaptoethanol just before buffer use), and vortex vigorously (*see* **Note 14**).
4. Add 60 μl of 20% sarkosyl. Incubate at 65 °C for 10 min and vortex several times.
5. The extract is passed through a lysate filter column of a RNA purification kit and combined with 300 μl of absolute ethanol (*see* **Note 15**).

6. 700 μl of the extract is used for the further purification steps (from RNA binding column) as indicated by the manufacturer.
7. Control the RNA quality and quantity by spectrophotometry (*see* **Note 16**).

3.2.5 Analysis of Virus RNA

Prior apple tree transformation, PCR reactions should be run to confirm recombinant virus clones, and measure the relative transcript abundance of virus RNA among total RNA extracted from quinoa.

1. cDNAs are synthesized using 1 μg of RNA, oligo dT18, and a reverse transcriptase, following manufacturer's instructions.
2. A PCR is performed on 0.5 μl of cDNA using Taq DNA polymerase, the GOI-specific forward primer used for cloning (e.g., rbcS forward primer), and the pEASL2R5L5 reverse primer.
3. Amplification is performed on a thermocycler, initiated by a denaturation step at 95 °C for 2 min followed by 35 amplification cycles of 20 s at 95 °C, 20 s at 52 °C, 30 s at 72 °C, and a final elongation step at 72 °C during 10 min.
4. PCR products are analyzed on 1.2% agarose TAE gel electrophoresis to detect amplicons of the expected size (*see* **Note 17**).
5. A qPCR is run on 0.5 μl cDNA using a SYBR Green Master Mix. Two couples of primers are used for each sample: pEASL2R5L5 forward and reverse primers to detect virus and CqEF1α forward and reverse primers as a reference gene. Assays were performed in triplicate.
6. The qPCR thermocycler is set up for 7 min at 95 °C followed by 40 cycles of 10 s at 95 °C and 20 s at 60 °C. SYBR signal is recorded at the end of each cycle. Amplification is followed by a melting curve analysis by increasing temperature at the rate of 0.5 °C every 5 s from 65 to 95 °C.
7. To establish calibration curves for virus genome and reference gene primers, qPCR is performed in the same conditions using from 10^3 to 10^8 copies of the purified pEASL2R5L5-GOI and a purified PCR product of CqEF1α, previously obtained from the cDNA using EF1α ORF forward and reverse primers (*see* **Note 18**).
8. Calibration curves are obtained by linear regression of Ct values (cycle threshold) by the log of copy number.
9. Transcript copy numbers of virus genome and reference gene in cDNA of infected quinoa samples are determined by using the obtained Ct values and calibration curves.
10. A ratio of virus copy number/EF1α copy number lower than 1 corresponds to a weak virus infection, whereas a ratio around 40 corresponds to a strong virus infection.

3.3 Apple Tree Transformation

3.3.1 Seed Germination

1. Surface seed sterilization is performed under a laminar hood.
2. Around 50 apple seeds are placed in a 50 ml Falcon tube and surface sterilized 12 min in 0.2% sodium hypochlorite (*see* **Note 19**).
3. Wash seeds three times with sterile water and remove a maximum of water.
4. Seeds are dropped on large petri dishes containing MS medium agar and separated using sterile forceps (Fig. 4a).
5. Plates are kept in dark at 4 °C for 1 month and then transferred at 21 °C in dark for few days until the emergence of almost 0.2–0.5 cm of the radicle (Fig. 4c) (*see* **Note 20**).
6. On the bench, maintain seeds with your fingers and use forceps to gently remove teguments (Fig. 4b). Take care of the root which is very breakable.
7. At this step, cotyledons are exposed, and apple seedlings can be stocked in the dark on a petri dish containing wet cotton for few hours at room temperature until the transformation can be done.

3.3.2 Biolistic Inoculation of Virus RNA

1. Around 30 mg of 0.6 μm gold particles are baked for 6 h at 220 °C and then homogenized in 1 ml of 70% ethanol for 5 min in sonication bath. Particles are collected by centrifugating 5 s at 16,000 × *g*, washed three times in water using sonication bath, and finally homogenized in 300 μl of water.
2. For coating, 25 μg of RNA are mixed with 1.3 mg of beads in 38 μl of RNAse-free water and precipitated using 12 μl of 5 M ammonium acetate and 220 μl isopropanol.
3. After 1-h incubation at −80 °C, beads are collected by centrifugation, washed five times in 200 μl of absolute ethanol, and finally homogenized in 40 μl absolute ethanol.
4. Transformation of apple tree seedling cotyledons can be performed with the Bio-Rad PDS1000/He system.
5. 10 μl of coated particles (around 0.3 mg) were dropped and dried on a microcarrier.
6. Twenty to twenty-five apple seedlings are placed 7.5 cm above the microcarrier stop screen, on wet cotton in a small petri dish, and maintained with a metal grind (Fig. 4e).
7. Transformation of cotyledons is conducted using a 1100 psi rupture disk under a vacuum pressure of 28 in. of Hg.
8. Four shoots are performed on each plant batch, two shoots for each cotyledons (seeds are turned using forceps after the two first shoots).

Fig. 4 Apple tree inoculation steps with recombinant ALSV RNAs. Seeds undergo vernalization 1 month at 4 °C in the dark and germinate on MS medium (**a**, **c**). Seed teguments are removed using forceps (**b**), and cotyledons (**d**) are inoculated with virus RNAs using biolistics. To limit damages, seeds are maintained on wet cotton with a metallic grid during the biolistic process (**e**). After few days of acclimation, inoculated seedlings (**f**) are transferred in pots (**g**). Differential leaf chlorosis of ALSV-rbcS inoculated plants helps to identify VIGS establishment and efficiency

3.3.3 Apple Seedling Acclimation and Phenotype Observation

1. After transformation, seeds are placed on wet cotton in a large petri dish and kept in the dark at 4 °C for 2 days.
2. Plates are transferred in greenhouse and covered with a paper (keep plates closed).
3. One day later, remove the paper (keep plates closed).
4. After 1 more day, cotyledons turn green. Open the plates in a small chamber containing water to maintain high humidity.
5. First leaves emerge within 4 days (Fig. 4f). Seedlings are transferred in pots and grown in greenhouse.
6. The first two leaves usually contain recombinant virus, but the targeted gene is not silenced. Heterogeneous silencing is observed in the third leaf, and finally, a full and stable silencing is measured from the fourth leaf (Fig. 4g).
7. Silencing analysis can be performed on six- to seven-leaf-stage apple tree. Leaves 4–7 are collected and pooled for further gene expression and metabolite analysis (*see* **Note 21**).

4 Notes

1. Three different sequences should be tested for each targeted gene to ensure at least one with proper virus stability in quinoa (recombinant virus is propagated in quinoa through the production of virus particles) and efficient silencing induction in apple tree.
2. RNA extraction on apple tree tissues can be done using commercial kit, but often results in poor quality and yield due to high flavonoids and sugar contents.
3. Do not allow samples to thaw before adding acetone to avoid RNAse activity.
4. Add almost four additional nucleotides on primer 5′-end to allow efficient digestion of PCR products for direct cloning in pEALSR2L5R5.
5. Confirm the insertion and the frame of positive clones by sequencing with pEASL2R5L5 forward primer.
6. Be careful of cross contamination when several recombinant virus constructs are inoculated in a row, especially if constructs for different targeted genes are produced at the same time. It is recommended to analyze the virus profile of every RNA produced as mention in Subheading 3.2.5 by using specific primers of each construct used in the same experiment set.
7. Instead of plasmid rub-inoculation, quinoa can be agrotransformed. To this aim, a different set of plasmids is needed. RNA1 and RNA2 genomes are available in Ti-plasmids

pCALSR1 and pCALSR2, respectively. However, since apple tree is not efficiently agro-transformed, the two-step virus amplification in quinoa is needed even with Ti-plasmids.

8. Rub-inoculations are performed into four leaves per quinoa plants of 25 days old, preferentially in the upper leaves (Fig. 3a). Use four plants per construct to ensure at least one efficiently infected. Do not rub too hard or leaves will wither in a day.
9. Depending on virus constructs and inoculum, the viral symptoms can be tricky to observe. This should be the case for plasmid rub-inoculation which leads to weak infection.
10. Collect upper leaves from every inoculated plants to prepare the virus extract for the second inoculation step even if symptoms are not observed.
11. Remind to grow a second series of quinoa 3–4 weeks after the first one, to have proper plants for the second inoculation step.
12. In contrary to the plasmid inoculation, the second inoculation using virus extracts should generate clear symptoms (Fig. 3d, e). If not, the recombinant virus is probably unstable and cannot be used to inoculate apple tree.
13. Excess of plant material has a negative impact on RNA extraction yield.
14. Do not allow samples to thaw before adding lysis buffer to avoid RNAse activity.
15. Use here an RNA extraction kit which includes two columns (a sample filter and a binding column). As example, this is the case for NucleoSpin RNA II (Macherey-Nagel) or RNeasy (Qiagen).
16. It is mandatory to extract high RNA quantities from infected quinoa samples. RNA amount and concentration are critical to obtain an efficient infection in apple tree.
17. It is recommended to analyze the virus profile of every RNA produced by using specific primers of each construct used in the same experiment set to detect cross contamination of virus.
18. PCR product of CqEF1α can be generated from any quinoa cDNA template.
19. Use 50 apple tree seeds or more to obtain around 20–25 transformable germinated seeds per construct (considering non-germinated seeds, seeds with too long roots, or seeds destroyed by experimentator during tegument removal).
20. Alternatively, seeds can be vernalized 3 months in vermiculite/water at 4 °C in dark and then transferred to 21 °C in the dark. However, germination is not synchronized, and 250 apple tree

seeds are needed to expect around 20–25 transformable germinated seeds.

21. Each inoculated apple tree has to be sampled and analyzed separately since the virus inoculation/silencing induction may not be homogeneous among plants.

References

1. Baulcombe DC (1999) Fast forward genetics based on virus-induced gene silencing. Curr Opin Plant Biol 2:109–113
2. Voinnet O (2001) RNA silencing as a plant immune system against viruses. Trends Genet 17:449–459
3. Vaucheret H, Béclin C, Fagard M (2001) Post-transcriptional gene silencing in plants. J Cell Sci 114:3083–3091
4. Ratcliff F, Montserrat A, Baulcombe DC (2001) Tobacco rattle virus as a vector for analysis of gene function by silencing. Plant J 25:237–245
5. Lange M, Yellina AL, Orashakova S, Becker A (2013) Virus-induced gene silencing (VIGS) in plants: an overview of target species and the virus-derived vector systems. Methods Mol Biol 975:1–14
6. Kant R, Dasgupta I (2019) Gene silencing approaches through virus-based vectors: speeding up functional genomics in monocots. Plant Mol Biol 100:3–18
7. Sasaki S, Yamagishi N, Yoshikawa N (2011) Efficient virus-induced gene silencing in apple, pear and Japanese pear using Apple latent spherical virus vectors. Plant Methods 7:15–25
8. Igarashi A, Yamagata K, Sugai T, Takahashi Y, Sugawara E, Tamura A, Yaegashi H, Yamagishi N, Takahashi T, Isogai M, Takahashi H, Yoshikawa N (2009) Apple latent spherical virus vectors for reliable and effective virus-induced gene silencing among a broad range of plants including tobacco, tomato, *Arabidopsis thaliana*, cucurbits, and legumes. Virology 386:407–416
9. Li C, Yamagishi N, Kasajima I, Yoshikawa N (2019) Virus-induced gene silencing and virus-induced flowering in strawberry (*Fragaria* × *ananassa*) using apple latent spherical virus vectors. Hortic Res 6:18–27
10. Li C, Sasaki N, Isogai M, Yoshikawa N (2004) Stable expression of foreign proteins in herbaceous and apple plants using apple latent spherical virus RNA2 vectors. Arch Virol 149:1541–1558
11. Li C, Yoshikawa N, Takahashi T, Ito T, Yoshida K, Koganezawa H (2000) Nucleotide sequence and genome organization of Apple latent spherical virus: a new virus classified into the family Comoviridae. J Gen Virol 81:541–547
12. Navarro Gallon SM, Elejalde-Palmett C, Daudu D, Liesecke F, Jullien F, Papon N, Dugé de Bernonville T, Courdavault V, Lanoue A, Oudin A, Glévarec G, Pichon O, Clastre M, St-Pierre B, Atehortua L, Yoshikawa N, Giglioli-Guivarc'h N, Besseau S (2017) Virus-induced gene silencing of the two squalene synthase isoforms of apple tree (*Malus* × *domestica* L.) negatively impacts phytosterol biosynthesis, plastid pigmentation and leaf growth. Planta 246:45–60

Chapter 15

Virus-Induced Gene Silencing for Functional Analysis of Flower Traits in Petunia

Shaun R. Broderick, Laura J. Chapin, and Michelle L. Jones

Abstract

Virus-induced gene silencing (VIGS) uses recombinant viruses to knock down the expression of endogenous plant genes, allowing for rapid functional analysis without generating stable transgenic plants. The *Tobacco rattle virus* (TRV) is a popular vector for VIGS because it has a wide host range that includes *Petunia* × *hybrida* (petunia), and it induces minimal viral symptoms. Using reporter genes like chalcone synthase (*CHS*) in tandem with a gene of interest (GOI; pTRV2-*PhCHS*-GOI), it is possible to visually identify silenced flowers so that phenotyping is more accurate. Inoculation methods and environmental conditions need to be optimized for each host plant-virus interaction to maximize silencing efficiency. This chapter will provide detailed protocols for VIGS in petunia, with an emphasis on the investigation of flower phenotypes.

Key words Corollas , Flowers, Flowering, *Petunia* × *hybrida*, Solanaceae, *Tobacco rattle virus* (TRV), VIGS

1 Introduction

Virus-induced gene silencing (VIGS) is a transcript suppression technique used for both forward and reverse genetic studies. The history of VIGS began with the discovery of posttranscriptional gene silencing (PTGS), referred to as co-suppression, in petunia [1]. This rapid, transient technique uses a viral delivery vector to introduce a short sequence of a target gene of interest (GOI) into plants without the need for stable transformation. PTGS results in the sequence-specific degradation of endogenous gene transcripts and allows for the identification of phenotypes associated with the downregulation of single genes or gene families. VIGS can be used to infect young plants, allowing for the characterization of embryonic and early development phenotypes that might be lethal in stable transgenic plants [2].

Vincent Courdavault and Sébastien Besseau (eds.), *Virus-Induced Gene Silencing in Plants: Methods and Protocols*, Methods in Molecular Biology, vol. 2172, https://doi.org/10.1007/978-1-0716-0751-0_15,

The uniformity of silencing throughout the plant and the severity of viral symptoms vary greatly in different host plant-virus combinations. The *Tobacco rattle virus* (TRV) has become a popular vector for VIGS because it has a wide host range, which includes petunia and other Solanaceae, and it induces minimal viral symptoms in these hosts. TRV also targets host RNAs efficiently throughout the plant, including the meristems. TRV is a bipartite (RNA1 and RNA2) positive-strand RNA virus [3]. TRV vectors designed for gene silencing include cDNA clones of TRV RNA1 (pTRV1) and TRV RNA2 (pTRV2), which are co-infiltrated into plants using *Agrobacterium tumefaciens.* pTRV2 contains the viral coat protein (CP) and a multiple cloning site (MCS) [3, 4]. Optimal TRV inoculation varies by species, but leaf infiltration with a needleless syringe and inoculation of mechanically wounded shoot apical meristems have been consistently successful techniques in petunia [5–8].

Uniform silencing throughout the plant is dependent on both the systemic spread of the virus-derived silencing signal and the successful movement of the virus within the plant [9]. Reporter or indicator genes are used to confirm the presence of silencing in different plant organs [10]. *Phytoene desaturase* (*PDS*), a gene involved in carotenoid biosynthesis, is often used as an indicator of silencing in vegetative tissues. The reduction in phytoprotective carotenoids leads to the photooxidation of chlorophyll, causing a whitening or photobleaching of the leaves [11]. Silencing *chalcone synthase* (*CHS*), a gene involved in anthocyanin production, causes purple-pigmented corollas to turn white [5]. These and other visual markers have been used in proof-of-concept experiments with various species, including petunia.

While *PDS* provides a clear visual indicator of silencing, downregulating *PDS* inhibits photosynthesis and significantly impairs normal plant growth. A modified pTRV2 vector that contains the green fluorescent protein (GFP) fused to the 3′-end of the coat protein provides an alternative reporter system that has not been shown to have a negative impact on plant growth [12]. When inoculated with pTRV2-*GFP-PDS*, the abundance of GFP protein in *Nicotiana benthamiana* plants is negatively correlated with the level of the *PDS* transcripts, suggesting that GFP can be used to estimate the presence of the virus and predict gene silencing [12, 13].

A disadvantage of VIGS is that sectors of plant tissue can escape PTGS. If the phenotypic effect of silencing a GOI is unknown or inconspicuous, then it becomes nearly impossible to ensure that the plant tissue analyzed or collected has undergone PTGS. To address this problem in flowers, Chen et al. [5] utilized a construct that included the visual reporter, *CHS*, that is cloned in tandem with a target GOI for functional analysis. Petunias exhibiting the phenotype that results from downregulating *CHS* also have downregulation of the GOI. This system can be used to visually identify flowers with GOI silencing, thereby making phenotyping more accurate. For example, when petunias were inoculated with pTRV2-*CHS-ACO* (ACC oxidase, the enzyme that converts ACC to ethylene), the resulting flowers were either entirely white, entirely purple, or sectored purple and white. White flowers had lower transcript levels of both *CHS* and the gene of interest (i.e., *ACO*). Additionally, they exhibited delayed senescence, reduced ACC-dependent ethylene production, and reduced ACO activity compared to purple flowers [5].

Uniform and consistent silencing depends on virus proliferation and plant growth, both of which can be improved by optimizing the growing environment. Temperature is one of the most important environmental factors controlling the efficiency of TRV-mediated silencing. Silencing phenotypes in *Solanum lycopersicum* (tomato) are best at temperatures at or below 21 °C, while temperatures around 25 °C are optimal for silencing in *Nicotiana benthamiana* [2]. VIGS experiments in petunia have mostly reported day/night temperatures of 25 °C/20 °C [5, 7] or 22 °C/18 °C [14]. Growth chamber (GC) experiments designed to determine the optimal daytime temperature for TRV-mediated silencing in petunia showed that the frequency (percent of plants with silencing) of *PDS* and *CHS* silencing was similar at 20, 23, or 26 °C (the nighttime temperature was set to 18 °C for each GC). *PDS* silencing effectiveness (percent of leaves showing photobleaching) was highest at 20 °C, while *CHS* silencing effectiveness was the same at all three daytime temperatures. When more precise calculations of silencing efficiency were obtained by quantifying the total white tissue that developed after inoculation, 20 °C resulted in significantly greater silencing in both leaves (*PDS*) and corollas (*CHS*) [6]. Additionally, utilizing an optimized VIGS protocol with consistent environmental conditions leads to earlier PTGS with higher efficiency. For example, in petunias, *PDS* silencing not only occurred 3 days sooner when the inoculum was applied to a wounded apical meristem, but it also increased the silencing efficiency by 13% [6].

Accurate flower phenotyping depends on comparison to the proper controls. The pTRV2 construct is often used as an empty vector control to account for any phenotypic effects of the virus. Viral symptoms of TRV constructs in Solanaceae are generally minimal, but symptoms of stunting or necrosis have been reported in *S. lycopersicum* [15], *S. nigrum* [16], *S. okadae* [17], *S. tuberosum* [18], and *P. hybrida* [6]. These symptoms can be reduced or eliminated by inserting a fragment of non-plant (*GUS* or *GFP*) or non-coding (a *PDS* intron) DNA into the MCS of pTRV2 [6, 15, 16]. Broderick and Jones [6] found that petunias inoculated with pTRV2 empty vector had stunted, necrotic shoots and delayed flowering, while minimal viral symptoms were observed in plants inoculated with the pTRV2-*NbPDS* or pTRV2-*PhCHS* constructs. Viral load as measured by gene expression of the coat protein was also higher in pTRV2 empty vector plants [6]. Because insert size in the pTRV2 vectors can affect viral replication and movement, it is also advisable to keep insert sizes of control vectors and experimental vectors similar. We, therefore, recommend the use of pTRV2-*PhCHS-GFP* constructs as negative controls for pTRV2-*PhCHS*-GOI functional analysis studies.

VIGS has been used to study many different genes in petunia, with the majority of experiments focusing on flower development, volatile production, and senescence [4, 5, 7, 8, 14, 19–23]. With the recent sequencing of the petunia genome [24], there is a need for large-scale functional analysis. VIGS provides a simple and rapid means of silencing petunia genes, and a pTRV2 gateway vector allows for fast and easy restriction enzyme- and ligation-free cloning of target genes for high-throughput analysis [19]. In this chapter, we provide a detailed protocol for TRV-mediated VIGS in petunia, with an emphasis on identifying flower phenotypes.

2 Materials

2.1 Petunia Cultivation

1. Seeds of *Petunia* × *hybrida* 'Picobella Blue' or another cultivar with high silencing efficiency (*see* Fig. 1 and **Note 1**).
2. Plug trays (128-cell) filled with moistened soil-less growing mix (e.g., Pro-Mix BX or similar).
3. Clear propagation domes for seedling germination.
4. Fertilizer such as 20-10-20 or 14-0-14.
5. 8 cm plastic pots filled with moistened soil-less growing mix.
6. Growth chambers set at 20 °C day/18 °C night, 150 μmol/m^2/s light (from metal halide lights) for 16 h days and 8 h nights. The target relative humidity is 40–50% (*see* Fig. 2 and **Note 2**).
7. Greenhouse data logger.

2.2 Vector Construction

1. *Tobacco rattle virus* (TRV)-based vectors.

   pTRV1 (pYL192; stock #CD3-1039; Arabidopsis Biological Resource Center).

   pTRV2-MCS (pYL156) contains the multiple cloning site to add the gene of interest or reporter gene fragments (stock #CD3-1040; Arabidopsis Biological Resource Center).

   pTRV2-*NbPDS* (*see* **Note 3**).

   pTRV2-*PhCHS* [5].

   pTRV2-*PhCHS-GFP* [6, 25].

   pTRV2-*GFP* (*see* **Note 4**) [6].

Fig. 1 Phenotypic effects of virus-induced gene silencing in 13 petunia cultivars using pTRV2-*NbPDS*, pTRV2-*PhCHS*, pTRV2-*GFP*, and a buffer-inoculated control. Petunias were grown under 20 °C day/18 °C night and inoculated using the apical meristem method 4 weeks after the seeds were sown

2. Taq DNA polymerase kit including 10 mM dNTP mixture.
3. 10 mM insert-specific forward and reverse primers that flank a region within the coding sequence for a GOI.
4. cDNA synthesized from petunia RNA or a plasmid containing the target gene of interest.
5. Autoclaved nanopure water.
6. Thermocycler.
7. PCR cleanup gel extraction kit.
8. Agarose gel electrophoresis equipment, agarose, and 1× Tris/borate/EDTA buffer (TBE).
9. NanoDrop spectrophotometer or similar.

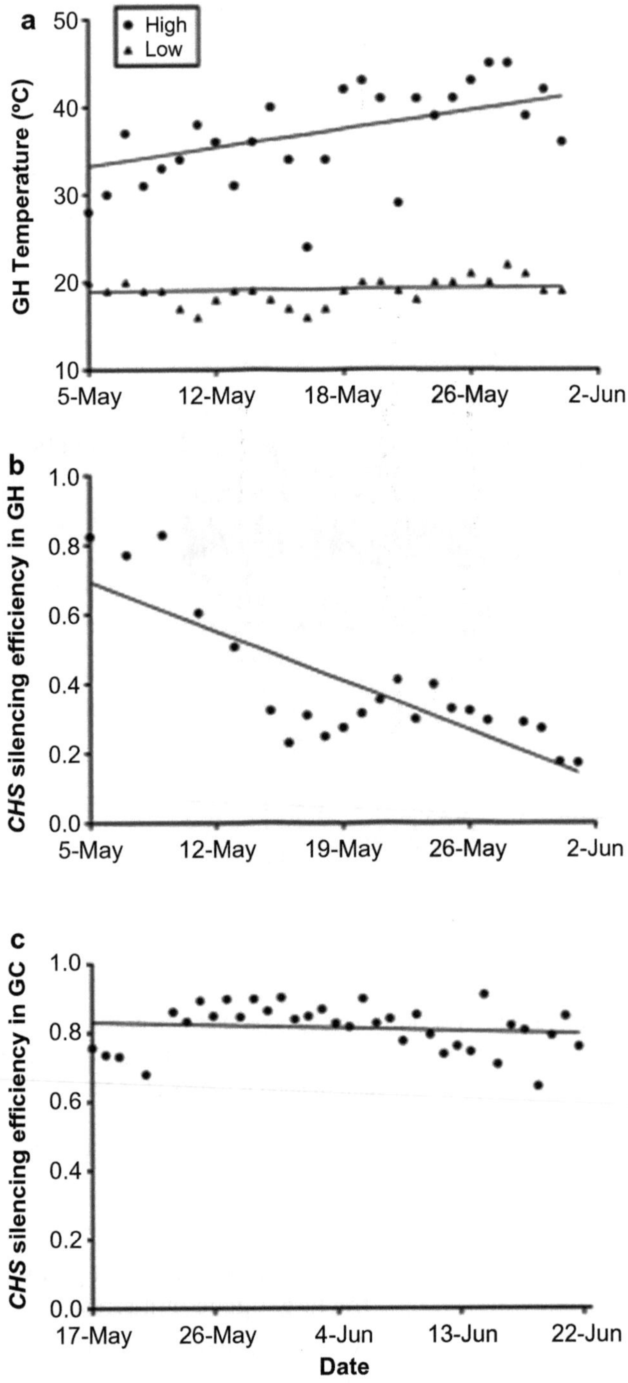

Fig. 2 *CHS* silencing efficiency and temperatures. (**a**) Daily springtime high and low temperatures in the greenhouse set at 20 °C day/18 °C night. (**b**) The decline of *CHS* silencing efficiency in petunia corollas over time when grown in the greenhouse (GH). (**c**) *CHS* silencing efficiency in petunia corollas grown in the growth chamber (GC) set at 20 °C day/18 °C night. Higher silencing efficiency occurred in the growth chamber where cooler temperatures could be better maintained

10. Restriction enzymes and buffers: *XbaI* and *BamHI* (or *EcoRI*, *StuI*, *NcoI*, *KpnI*, *SacI*, *MluI*, *XhoI*, *SrfI*, or *SmaI* depending on the site chosen for cloning in the MCS).
11. T4 DNA ligase kit.
12. *Escherichia coli* competent cell strain DH5α (*see* **Note 5**).
13. Luria-Bertani (LB) sterile liquid media and plates containing 50 μg/ml kanamycin.
14. Water bath set at 42 °C and shaking incubator.
15. Plasmid purification kit.
16. pTRV2 MCS-specific primers: forward primer 5′-CTCAAGG AAGCACGATCAGC-3′ and reverse primer 5′-TTGAACCT AAAACTTCAGACACG-3′.
17. *Coat protein* (*CP*) primers: forward primer 5′-CTGGGTTACT AGCGGCACTGAATA-3′ and reverse primer 5′-TCCACC AAACTTAATCCCGAATAC-3′ [26].
18. Sterile screw cap cryovials.
19. 30% sterile glycerol.
20. Liquid N_2.
21. −80 °C freezer.

2.3 Agrobacterium Preparation

1. Electrocompetent cells of *Agrobacterium tumefaciens* strain GV3101 stored at −80 °C.
2. LB sterile liquid media and agar plates containing 25 μg/ml gentamicin, 50 μg/ml kanamycin, and 10 μg/ml rifampicin (*see* **Note 6**).
3. Electroporator that can deliver 1500 volts.
4. Culture LB-MES media: LB containing 10 mM MES, pH 5.7, 25 μg/ml gentamicin, 50 μg/ml kanamycin, and 10 μg/ml rifampicin.
5. 15-ml capped glass culture tubes.
6. Subculture LB-MES media: LB containing 10 mM MES, pH 5.7, 20 μM acetosyringone, 25 μg/ml gentamicin, 50 μg/ml kanamycin, and 10 μg/ml rifampicin (*see* **Note 7**).
7. Shaking incubator.
8. Spectrophotometer (to measure OD_{600}).
9. Refrigerated centrifuge (4000 × *g*).
10. Agroinduction media: LB containing 10 mM MES, pH 5.7, 500 mM $MgCl_2$, and 200 μM acetosyringone.

2.4 Apical Meristem Inoculation

1. Forceps.
2. Disposable dropper.
3. Sterile 15 ml Falcon tubes.

2.5 Phenotyping VIGS Petunias

1. Camera or scanner, software (e.g., WinCAM; *see* **Note 8**).
2. String tags to assist in marking flowers.
3. Forceps.

2.6 Confirming GOI Downregulation

1. RNA extraction reagent (e.g., TRIzol Reagent, Invitrogen).
2. RNase-free DNase.
3. cDNA Synthesis Kit including oligo dT and/or random primers.
4. 2× SYBR Green Master Mix.
5. qPCR thermocycler.
6. PCR primers. TRV *CP* forward primer 5′-GCGTTCAAGAGACGAGCTGAT-3′ and reverse primer 5′-TCGTCGTAACCGTTGTGTTTG-3′. Reference genes to include *PhActin* forward primer 5′-AGCCAACAGAGAGAAGATGACCCA-3′ and reverse primer 5′-ACACCATCACCAGAGTCCAACACA-3′; *PhSAND* forward primer 5′-CTTACGACGAGTTCAGATGCC-3′ and reverse primer 5′-TAAGTCCTCAACACGCATGC-3′; *PhRPS13* forward primer 5′-CAGGCAGGTTAAGGCAAAGC-3′ and reverse primer 5′-CTAGCAAGGTACAGAAACGGC-3′; and *PhEF1α* forward primer 5′-CCTGCTCAAATTGGAAACGG-3′ and reverse primer 5′-CAGATCGCCTGTCAATCTTGG-3′. The primers for *PhSAND*, *PhRPS13*, and *PhEF1α* are described in [27]. Primers for GOI and other markers (i.e., senescence marker gene like *PhCP10* if evaluating flower senescence; *see* **Note 9**).

3 Methods

An overview of the VIGS protocol for petunia can be found in Fig. 3.

3.1 Petunia Cultivation

1. Sow seeds of a *Petunia* × *hybrida* cultivar like 'Picobella Blue' that exhibits efficient gene silencing. Place seeds on top of moistened soil-less mix in clean, 128-cell plug trays (*see* Fig. 1 and **Note 10**) [2].
2. Leave the seeds exposed on the soil's surface, and cover the plug tray with a dome lid to maintain high humidity for germination (*see* **Note 11**).
3. Grow seedlings in a growth chamber under 150 μmol/m^2/s light (from metal halide lights) for 16 h days and 8 h nights. The target relative humidity is 40–50% with a temperature of 20 °C (*see* Fig. 2).
4. After the cotyledons have expanded, begin fertilizing plants with 100 ppm N from a 20-10-20 fertilizer. Let the media

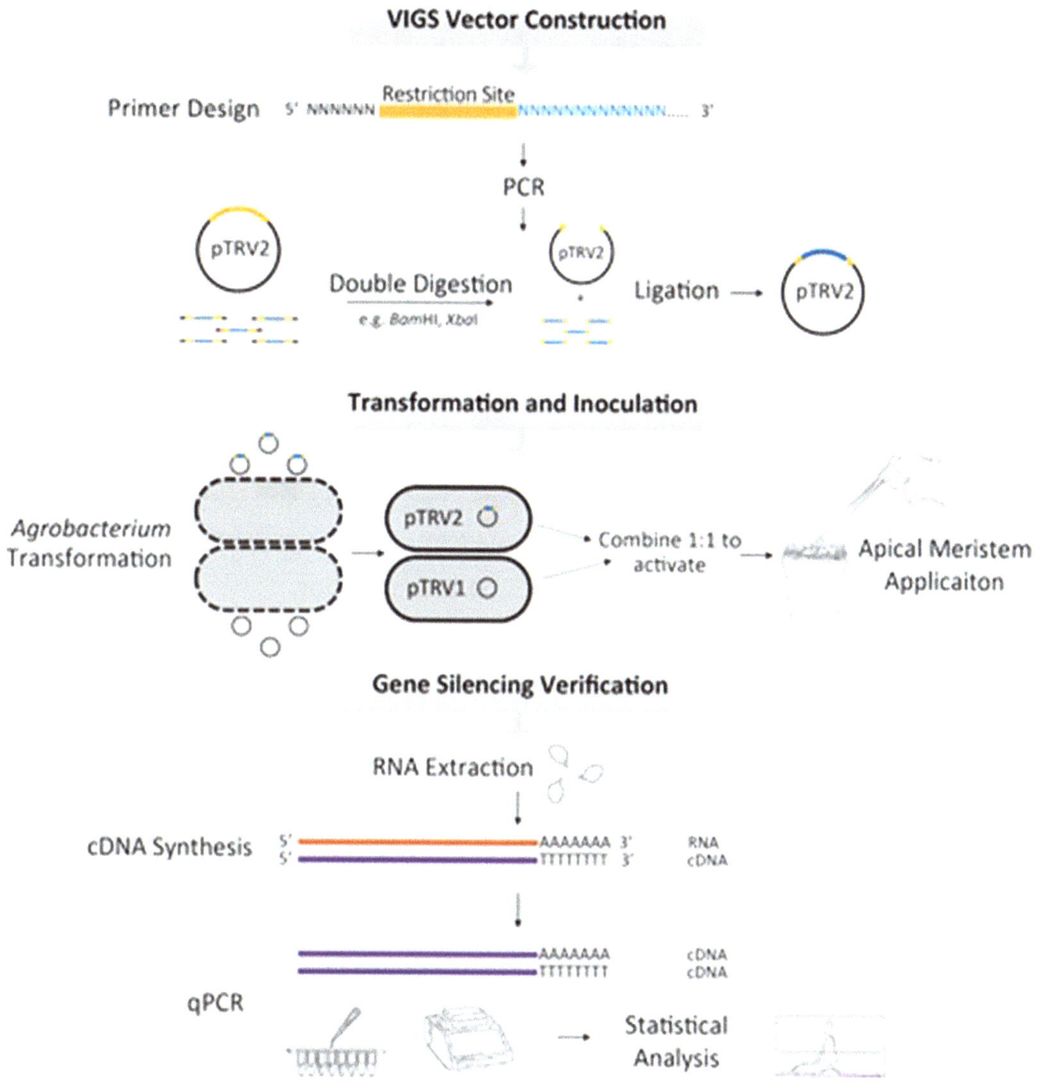

Fig. 3 Overview of virus-induced gene silencing (VIGS) protocol in petunia

dry between watering. When true leaves emerge, increase fertilizer to 200–250 ppm N and maintain a soil pH of 5.5–6.3 (*see* **Note 12**).

5. Three to four weeks after sowing the seeds, plants should be transplanted into 8 cm plastic pots. At 4 weeks after sowing, the petunias will have about four or five true leaves and will be ready for inoculation (*see* Fig. 4 and **Note 13**).

Tobacco rattle virus (TRV) is a bipartite virus, and the two parts have been used to create plasmids pTRV1 and pTRV2. pTRV2 contains the MCS for cloning the fragment of the GOI.

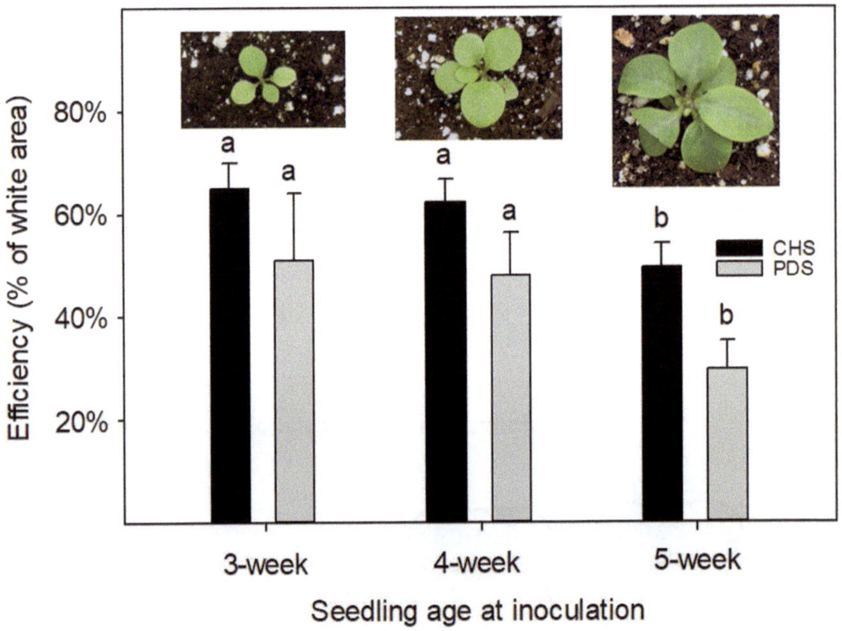

Fig. 4 *CHS* and *PDS* silencing efficiency decreases in 5-week-old petunia plants. Three- to four-week-old petunia plants have higher silencing efficiency than 5-week-old plants. The petunia seedling photos are scaled to size and representative of seedlings at the age specified on the *x*-axis. Petunias were grown under 20 °C day/18 °C night and inoculated using the apical meristem method. *CHS* silencing was measured at the onset of flowering by analyzing photos of each petunia corolla over 4 weeks. *PDS* silencing was measured 4 weeks after inoculation. Seed sowing dates were offset so the seedlings could be inoculated on the same day. Letters above the bars were used to signify statistical significance, but the letters pertain specifically to *PDS* and *CHS* separately (i.e., letters from the *CHS* black bars should not be compared to the letters of the *PDS* gray bars)

3.2 Vector Construction for Silencing Genes of Interest (GOI)

1. Determine if you want to use pTRV2 alone or with a reporter gene that will give a visual indicator of silencing in the leaves (pTRV2-*NbPDS*; *see* **Note 14**) or the corollas (pTRV2-*PhCHS*; *see* **Note 15**).
2. Analyze the GOI sequence for target regions that are amenable to silencing using a siRNA selection program. The optimal fragment length is from 200 to 400 bp. If the CHS-*GFP* vector is used as an experimental control vector, ensure that the GOI fragment is about 256 bp to correspond to the length of the *GFP* insert in the control vector (*see* **Note 16**).
3. Identify restriction enzymes from pTRV2's MCS (*EcoRI*, *XbaI*, *StuI*, *NcoI*, *BamHI*, *KpnI*, *SacI*, *MluI*, *XhoI*, *SrfI*, and *SmaI*) that have a unique restriction site that is not found in the GOI target sequence identified above (*see* **Note 17**).
4. Design forward and reverse primers to flank the target region identified in **step 2** (Subheading 3.2) using a primer design program of your choice. Add one of the restriction site

sequences to the 5′-end of the forward primer, and add the other restriction site sequence to the 5′-end of the reverse primer (*see* **Note 18**).

5. Set up three, 50 μl PCR reactions using *Taq* polymerase with the restriction-site-containing primer pairs for each GOI (*see* **Note 19**). The template for amplification can be cDNA synthesized from petunia RNA or a plasmid containing the target GOI. Two microliters of template with a final concentration of less than 250 ng should be used for each PCR.
6. Prepare a 1% agarose gel for electrophoresis using 1× TBE. Mix each PCR reaction with a stain that can be visualized under a UV light. Run the gel to separate the DNA products. Excise the bands of interest from the gel using a sharp scalpel, and purify the products using the DNA gel purification kit. Quantify the DNA using a NanoDrop spectrophotometer.
7. Set up a double digest reaction for each purified PCR product (approximately 500 ng) as well as the pTRV2-MCS or pTRV2-*PhCHS* vector (3 μg). Stop the reaction after the manufacturer's recommended time (*see* **Note 20**).
8. Set up a ligation reaction in a sterile PCR tube by combining the digested GOI fragment and the digested pTRV2-MCS or pTRV2-*PhCHS* vector in a 3:1 molar ratio of insert/vector. Add the ligation buffer and T4 DNA ligase to the reaction as per the manufacturer's instructions (*see* **Note 21**).

   Ligation formula:

$$\text{Insert (ng)} = \frac{\text{vector (ng)} \times \text{insert size (Kbp)}}{\text{vector size (Kbp)} \times \text{insert : vector ratio)}}$$

9. To transform chemically ($CaCl_2$) competent *E. coli* strain DH5α cells with the pTRV2-GOI or pTRV2-*PhCHS*-GOI vectors, thaw the cells from the −80 °C freezer on ice. Add 5 μl of a ligation reaction to 50 μl of *E. coli* cells. Do this for each ligation reaction. Place the tubes on ice for 10–30 min. Move the tubes to a water bath at 42 °C for 90 s. Immediately return them to ice. Add 1 ml of LB to each tube, and incubate at 37 °C while gently shaking the tubes for 1 h. Transfer 50–100 μl of each transformed bacterial culture onto selective LB plates containing 50 μg/ml kanamycin. Wrap each plate in parafilm and incubate the plate overnight at 37 °C (*see* **Note 22**).
10. Select about five individual colonies from each plate using a sterile pipette tip. Gently touch the tip to a reference plate of LB containing 50 μg/ml kanamycin, and then swish the tip in a 0.2 ml PCR tube containing 10 μl of sterile nanopure water (the water should become slightly cloudy). Cap each tube and incubate for 10 min at 85–95 °C degrees. Use 1 μl of the

heated water as the PCR template to screen for the presence of the ligated vectors with pTRV2 MCS-specific primers (*see* **Note 23**). Run the PCR products on an agarose gel to verify the presence and size of the bands.

11. Validated colonies are also used to generate glycerol stocks for long-term storage. To do this, combine 500 μl of each culture with 500 μl of autoclaved 30% glycerol in cryotubes. Flash freeze them in liquid nitrogen and store them in the −80 °C freezer for future use.

3.3 *Agrobacterium Transformation*

1. After PCR verification, remove the reference plate and select the colonies that tested positive for the insert. Use a sterile pipette tip to transfer the colony to 5 ml of selective LB containing 50 μg/ml kanamycin, and grow overnight in a shaking incubator at 37 °C and 200 rpm.
2. Isolate and purify the plasmid DNA using a purification kit (*see* **Note 24**).
3. To transform the plasmid DNA into *A. tumefaciens* strain GV3101 electrocompetent cells, thaw the cells on ice, and add 1 μl of plasmid DNA to 50 μl of thawed cells. Transfer to a sterile, chilled 0.1 cm electroporation cuvette, and avoid the formation of air bubbles. Electroporate at 1500 V. Add 1 ml of LB to the cuvette and then transfer the culture from the cuvette to a culture tube. Incubate at 28 °C for 2 to 3 h while gently shaking the culture. Plate 100 μl of the recovered culture onto selective LB agar containing 25 μg/ml gentamicin, 50 μg/ml kanamycin, and 10 μg/ml rifampicin. Grow plates for 48 h at 28 °C (*see* **Note 25**).
4. To validate the *Agrobacterium* clones, select two or three colonies using a sterile pipette tip. Touch the tip to a selective LB reference plate containing 25 μg/ml gentamicin, 50 μg/ml kanamycin, and 10 μg/ml rifampicin and to the inside of a 0.2 ml PCR tube. Then expel the pipette tip into a culture tube containing 3 ml of selective LB. Add 10 μl of sterile nanopure water to each PCR tube and incubate at 85–95 °C for 10 min. Use the heated sample as a template for PCR validation using the pTRV2 MCS primer set as outlined in **step 10** (Subheading 3.2).
5. Make a glycerol freezer stock for each colony with successful GOI ligation into pTRV2-MCS or pTRV2-*PhCHS*-GOI by combining 500 μl of an overnight culture with 500 μl of autoclaved 50% glycerol in a cryotube. Flash freeze the cultures in liquid N_2 and transfer them to a −80 °C freezer.

3.4 *Apical Meristem Inoculation*

After comparing multiple methods of inoculation, it was determined that apical meristem inoculation resulted in more efficient silencing and required the least amount of inoculum. Additionally, using this method, *PDS* silencing can be observed 10 days after inoculation in the youngest leaves, and *CHS* silencing is usually observed at the onset of flowering [6]. Silencing phenotypes will continue for several months (and even years [28]) if optimal environmental conditions are maintained.

1. Streak out each pTRV2-GOI or pTRV2-*PhCHS*-GOI along with pTRV1 and the control vectors (e.g., empty pTRV2, pTRV2-*GFP*, pTRV2-*PhCHS-GFP*) from −80 °C *Agrobacterium* GV3101 freezer stocks onto selective LB plates containing 25 μg/ml gentamicin, 50 μg/ml kanamycin, and 10 μg/ml rifampicin, and grow for 36–48 h at 28 °C (*see* **Note 26**).
2. Transfer a single colony with a sterile pipette tip or toothpick to 3 ml of selective LB-MES containing 25 μg/ml gentamicin, 50 μg/ml kanamycin, and 10 μg/ml rifampicin, and grow for 24–30 h at 28 °C in a shaking incubator at 250 rpm.
3. Inoculate 5 ml LB-MES media with 250 μl of each culture. Incubate for 6–8 h at 28 °C in a shaking incubator at 250 rpm. Measure the optical density (OD_{600}). The target OD_{600} is between 0.8 and 1.0 (*see* **Note 27**).
4. Harvest the cells by centrifuging for 15 min at 2000 × *g*. Pour off as much supernatant as possible without disturbing the pellet.
5. Resuspend cell pellet in an appropriate volume of agroinduction media so that a final OD_{600} of about 1.0–1.1 is obtained. Incubate in the dark at room temperature (~25 °C) for at least 2–3 h to overnight (*see* **Note 28**).
6. Combine each individual pTRV2 vector with pTRV1 in a 1:1 ratio by volume.
7. Remove the apical meristem from each petunia using forceps and apply 3–5 drops (~150 μl) of inoculum from a disposable pipette (*see* **Note 29**).
8. Place the plants in a growth chamber set at 20/18 °C day/night temperatures with a 16 h day. Increase the humidity and darkness after inoculation by placing clear domes over pots and turning off the lights for 24 h.
9. Include non-inoculated control plants. (These may also be called mock-inoculated or buffer-inoculated controls.) Remove the apical meristem and apply 3–5 drops of agroinduction media (no *Agrobacterium*). These plants serve as a control for viral symptoms and are especially important when investigating new constructs or new environmental conditions (*see* **Note 30**).

3.5 Experimental Design for VIGS Petunia Experiments

The extent of silencing and the severity of associated phenotypes are largely determined by having growing conditions that are conducive to viral infection and replication (*see* Fig. 2) [6]. Growth chambers are highly advantageous for VIGS experiments because of their ability to maintain specific temperatures that are more optimal for virus infection, but they are not completely uniform and do not eliminate environmental variability [29]. For example, light intensity decreases as bulbs age. Therefore, some important considerations and guidelines are listed below to improve phenotyping success for VIGS petunias:

1. Consult a statistician during the planning stages of each experiment to help you identify the best design for your experiment.
2. Good cultural practices should be utilized during the growth of petunias in growth chamber as phenotypes can be hidden or exaggerated by poor plant care. Maintain consistent irrigation and fertilization during growth (*see* **Note 31**).
3. Use data loggers to monitor and record temperature, light intensity, and humidity during the experiment.
4. Especially when one of your test variables is an environmental parameter, repeat the experiment in the same chamber over time or across multiple chambers.
5. If multiple chambers are required to accommodate all the treatments, each chamber must include control plants, including the mock-inoculated control.
6. Randomly assign plants to a specific treatment (e.g., a VIGS GOI construct), and randomize treatments within the chambers to reduce environmental and genetic variability.
7. Use blocking if known (measurable) sources of environmental or genetic variation exist in the experiment. For example, if one corner of the growth chamber is dimmer or a different temperature, blocking should be used. If no known variation exists, utilize a completely randomized design.
8. Because of the variability of gene silencing that results from VIGS, collect data from many flowers (or other tissues) per plant and from many plants per treatment to obtain an accurate estimate of the mean and identify significant differences between your treatments.

3.6 Recording Phenotypic Data

1. For determining accelerated and/or delayed senescence phenotypes as an example, remove the anthers (i.e., deanther) from flowers with forceps on the day of flower opening (anthesis, day 0) (*see* **Note 32**). To determine flower longevity, record the day the first visual symptoms of senescence (i.e., wilting or abscission) are observed.

Fig. 5 Image analysis of a petunia cv 'Sophistica Blue Morn' corolla using WinCAM after successful color training. The orange line was manually added to aid in visually separating the original image from the WinCAM overlay. The legend in the bottom-right corner displays the color classes

2. Quantify the phenotypic effects of VIGS silencing (e.g., white flowers or white leaves) using the following formulae (*see* **Note 33**):

$$100 \times (\text{silencing frequency}) = \frac{\text{plants with visible silencing}}{\text{total plants treated}}$$

$$100 \times (\text{silencing effectiveness}) = \frac{\text{leaves or flowers with visible silencing}}{\text{total leaves or flowers}}$$

$$100 \times (\text{silencing efficiency}) = \frac{\text{white colored pixels}}{\text{total pixels} - \text{background pixels}}$$

3. To obtain the silencing efficiency of *CHS*, photograph each flower from each plant for the duration of the experiment using a background with a contrasting color. For *PDS*, scan an image of the leaves from each plant by removing all of them and placing them on a flatbed scanner with a contrasting background provided by a colored piece of paper or plastic.
4. Train WinCAM to assign the pixels in each photo that are of a similar hue into one of the color classes: background, silencing phenotype, and wild type. After pixels have been selected, perform an image analysis on a few photos to verify that the training step will correctly assign the pixels to specific color groups. Save the color training file for future analyses (*see* Fig. 5 and **Note 34**).
5. Run a batch analysis on the photos to obtain the ratio of pixels assigned to one of the three color classes. Use the silencing efficiency formula to calculate the percentage of silencing.

3.7 Confirming GOI Downregulation

Quantitative PCR is used to confirm the downregulation of the GOI in pTRV2-*PhCHS*-GOI tissues compared to controls (pTRV2-*PhCHS-GFP* and uninoculated). At least three biological replicates should be quantified to obtain the expression level.

1. Collect leaf or flower tissue (*see* **Note 35**).
2. Collect individual corollas or leaves in 2-ml capped tubes, flash freeze in liquid N_2, and store at −80 °C until RNA extraction. At least four biological replicates should be collected and extracted. This allows for 3 replicates for qPCR and provides one backup sample if needed.
3. Extract the total RNA using TRIzol Reagent or equivalent (*see* **Note 36**). Treat extracted RNA with RNase-free DNase to remove any genomic DNA contamination. Determine the integrity and purity of the RNA. RNA purity can be monitored with the $OD_{260/280}$ ratio of 1.8–2.0, which indicates that the RNA has been successfully separated from protein or phenol contamination. Additionally, RNA can be run on a denaturing gel to visualize the 28S and 18S rRNA bands that will be in a 2:1 ratio in high-quality RNA.
4. Synthesize cDNA using a cDNA Synthesis Kit. One-microgram RNA can be used as a starting RNA concentration for the synthesis of cDNA. Follow the manufacturer's protocol. Throughout the experiment, use the same starting RNA concentration for cDNA synthesis. Store cDNA at −20 °C.
5. Design primers using primer design software. Primers for qPCR should be between 18 and 30 bp long and have a GC content of 46–60%. The PCR products should be between 100 and 150 bp. Determine the optimal annealing temperature for each primer by running a gradient PCR. Select an annealing temperature that is adequate for all primer sets in an experimental study. Primer specificity should be analyzed for every qPCR through melt curve analysis performed at the end of the qPCR. Confirm the amplicon size for at least one sample per primer set by visualization on an agarose gel.
6. Reference genes (two minimum) should be used to accurately calculate relative gene expression. These reference genes should be confirmed to be stable in the experimental tissue. The M-value can be calculated with the geNorm method [30]. To account for variability and ensure reproducibility, at least three biological replicates should be included in each gene expression experiment, and two technical replicates should be run for each sample. You should confirm that your reference genes are also stable in tissue infected with TRV (*see* Fig. 6).

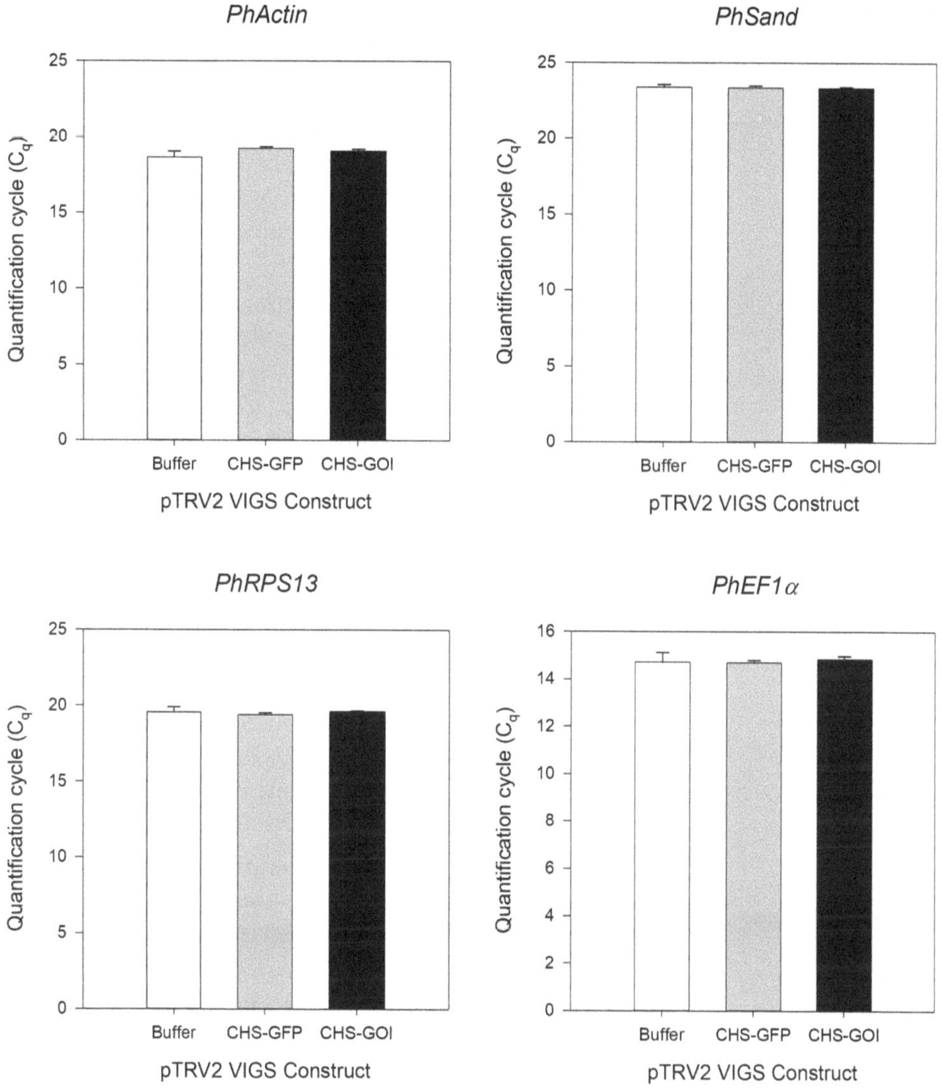

Fig. 6 Stable expression of petunia reference genes in corollas from plants inoculated with TRV constructs. *PhACTIN*, *PhSAND*, *PhRPS13*, and *PhEF1α* serve as good reference genes for petunia corollas [27, 41]. Quantitative PCR was conducted to confirm that the patterns of gene expression were still stable in petunia cv 'Picobella Blue' corollas from plants inoculated with pTRV2-*PhCHS-GFP* and pTRV2-*PhCHS*-GOI compared to buffer (non-inoculated) controls. In the example above, the GOI is petunia phosphate transporter 1 (*PhPT1*) [41]

7. Calculate the gene expression using the $\Delta\Delta C_q$ method using at least two reference genes [31]. qPCR efficiency should be evaluated for each primer set. Generate a standard curve with experimental cDNA serial dilution. Conduct qPCR on the known dilution series, and evaluate the efficiency of the PCR. Target efficiency should be between 90% and 110% (*see* **Note 37**).

4 Notes

1. Horticulture companies sometimes replace plant material while still selling it under the same trade name; therefore, it is advisable to communicate with suppliers to make sure the cultivar is the same as a previous purchase. Make sure enough seed has been purchased to complete an entire experiment and record the seed lot.
2. Plants can be grown in the greenhouse in moderate climates or when outside temperatures allow for lower greenhouse temperatures to be maintained. High temperatures reduce silencing efficiency and lower viral titer (*see* Fig. 2).
3. *Nicotiana tabacum NtPDS* (YL124) and *Arabidopsis thaliana AtPDS* (YL154) are available from Arabidopsis Biological Resource Center (stock #CD3-1045 and CD3-1047, respectively).
4. pTRV2-*GFP* and pTRV2-*PhCHS-GFP* have 265 bp fragments of *green fluorescent protein* (*GFP*). The *GFP* fragment came from pFELV [25]. It does not lead to silencing or the production of fluorescent GFP protein. Instead, it minimizes viral symptoms caused by pTRV2-MCS inoculation and allows pTRV2-*PhCHS-GFP* and pTRV2-*PhCHS*-GOI constructs to be of similar size. The GFP fragment in tandem with *PhCHS* serves as a control for pTRV2-*GFP*-GOI.
5. Other *E. coli* competent cell strains can be used for pTRV transformation; however, they should not be kanamycin resistant.
6. Strain GV3101 is resistant to gentamicin and rifampicin. Kanamycin resistance is conferred by pTRV vectors.
7. Prepare the 0.1 M acetosyringone stock solution with DMSO.
8. Other software capable of analyzing photo colors such as Image Color Extract [32] may be used.
9. PCR primers to amplify *CHS* or any GOI should be designed outside the region that was targeted for gene silencing to avoid amplification of virus RNA from the pTRV2 construct. Ye et al. [33] provide information and a web-based tool for designing primers for PCR and qPCR applications. Yuan et al. [34] developed a web-based tool to aid in identifying regions of target genes that are more amenable to RNAi silencing.
10. Cultivar selection is important. Not all petunia cultivars have efficient silencing (*see* Fig. 1). A cultivar such as 'Picobella Blue' (Syngenta Flowers, Gilroy, CA, USA) is a good choice. It has a high degree of silencing efficiency compared to other commercial cultivars, and its compact size is more amenable to growth

chambers [2]. In contrast, the petunia model cv 'Mitchell Diploid' is much larger, and its white flowers provide no visual cues for gene silencing in floral tissue when using *CHS*. Carefully consider other characteristics specific and/or relevant to the study such as senescence patterns, flower size, and plant architecture.

11. Many commercially available petunia seeds are pelletized. Apply water gently (such as with a spray bottle) until the pellets break down to promote faster and consistent germination. Petunia seeds germinate best under highly humid conditions such as those provided with a clear propagation dome, but do not oversaturate the media with water.
12. Fertilization can be modified for nutrient stress studies, but for most phenotypic evaluations, plants should not be under nutrient stress.
13. Inoculating plants 5 weeks after sowing results in a significant decline in silencing efficiency compared to 3 or 4 weeks (*see* Fig. 4) [2].
14. Whitening (i.e., photobleaching) of the leaves is a visible indicator of PTGS of *phytoene desaturase* (*PDS*). *PDS* from *N. benthamiana* (*NbPDS*) is 95% identical to *P. hybrida's PDS* at the nucleotide level. *NbPDS* efficiently silences *PDS* and induces photobleaching in petunia and many other solanaceous species (*see* Fig. 1) [26, 35].
15. Silencing *chalcone synthase* (*CHS*) results in the breakdown of anthocyanin pigments in the corolla, and petal tissue that is colored becomes white. This is seen as entirely white corollas or sectors of a corolla that are white (*see* Fig. 1). If *CHS* is used as a visual indicator of gene silencing in a flower, the GOI insert should be cloned into pTRV2-*PhCHS*-GOI. Alternatively, pTRV2 vectors containing full-length GFP fused to the coat protein can be used to track the location of the virus within the plant [12]. This is a superior indicator of gene silencing in the leaves compared to *PDS* because it has less impact on plant growth than inhibiting chlorophyll production.
16. Inserts larger than 400 bp can increase the chance of off-target silencing [36]. Thermo Fisher Scientific has an online resource to identify the best regions for RNAi silencing (https://rnaidesigner.thermofisher.com/rnaiexpress/design.do). Use mRNA/cDNA as your input and select 50% as the maximum GC content. Also, Sol Genomics Network (SGN) has developed the SGN VIGS Tool that can be used to identify regions for VIGS in the petunia genome while reducing off-target silencing (https://solgenomics.net/) [37]. Conserved regions within gene families can be used to silence more than one gene from a single VIGS construct. Alternatively, it may be possible

to utilize unique regions of a specific gene or untranslated regions to exclude the downregulation of related genes. VIGS constructs can also be designed to contain fragments from multiple genes to silence more than one gene with one construct [38].

17. It is helpful to choose two restriction enzymes that can be digested together using the same buffer (i.e., double digestion).

18. Follow the restriction enzyme manufacturer's instructions, and add an appropriate number of nucleotides (usually 4–6) to the 5′-end of the restriction site to improve the efficiency of restriction enzyme digestion. Ensure that the added sites will not produce primer-dimers or inhibitory secondary structures.

19. Amplification with the restriction-site primers may result in off-target amplification or poor amplification. In these cases, primers without the restriction sites should be used. The restriction sites can be added afterward in a second round of PCR.

20. The reaction can be heat inactivated if both enzymes are heat sensitive. If not, the reaction can be stopped by DNA purification using ethanol precipitation. This may also reduce the false-positive colonies that can develop. This can be done by mixing 0.1 volume of 3.0 M sodium acetate (pH 5.2) to each reaction. Then add 2 volumes of room temperature 95% ethanol, and incubate at room temperature for 2 h. Centrifuge for 15 min at maximum speed in a microcentrifuge at room temperature, and immediately pipette off the supernatant being careful to keep the tip away from the glassy pellet. (It may not be visible.) If more liquid remains, briefly spin down the tubes again (1 min) and remove any remaining liquid. Wash with 200–500 μl of cold (<10 °C) 70% ethanol. Re-centrifuge at 4 °C for 5 min and decant as outlined above. Resuspend the pellet in an appropriate amount of water (such as 10–50 μl). Quantify the DNA using a NanoDrop spectrophotometer.

21. Rapid ligation buffer can be used for successful ligation, which greatly reduces the ligation time.

22. LB or SOC can be used for recovering the cells after the heat shock treatment.

23. The pTRV2 MCS-specific primer set may be enough to verify that the ligation and transformation steps were successful (expected product size if empty is 192 bp). However, the GOI primer set can be used to ensure that the insert from the GOI is transformed into the MCS in pTRV2. The *coat protein* (*CP*) primer set (expected product size 401 bp [26]) alone does not verify that the GOI insert was successfully cloned

into the MCS, but the forward *CP* primer and the reverse MCS primer could also be used for verification.

24. Plasmid DNA samples should be sent for sequence verification.
25. LB or SOC can be utilized for recovery after electroporation. A serial dilution may be necessary to obtain single *Agrobacterium* colonies if the transformation is highly efficient.
26. Inoculation of petunia (and other solanaceous crops) with the pTRV2 empty vector as a control results in viral symptoms which vary in intensity depending on the cultivar. Symptoms can be so severe that empty pTRV2 cannot be used as a suitable control [2]. The pTRV2-*GFP* construct should be used in place of the empty vector control. It should have an insert that is similarly sized with the GOI so that viral replication rates are similar. This construct has a 265 bp fragment of *GFP* (no functional fluorescent protein is produced), and there are no endogenous sequences to target for silencing. *GFP* or any other fragment can be cloned into the pTRV2-MCS following the protocol in Subheading 3.2. Other DNA inserts could be utilized. For example, *β-glucuronidase* from *E. coli* has been successfully used to minimize viral symptoms [39]. If *CHS* is used as a silencing reporter gene in combination with the gene of interest, the control construct should be pTRV2-*PhCHS-GFP*.
27. It is important that enough pTRV1 inoculum is prepared because it will need to be co-inoculated with pTRV2 and all pTRV2 derivatives.
28. Overnight is preferred because it gives more time for the *Agrobacterium* to respond to the acetosyringone in the media.
29. Removing the meristem too far in advance may reduce the efficiency of the inoculation.
30. Plants infected with TRV must be grown under BL2-P containment according to NIH guidelines [40]. Protocols should be approved by your organization's Institutional Biosafety Committee (IBC) or equivalent. This protocol will detail the safety precautions required to prevent the release of the modified TRV. Do not start experiments without obtaining appropriate approvals.
31. Petunias growing in growth chambers may need higher fertilizer rates and modified irrigation scheduling to maintain healthy growth compared to petunias grown in the greenhouse.
32. Removal of the anthers, while buds are just beginning to open, will prevent self-pollination and premature senescence. After anther removal, tags can be strung on the peduncle to mark them with the date and sample number.

33. Silencing efficiency is the most accurate and statistically powerful method of calculating VIGS silencing differences, but it also requires more time. Silencing frequency is easily obtained by counting the plants that exhibit silencing phenotypes and dividing that by the total plants in a treatment, but it is not as powerful statistically and doesn't provide information about the extent of silencing on each plant. Silencing effectiveness requires a little more time to collect than silencing frequency, but it is more statistically powerful and provides more details about the extent of silencing without relying on image analysis.
34. WinCAM's default color classes are labeled as "background," "diseased," and "healthy" (*see* Fig. 5).
35. Due to the variability in silencing throughout the plant, it is best if you collect tissue that has a visual reporter indicating the presence of the TRV construct. For example, collect white (not purple) flowers from petunia cv 'Picobella Blue' that have been inoculated with pTRV2-*PhCHS*-GOI.
36. Other methods of RNA extraction, such as commercially available kits, may be used if sufficient amounts of RNA can be obtained; however, certain tissues, such as petunia corollas, generally do not yield as much RNA as leaves when using kits.
37. Consult the MIQE guidelines for further details about qPCR experimental design and parameters [31].

Acknowledgments

The authors would like to acknowledge the American Floral Endowment, the CFAES Research Competitive Grants Program, and The Ohio State University D.C. Kiplinger Floriculture Endowment for funding support. Salaries and research support were provided in part by state and federal funds appropriated to the OARDC, The Ohio State University, and by USDA Hatch MIS-212060, Mississippi State University.

References

1. Napoli C, Lemieux C, Jorgensen R (1990) Introduction of a chimeric chalcone synthase gene into petunia results in reversible co-suppression of homologous genes *in trans*. Plant Cell 2:279–289. https://doi.org/10.1105/tpc.2.4.279
2. Burch-Smith TM, Anderson JC, Martin GB, Dinesh-Kumar SP (2004) Applications and advantages of virus-induced gene silencing for gene function studies in plants. Plant J 39:734–746. https://doi.org/10.1111/j.1365-313X.2004.02158.x
3. Ratcliff F, Martin-Hernandez AM, Baulcombe DC (2001) Tobacco rattle virus as a vector for analysis of gene function by silencing. Plant J 25:237–245. https://doi.org/10.1046/j.0960-7412.2000.00942.x
4. Liu Y, Schiff M, Marathe R, Dinesh-Kumar SP (2002) Tobacco *Rar1*, *EDS1* and *NPR1/NIM1* like genes are required for *N*-mediated resistance to tobacco mosaic virus. Plant J

30:415–429. https://doi.org/10.1046/j.1365-313X.2002.01297.x
5. Chen J-C, Jiang C-Z, Gookin T et al (2004) Chalcone synthase as a reporter in virus-induced gene silencing studies of flower senescence. Plant Mol Biol 55:521–530. https://doi.org/10.1007/s11103-004-0590-7
6. Broderick SR, Jones ML (2014) An optimized protocol to increase virus-induced gene silencing efficiency and minimize viral symptoms in petunia. Plant Mol Biol Rep 32:219–233. https://doi.org/10.1007/s11105-013-0647-3
7. Spitzer B, Zvi MMB, Ovadis M et al (2007) Reverse genetics of floral scent: application of tobacco rattle virus-based gene silencing in petunia. Plant Physiol 145:1241–1250. https://doi.org/10.1104/pp.107.105916
8. Noor SH, Ushijima K, Murata A et al (2014) Double flower formation induced by silencing of C-class MADS-box genes and its variation among petunia cultivars. Sci Hortic 178:1–7. https://doi.org/10.1016/j.scienta.2014.07.029
9. Becker A, Lange M (2010) VIGS – genomics goes functional. Trends Plant Sci 15:1–4. https://doi.org/10.1016/j.tplants.2009.09.002
10. Robertson D (2004) VIGS vectors for gene silencing: many targets, many tools. Annu Rev Plant Biol 55:495–519. https://doi.org/10.1146/annurev.arplant.55.031903.141803
11. Kumagai MH, Donson J, della-Cioppa G et al (1995) Cytoplasmic inhibition of carotenoid biosynthesis with virus-derived RNA. Proc Natl Acad Sci U S A 92:1679–1683. https://doi.org/10.1073/pnas.92.5.1679
12. Tian J, Pei H, Zhang S et al (2014) TRV–GFP: a modified *tobacco rattle virus* vector for efficient and visualizable analysis of gene function. J Exp Bot 65:311–322. https://doi.org/10.1093/jxb/ert381
13. Cheng C, Gao J, Ma N (2018) Investigation of petal senescence by TRV-mediated virus-induced gene silencing in rose. In: Guo Y (ed) Plant senescence: methods and protocols. Springer New York, New York, pp 49–63
14. Liang Y-C, Reid MS, Jiang C-Z (2014) Controlling plant architecture by manipulation of gibberellic acid signalling in petunia. Hortic Res 1:14061. https://doi.org/10.1038/hortres.2014.61
15. Wu C, Jia L, Goggin F (2011) The reliability of virus-induced gene silencing experiments using tobacco rattle virus in tomato is influenced by the size of the vector control. Mol Plant Pathol 12:299–305. https://doi.org/10.1111/j.1364-3703.2010.00669.x
16. Hartl M, Merker H, Schmidt DD, Baldwin IT (2008) Optimized virus-induced gene silencing in *Solanum nigrum* reveals the defensive function of leucine aminopeptidase against herbivores and the shortcomings of empty vector controls. New Phytol 179:356–365. https://doi.org/10.1111/j.1469-8137.2008.02479.x
17. Brigneti G, Martín-Hernández AM, Jin H et al (2004) Virus-induced gene silencing in *Solanum* species. Plant J 39:264–272. https://doi.org/10.1111/j.1365-313X.2004.02122.x
18. Ghazala W, Varrelmann M (2007) *Tobacco rattle virus* 29K movement protein is the elicitor of extreme and hypersensitive-like resistance in two cultivars of *Solanum tuberosum*. Mol Plant Microbe Interact 20:1396–1405. https://doi.org/10.1094/MPMI-20-11-1396
19. Liu Y, Schiff M, Dinesh-Kumar SP (2002) Virus-induced gene silencing in tomato. Plant J 31:777–786. https://doi.org/10.1046/j.1365-313X.2002.01394.x
20. Sun D, Li S, Niu L et al (2017) *PhOBF1*, a petunia ocs element binding factor, plays an important role in antiviral RNA silencing. J Exp Bot 68:915–930. https://doi.org/10.1093/jxb/erw490
21. Sun D, Nandety RS, Zhang Y et al (2016) A petunia ethylene-responsive element binding factor, *PhERF2*, plays an important role in antiviral RNA silencing. J Exp Bot 67:3353–3365. https://doi.org/10.1093/jxb/erw155
22. Wang H, Chang X, Lin J et al (2018) Transcriptome profiling reveals regulatory mechanisms underlying corolla senescence in petunia. Horticult Res 5:16. https://doi.org/10.1038/s41438-018-0018-1
23. Cna'ani A, Spitzer-Rimon B, Ravid J et al (2015) Two showy traits, scent emission and pigmentation, are finely coregulated by the MYB transcription factor PH4 in petunia flowers. New Phytol 208:708–714. https://doi.org/10.1111/nph.13534
24. Bombarely A, Moser M, Amrad A et al (2016) Insight into the evolution of the Solanaceae from the parental genomes of *Petunia hybrida*. Nat Plants 2:16074. https://doi.org/10.1038/nplants.2016.74
25. Hernandez-Garcia CM, Bouchard RA, Rushton PJ et al (2010) High level transgenic expression of soybean (*Glycine max*) GmERF and Gmubi gene promoters isolated by a novel promoter analysis pipeline. BMC Plant Biol

10:237. https://doi.org/10.1186/1471-2229-10-237

26. Ryu C-M, Anand A, Kang L, Mysore KS (2004) Agrodrench: a novel and effective agroinoculation method for virus-induced gene silencing in roots and diverse *Solanaceous* species. Plant J 40:322–331. https://doi.org/10.1111/j.1365-313X.2004.02211.x
27. Mallona I, Lischewski S, Weiss J et al (2010) Validation of reference genes for quantitative real-time PCR during leaf and flower development in *Petunia hybrida*. BMC Plant Biol 10:4. https://doi.org/10.1186/1471-2229-10-4
28. Senthil-Kumar M, Mysore KS (2011) Virus-induced gene silencing can persist for more than 2 years and also be transmitted to progeny seedlings in *Nicotiana benthamiana* and tomato. Plant Biotechnol J 9:797–806. https://doi.org/10.1111/j.1467-7652.2011.00589.x
29. Hammer PA, Hopper DA (1997) Experimental design. In: Langhans R, Tibbitts T (eds) Plant growth chamber handbook. Iowa State University, Ames, IA, pp 177–187
30. Hellemans J, Mortier G, Paepe AD et al (2007) qBase relative quantification framework and software for management and automated analysis of real-time quantitative PCR data. Genome Biol 8:r19. https://doi.org/10.1186/gb-2007-8-2-r19
31. Bustin SA, Benes V, Garson JA et al (2009) The MIQE guidelines: *Minimum Information* for publication of *Quantitative* real-time PCR *Experiments*. Clin Chem 55:611–622. https://doi.org/10.1373/clinchem.2008.112797
32. Gelotte K (2007) Image color extract. Cool PHP tools. https://www.coolphptools.com/color_extract
33. Ye J, Coulouris G, Zaretskaya I et al (2012) Primer-BLAST: a tool to design target-specific primers for polymerase chain reaction. BMC Bioinformatics 13:134. https://doi.org/10.1186/1471-2105-13-134
34. Yuan B, Latek R, Hossbach M et al (2004) siRNA Selection Server: an automated siRNA oligonucleotide prediction server. Nucleic Acids Res 32:W130–W134. https://doi.org/10.1093/nar/gkh366
35. Senthil-Kumar M, Hema R, Anand A et al (2007) A systematic study to determine the extent of gene silencing in *Nicotiana benthamiana* and other Solanaceae species when heterologous gene sequences are used for virus-induced gene silencing. New Phytol 176:782–791. https://doi.org/10.1111/j.1469-8137.2007.02225.x
36. Tuttle JR, Haigler CH, Robertson D (2015) Virus-induced gene silencing of fiber-related genes in cotton. In: Mysore KS, Senthil-Kumar M (eds) Plant gene silencing: methods and protocols. Springer New York, New York, pp 219–234
37. Fernandez-Pozo N, Rosli HG, Martin GB, Mueller LA (2015) The SGN VIGS tool: user-friendly software to design virus-induced gene silencing (VIGS) constructs for functional genomics. Mol Plant 8:486–488. https://doi.org/10.1016/j.molp.2014.11.024
38. Purkayastha A, Dasgupta I (2009) Virus-induced gene silencing: a versatile tool for discovery of gene functions in plants. Plant Physiol Biochem 47:967–976. https://doi.org/10.1016/j.plaphy.2009.09.001
39. Gargul JM, Mibus H, Serek M (2015) Manipulation of *MKS1* gene expression affects *Kalanchoë blossfeldiana* and *Petunia hybrida* phenotypes. Plant Biotechnol J 13:51–61. https://doi.org/10.1111/pbi.12234
40. Wilson DE, Chosewood LC (2009) Biosafety in microbiological and biomedical laboratories, 5th edn. U.S. Department of Health and Human Services, Washington, DC
41. Chapin LJ, Jones ML (2009) Ethylene regulates phosphorus remobilization and expression of a phosphate transporter (*PhPT1*) during petunia corolla senescence. J Exp Bot 60:2179–2190. https://doi.org/10.1093/jxb/erp092

Chapter 16

Virus-Induced Gene Silencing in Rose Flowers

Huijun Yan, Zhao Zhang, Jean-Louis Magnard, Benoît Boachon, Sylvie Baudino, and Kaixue Tang

Abstract

Virus-induced gene silencing (VIGS) is a favorable method to study gene function by posttranscriptional gene silencing in plants. Here we describe a methodology of graft-accelerated VIGS in rose aimed at obtaining posttranscriptional gene silencing in the flower. The resulting phenotype can be observed within 5–6 weeks post infiltration. By using this method, we successfully silenced the expression of several genes involved in processes such as scent production, petal coloration, or flower architecture. We showed that graft-accelerated VIGS was faster, more efficient, and more convenient than conventional methods previously developed in rose such as agroinfiltration of young plantlets and in vitro cultured tissues or seeds.

Key words VIGS, Rose, Vacuum infiltration, Graft, Posttranscriptional gene silencing

1 Introduction

Virus-induced gene silencing is a rapid approach to study gene function by knocking down gene expression in plants [1, 2]. VIGS is based on posttranscriptional gene silencing (PTGS), in which a sequence-specific RNA degradation is triggered by double-stranded RNA [1]. In VIGS, a recombinant virus carrying a partial sequence of the target gene is used to infect the plant. When the virus replicates and spreads throughout the whole plant, the viral gene transcripts are recognized and degraded together with those of the plant target gene by plant endogenous PTGS, and the expression of the target gene is knocked down [3, 4].

Nowadays, VIGS has been broadly used for functional characterization of genes in different plant organs, including leaves of eggplant [5]; leaves and tubers of potato [6]; cornels of *Gladiolus hybridus* [7]; roots of *Nicotiana benthamiana*, *Arabidopsis thaliana*, and *Lycopersicon esculentum* [8]; fruits of tomato [9, 10]; as well as flowers of *Petunia hybrida* [11].

Vincent Courdavault and Sébastien Besseau (eds.), *Virus-Induced Gene Silencing in Plants: Methods and Protocols*, Methods in Molecular Biology, vol. 2172, https://doi.org/10.1007/978-1-0716-0751-0_16,

Regarding rose, a VIGS approach using young plantlets issued from in vitro cultivated rose cuttings was established by Tian et al. [2]. The plantlets were submerged in an *Agrobacterium tumefaciens* suspension carrying the vectors encoding the recombinant *Tobacco rattle virus* (TRV). Vacuum infiltration was used to transform the plant material. Using this procedure, the infected plantlets flowered about 2 months post inoculation, and efficiency of the transformation procedure could be evaluated by observing alteration in flower coloration [2] or petal abscission [12]. However, the long delay between agro-inoculation and flowering often leads to attenuated efficiency of gene silencing. Furthermore, when the transformed plantlets are issued from in vitro culture, rooting may be difficult for some rose varieties (e.g., *R. odorata* var. *gigantea*). In addition, when plantlets are obtained directly from seeds, cross pollination can also lead to trait segregation [13, 14].

Recently, an alternative approach based on recombinant TRV inoculation in detached rose petals proved to be able to induce gene silencing [15]. An observable phenotype and a measurable down expression of the target gene were detected within a week. This method has been applied to study several processes linked to petal physiology, namely, dehydration tolerance [15], petal expansion [16], as well as senescence [17]. However, this method has limitations. For instance, detached rose petals can obviously not be used to study flower development or architecture, as well as petal abscission.

Some of the limitations of the previously described procedures could be overcome thanks to the recent development and use of a novel method for rapid gene silencing in complete rose flowers, called "graft-accelerated VIGS" [18]. In our hands graft-accelerated VIGS proved to be able to generate silenced flowers in 30–35 days with an improved efficiency: 46% silenced flowers versus 10–30% for previously developed agroinfiltration methods. We previously demonstrated that this procedure could be used for rapid and efficient studies of gene function in rose flowers, such as genes involved in fragrance production, flower architecture, as well as petal coloration [18, 19]. The rose genome has recently been uncovered [20, 21], thus enabling to study the function of genes involved in many processes of interest such as fragrance biosynthesis or development. This brings a real need to develop reliable methods for reverse genetic studies. We propose that graft-accelerated VIGS can be extensively used in reverse genetics of rose flowers.

2 Materials

2.1 Media and Buffers

1. YEB medium: 1 g/L yeast extract, 5 g/L tryptone (*see* **Note 1**), 5 g/L beef extract, 5 g/L sucrose, 2 g/L $MgSO_4$, 10 mM MES, pH 7.0, 20 mM acetosyringone. Solid YEB medium contained 15 g/L agar.
2. 50 mg/mL kanamycin stock solution in water and 25 mg/mL rifampicin stock solution in DMSO.
3. *Agrobacterium* infiltration buffer: 10 mM $MgCl_2$, 10 mM MES, pH 5.6, 150 mM acetosyringone.

2.2 Vectors and Constructs

1. The binary pTRV vectors, pTRV1 and pTRV2, were used for VIGS experiments [3]. The pTRV1 plasmid contains genes for virus replication and spread, while pTRV2 encodes virus coat proteins and carries a fragment of the VIGS-targeted gene of interest (*see* **Note 2**). Both vectors carry duplicated CaMV 35S promoter (2 × 35S promoter) and the selection marker for kanamycin resistance. The gene of interest fragment sequence to be inserted in the pTRV2 vector should be 200–400 bp and specific to the gene to be silenced. The SGN VIGS Tool (http://vigs.solgenomics.net/) [22] now includes the genome of *Rosa chinensis* 'Old Blush' and can be used to design the fragment sequence and avoid off-target silencing. The gene fragment can be inserted in the pTRV2-gateway vector through Gateway LR cloning using the attR1 and attR2 integration sites located on the vector (*see* **Note 3**).

2.3 Plant Material

1. *Rosa chinensis* 'Old Blush' plants.
2. Greenhouse: 13 h/11 h day/night cycle, 28/20 °C day/night temperature, the relative humidity was 75–80% (*see* **Note 4**).

2.4 Equipments

1. Spectrophotometer.
2. Bacterial incubator.
3. Vacuum pump and vacuum bell jar.
4. Centrifuge.
5. Polyethylene plastic gloves.
6. Parafilm.
7. Grafting knife.
8. Thermocycler.

3 Methods

All procedures are carried out at room temperature (*see* **Note 5**).

3.1 Preparation of the Scions

1. The grafted organ was taken from a stem with axillary sprouts. It was cut from the plant as a scion. The scion must be sprouting (not dormant), with one axillary vegetative bud located approximately in its middle, and should be about 4 cm long (Fig. 1a) (*see* **Notes 6** and 7).
2. The scions were kept in the water until grafting (Fig. 1b).

3.2 Vacuum Agroinfiltration

1. The *Agrobacterium* strain EHA105 (*see* **Note 8**) was transformed with pTRV1 or with recombinant pTRV2 carrying a

Fig. 1 Preparation of grafting scions and vacuum infiltration. (**a**) Sampling of the scion used for vacuum infiltration. (**b**) The scions in water prior to vacuum infiltration. (**c**) Vacuum bell jar and pump used for *Agrobacterium* infiltration

fragment of the gene of interest (pTRV2-GOI). The resulting agrobacteria were grown at 28 °C on YEB solid medium containing 50 μg/mL kanamycin for selecting the pTRV1 and pTRV2 plasmids and 25 μg/mL rifampicin for selection of the *Agrobacterium* strain.

2. 5 mL YEB medium containing 50 μg/mL kanamycin and 25 μg/mL rifampicin was inoculated with colonies issued from the pTRV1 or pTRV2-GOI transformations and grown overnight at 28 °C under continuous agitation (180 rpm).
3. The next day, 100 μL of the overnight *Agrobacterium* cultures were inoculated into 50 mL YEB liquid medium containing 50 μg/mL kanamycin and 25 μg/mL rifampicin, and cultured in a shaker at 28 °C and 180 rpm until OD_{600} reaches a value of 0.3–0.5.
4. The agrobacteria containing pTRV1 or recombinant pTRV2-GOI were harvested by centrifugation (500 × *g*, 15 min), resuspended in *Agrobacterium* infiltration buffer, and mixed in a ratio of 1:1 (v/v).
5. The cellular density was finally adjusted to an OD_{600} of 1.0 with infiltration buffer (*see* **Note 9**). The resulting suspension was kept in the dark at room temperature for 2 h prior to vacuum infiltration for virulence gene activation with acetosyringone.
6. Twenty rose scions per gene/construct were submerged in 80 mL *Agrobacterium* suspension in a 100 mL glass beaker. The beaker was transferred into a vacuum bell jar (Fig. 1c). A 25 kPa vacuum was applied to the preparation for 60 s before being released slowly. The procedure was repeated a second time in identical conditions.
7. The treated scions were then washed for three seconds in distilled water and grafted onto appropriate rootstocks (*see* **Note 10**).

3.3 Grafting

1. Using a grafting knife, a wedge-shaped cut was performed at the basal extremity of the infiltrated scions to be grafted onto the rootstocks (Fig. 2a).
2. The plant to be used as rootstock was from the same cultivar as the scion. Branches 0.5–1 cm in diameter were used as rootstocks. The branch that was chosen to receive the graft was cut near the middle of the stem, between two nodes. Then, buds and leaves were removed from the stem (Fig. 2b) (*see* **Note 7**). Finally, a grafting knife was used to split the rootstock in two equal parts over a height of about 1.5 cm.
3. The scion has to be inserted immediately after preparing the rootstock. It should be pushed all the way down into the cut, keeping the cambium layers of scion and rootstock in contact (Fig. 2c). Finally, a stretchy material such as Parafilm laboratory

Fig. 2 Grafting procedure. (**a**) Wedge-shaped section on transformed scions. (**b**) Rose branch used as a rootstock. (**c**) Grafting of an infiltrated scion onto a rootstock. (**d**) Fixation of the graft using Parafilm laboratory film. (**e**) PE plastic glove fingers. (**f**) Positioning of PE plastic gloves on grafted material. (**g**) Blooming of the first flower on a grafted scion. The arrow points to the junction between the scion and rootstock

film was wrapped around the junction to hold the scion in place (Fig. 2d) (*see* **Notes 11** and **12**).

4. Grafted material was kept in humid conditions for 7–10 days by glove fingers cut in polyethylene plastic gloves (Fig. 2e, f) (*see* **Note 13**).

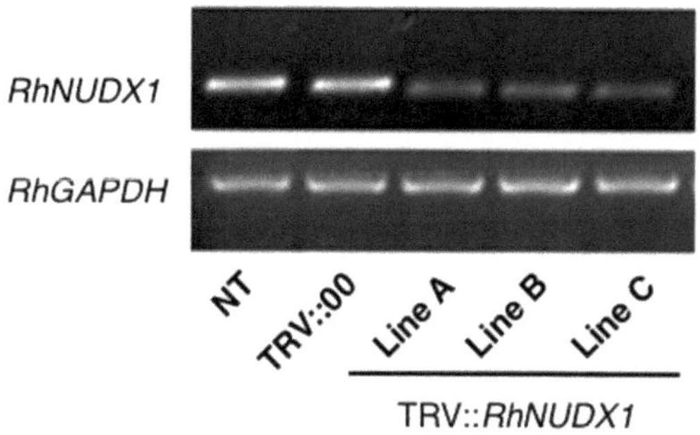

Fig. 3 Silencing of *RhNUDX1* in *R. chinensis* 'Old Blush'. Expression of *RhNUDX1* [23] in non-inoculated flowers (NT), flowers transformed by an empty pTRV2 vector (TRV::00) and by a vector containing a 331 bp fragment of the *RhNUDX1* gene (TRV::*RhNUDX1*). Lines A, B, and C represent three independent *RhNUDX1* silenced flowers. Semiquantitative RT-PCR was performed with the *RhNUDX1*-specific primers RhNUDX1-F (5′-TGTTGAGAGGCCAAAACGTG-3′) and RhNUDX1-R (5′-GTACTGCGATGGTTTGGCTT-3′). Glyceraldehyde phosphate dehydrogenase (*RhGAPDH*) amplified with primers RhGAPDH-F (5′-ATCCATTCATCACCACCGACTA CA-3′) and RhGAPDH-R (5′-GCATCCTTACTTGGGGCAGAGA-3′) was used as a reference gene

Table 1
Geraniol emission in control and *RhNUDX1* silenced flowers

	NT	TRV::00	Line A	Line B	Line C
Relative content (%)	4.19 ± 0.20	3.73 ± 0.35	1.52 ± 0.38	2.16 ± 0.52	1.51 ± 0.27

Rose flowers (*R. chinensis* 'Old Blush') were harvested at the fully open stage of development. Floral volatiles were collected by solid-phase micro-extraction (SPME) at 30 °C for 45 min using a 75 μm CAR/PDMS SPME fiber (Supelco, PA, USA) inserted into a 100 mL sample vial containing 1 g of petals. The SPME fiber was then inserted into the front inlet of a gas chromatograph (GC, HP 6890A) connected to a mass spectrometer (MS, HP 5973, GC/MS: Agilent Technologies, USA) and desorbed at 250 °C for 2 min. The temperatures of the transfer line and ion trap were 250 and 300 °C, respectively. Values correspond to the emission of geraniol as a percentage of total volatiles and are the means ± standard deviations of three independent measurements. Tukey's test indicates significant differences ($p < 0.01$) between silenced lines (A, B, or C) and both not transformed and TRV::00 transformed plants. *See* Fig. 3 for sample nomenclature

5. The grafted organ bloomed 30–40 days after grafting (*see* **Note 14**) (Fig. 2g). In our hands the percent survival of the graft was 90%.
6. The efficiency of the VIGS procedure was assessed by RT-PCR measurement of down expression of the target gene in flowers and by apparition of a knockdown phenotype. As an example, targeting of the *Nudix hydrolase 1* gene (*RhNUDX1*) (GenBank accession number JQ820249) [23], involved in synthesis of the scent compound geraniol, yielded significant decrease in both gene expression (Fig. 3) and geraniol emission (Table 1)

in silenced plants compared to control. Among the surviving grafted organs, the percent of material, which was actually transformed and presented a typical phenotype, was about 46% [18, 19].

4 Notes

1. Bacteriological tryptone can be replaced by bacteriological peptone.
2. The pTRV1 plasmid can be ordered from the Arabidopsis Biological Resource Center (ABRC): stock no. CD3-1039, vector name YL192.

   The Gateway-compatible pTRV2 vector corresponds to ABRC stock no. CD3-1041, vector name YL279.
3. Following cloning the pTRV2 will contain the antisense sequence of the gene of interest. Cloning into pTRV2 can also be performed by conventional restriction digestion-based cloning or ligation independent cloning (LIC) [24].
4. Culture conditions of the plants both prior and after grafting should be carefully monitored. The temperature is a key factor for the success of this procedure, especially for the successful development of the grafted organ since callus development is favored in the indicated conditions.
5. All genetically modified organisms should be cultured, grown, transported, or used in a proper way according to requirement of local legislation.
6. For grafting, the terminal bud of the scion was removed.
7. Ideally the diameters of the scion and rootstock should be the same (0.5–1 cm).
8. *Agrobacterium* strain GV3101 can be used instead of EHA105.
9. The volume of infiltration mixture to prepare should be about 80 mL for 20 scions.
10. Following infiltration, the scions are kept in water until grafting. It takes about 1 h for a trained experimenter to graft 20 scions.
11. Parafilm film can be replaced by cling film.
12. Several scions can be grafted on different branches of the same plant as long as they are transformed by the same construct.
13. The key point is to have a fast growing tissue to favor wound healing between the scion and the rootstock.

14. The blooming time depends on temperature. The optimal day temperature is 25–28 °C, and night temperature should be around 18–20 °C. Lowering temperature below 24 °C delayed blooming by more than 10 days.

Acknowledgments

This work was supported by the National Natural Science Foundation of China (Grants numbers 31660579, 31872144, and 31501791). This research was funded in part by the Beijing Natural Science Foundation (6162017).

References

1. Senthil-Kumar M, Mysore KS (2011) New dimensions for VIGS in plant functional genomics. Trends Plant Sci 16:656–665
2. Tian J, Pei H, Zhang S, Chen J, Chen W, Yang R, Meng Y, You J, Gao J, Ma N (2014) TRV–GFP: a modified Tobacco rattle virus vector for efficient and visualizable analysis of gene function. J Exp Bot 65:311–322
3. Liu Y, Schiff M, Dinesh-Kumar S (2002) Virus-induced gene silencing in tomato. Plant J 31:777–786
4. Burch-Smith TM, Anderson JC, Martin GB, Dinesh-Kumar SP (2004) Applications and advantages of virus-induced gene silencing for gene function studies in plants. Plant J39:734–746
5. Liu H, Fu D, Zhu B, Yan H, Shen X, Zuo J, Zhu Y, Luo Y (2012) Virus-induced gene silencing in eggplant (*Solanum melongena*). J Integr Plant Biol 54:422–429
6. Faivre-Rampant O, Gilroy EM, Hrubikova K, Hein I, Millam S, Loake GJ, Birch P, Taylor M, Lacomme C (2004) Potato virus X-induced gene silencing in leaves and tubers of potato. Plant Physiol 134:1308–1316
7. Zhong X, Yuan X, Wu Z, Khan MA, Chen J, Li X, Gong B, Zhao Y, Wu J, Wu C (2014) Virus-induced gene silencing for comparative functional studies in *Gladiolus hybridus*. Plant Cell Rep 33:301–312
8. Valentine T, Shaw J, Blok VC, Phillips MS, Oparka KJ, Lacomme C (2004) Efficient virus-induced gene silencing in roots using a modified tobacco rattle virus vector. Plant Physiol 136:3999–4009
9. Fu DQ, Zhu BZ, Zhu HL, Jiang WB, Luo YB (2005) Virus-induced gene silencing in tomato fruit. Plant J 43:299–308
10. Zhu HL, Zhu BZ, Shao Y, Wang XG, Lin XJ, Xie YH, Li YC, Gao HY, Luo YB (2006) Tomato fruit development and ripening are altered by the silencing of *LeEIN2* gene. J Integr Plant Biol 48:1478–1485
11. Chen J, Jiang C, Gookin T, Hunter D, Clark D, Reid M (2004) Chalcone synthase as a reporter in virus-induced gene silencing studies of flower senescence. Plant Mol Biol 55:521–530
12. Gao Y, Liu C, Li X, Xu H, Liang Y, Ma N, Fei Z, Gao J, Jiang CZ, Ma C (2016) Transcriptome profiling of petal abscission zone and functional analysis of an Aux/IAA family gene RhIAA16 involved in petal shedding in rose. Front Plant Sci 7:1375
13. Sassa H, Nishio T, Kowyama Y, Hirano H, Koba T, Ikehashi H (1996) Self-incompatibility (S) alleles of the Rosaceae encode members of a distinct class of the T 2/S ribonuclease superfamily. Mol Gen Genet 250:547–557
14. Serge G (2001) Rose breeding technologies. Acta Hortic 547:23–26
15. Dai F, Zhang C, Jiang X, Kang M, Yin X, Lü P, Zhang X, Zheng Y, Gao J (2012) RhNAC2 and RhEXPA4 are involved in the regulation of dehydration tolerance during the expansion of rose petals. Plant Physiol 160:2064–2082
16. Luo J, Ma N, Pei H, Chen J, Li J, Gao J (2013) A DELLA gene, *RhGAI1*, is a direct target of EIN3 and mediates ethylene-regulated rose petal cell expansion via repressing the expression of *RhCesA2*. J Exp Bot 64:5075–5084
17. Lü P, Zhang C, Liu J, Liu X, Jiang G, Jiang X, Khan MA, Wang L, Hong B, Gao J (2014) RhHB1 mediates the antagonism of gibberellins to ABA and ethylene during rose (*Rosa hybrida*) petal senescence. Plant J 78:578–590

18. Yan HJ, Shi S, Ma N, Cao X, Zhang H, Qiu X, Wang Q, Jian H, Zhou N, Zhang Z, Tang KX (2018) Graft-accelerated virus-induced gene silencing facilitates functional genomics in rose flowers. J Integr Plant Biol 60:34–44
19. Yan H, Baudino S, Caissard JC, Nicolè F, Zhang H, Tang K, Li S, Lu S (2018) Functional characterization of the eugenol synthase gene in rose. Plant Physiol Biochem 129:21–26
20. Raymond O, Gouzy J, Just J, Badouin H, Verdenaud M, Lemainque A, Vergne P, Moja S, Choisne N, Pont C, Carrère S, Caissard JC, Couloux A, Cottret L, Aury JM, Szecsi J, Latrasse D, Madoui MA, François L, Fu XP, Yang SH, Dubois A, Piola A, Larrieu A, Perez M, Labadie K, Perrier L, Govetto B, Labrousse Y, Villand P, Bardoux C, Boltz V, Lopez-Roques C, Heitzler P, Vernoux T, Vandenbussche M, Quesneville H, Boualem A, Bendahmane A, Liu C, Le Bris M, Salse J, Baudino S, Benhamed M, Wincker P, Bendahmane M (2018) The *Rosa* genome provides new insights into the domestication of modern roses. Nat Genet 50:772–777
21. Hibrand Saint-Oyant L, Ruttink T, Hamama L, Kirov I, Lakhwani D, Zhou NN, Bourke PM, Daccord N, Leus L, Schulz D, Van de Geest H, Hesselink T, Van Laere K, Debray K, Balzergue S, Thouroude T, Chastellier A, Jeauffre J, Voisine L, Gaillard S, Borm TJA, Arens P, Voorrips RE, Maliepaard C, Neu E, Linde M, Le Paslier MC, Bérard A, Bounon R, Clotault J, Choisne N, Quesneville H, Kawamura K, Aubourg S, Sakr S, Smulders MJM, Schijlen E, Bucher E, Debener T, De Riek J, Foucher F (2018) A high-quality genome sequence of *Rosa chinensis* to elucidate ornamental traits. Nat Plants 4:473–484
22. Fernandez-Pozo N, Rosli HG, Martin GB, Mueller LA (2015) The SGN VIGS tool: user-friendly software to design virus-induced gene silencing (VIGS) constructs for functional genomics. Mol Plant 8:486–488
23. Magnard JL, Roccia A, Caissard JC, Vergne P, Sun P, Hecquet R, Dubois A, Hibrand-Saint Oyant L, Jullien F, Nicolè F (2015) Biosynthesis of monoterpene scent compounds in roses. Science 349:81–83
24. Dong Y, Burch-Smith TM, Liu Y, Mamillapalli P, Dinesh-Kumar SP (2007) A ligation-independent cloning tobacco rattle virus vector for high-throughput virus-induced gene silencing identifies roles for NbMADS4-1 and -2 in floral development. Plant Physiol 145:1161–1170

Chapter 17

Virus-Induced Gene Silencing (VIGS) in Flax (*Linum usitatissimum* L.) Seed Coat: Description of an Effective Procedure Using the *transparent testa 2* Gene as a Selectable Marker

Christophe Hano, Samantha Drouet, and Eric Lainé

Abstract

Virus-induced gene silencing (VIGS) has been successfully applied for functional analysis of genes in many plant species. Many protocols have been established but mainly for gene expression study in vegetative tissue. Here, we present the critical steps of an optimized procedure of VIGS in flax (*Linum usitatissimum* L.) seed coat using the *transparent testa 2* gene as a selectable marker. The present protocol may serve as an effective tool for functional characterization of genes involved in seed coat formation and/or biological functions.

Key words Flax, Flavonoids, Functional genetic, Seed coat, Virus-induced gene silencing

1 Introduction

Seed biology is now quite well understood and characterized in many models as well as crop plant species. However, there are still many unclear fundamental questions about the development of the integument (aka seed coat), a maternal tissue surrounding the embryo and albumen [1]. This tissue is recognized to play essential roles such as the seed protection against biotic and abiotic stress, the seed dispersal strategies, and the seed longevity and persistence in the soil and its dormancy or germination [1]. The seed coat is an important source of valuable compounds such as flavonoids, lignin, lignans, polysaccharides, lipid polyesters, and waxes that could find many applications in the industry but are still underexploited [1–3]. It is also the site of accumulation of anti-nutritional factors that would require modifications to improve the quality of meals [1, 4]. Surprisingly, few targeted developments have been made to improve the nutritional and health benefits associated with their

Vincent Courdavault and Sébastien Besseau (eds.), *Virus-Induced Gene Silencing in Plants: Methods and Protocols*, Methods in Molecular Biology, vol. 2172, https://doi.org/10.1007/978-1-0716-0751-0_17,

consumption. This could be explained by our still too fragmentary current knowledge of the seed coat biology, in comparison with our understanding of the accumulation processes of important reserves like fatty acids or proteins in the embryo. Indeed, most of this knowledge comes from functional genetics data obtained using model plants such as *Arabidopsis thaliana* and still too few coming from crops [1, 4]. This is the case of lignans, specialized metabolites with an undeniable high potential for the pharmaceutical or cosmetics, accumulated in the seed coat of flax, because *A. thaliana* is not prone to produce these compounds [3, 4]. In addition to lignans, flax seed coat also accumulates a wide range of attractive compounds including mucilage, flavonoids, and hydroxycinnamic acids [5]. However, functional genomics in flax genes involved in seed(coat) development is quite challenging, since stable transformation using *Agrobacterium tumefaciens* is a long process with a low transformation efficiency [6]. Therefore, virus-induced gene silencing (VIGS) could represent an interesting alternative. This technique has been successfully employed for gene manipulation in many crop species on a wide range of plant tissues including seeds [7]. Recently, in stem tissue of flax, susceptible to *Tobacco rattle virus* (TRV), the TRV-derived viral vectors, TRV1 and TRV2, were used to silence cellulose synthase genes [8]. Knockdown of genes leading to coloration changes such as the *phytoene desaturase* (*PDS*) for vegetative parts or *chalcone synthase* (CHS) for reproductive parts (flowers or seeds) provides a robust means to observe the gene silencing response [7]. In *A. thaliana*, the brown coloration of the seed coat is controlled by the specific cell expression of late biosynthetic genes involved in flavonoid biosynthesis leading to a subsequent accumulation of proanthocyanidins essentially determined by the R2R3 MYB transcription factor *transparent testa 2* (*TT2*) [9]. For this reason, we have developed a VIGS protocol to silence genes involved in seed coat development using the TRV-derived viral vectors and used *TT2* as a reporter gene to observe the gene silencing response. This protocol is based on the exogenous inoculation of maturing flaxseed bolls with their peduncle soaking in a sap extract from TRV1- and TRV2-*TT2*-transformed tobacco leaves.

2 Materials

2.1 Chemicals, Media, and Apparatus

All the chemicals used were of analytical or molecular biology grade. Molecular biology grade double distilled water was used for the preparation of buffers and media.

1. 1 M acetosyringone stock solution in DMSO.
2. Antibiotic stock solutions: kanamycin, rifampicin, and gentamicin are prepared separately at 50 mg/mL in water, dispensed

into 1 mL aliquot, and stored at −20 °C. Dispense this stock solution into 1 mL aliquots and store at −20 °C.

3. 50 mM sodium phosphate buffer, pH 7.2: autoclaved, store at room temperature and use aseptically under laminar flow hood.
4. YEB medium: 5 g/L beef extract, 1 g/L yeast extract, 5 g/L peptone, 5 g/L sucrose, and 0.5 g/L $MgCl_2$. The pH was adjusted with 5 M NaOH to 7.0 before autoclaving (20 min, 121 °C). For solid medium 15 g/L of agar was added.
5. Agroinduction media: YEB medium containing 10 mM MES, 10 mM $MgCl_2$, and 200 μM acetosyringone.
6. Digital shaking water bath.
7. UV-visible spectrophotometer.
8. Centrifuge for 15 and 50 mL conical centrifuge tubes.
9. Microcentrifuge for 1.5 and 2.0 mL microtubes.
10. Assay 96-well microplates in polystyrene.
11. Tissue culture-treated 6-well microplates.
12. Mortar and pestle.

2.2 Plant Materials and Growth Conditions

1. Flax (*Linum usitatissimum* L.) seeds cultivar 'Barbara'.
2. Greenhouse at 25 °C under a 16 h photoperiod, 30 μmol/m^2/s photosynthetically active radiation conditions [10].
3. 7 × 7 cm and 8 cm depth pots.
4. Commercial garden soil (N, 250 g/m^3; P_2O_5, 120 g/m^3; K_2O, 80 g/m^3; 37% of dry matter; 65% of organic matter; pH, 6.2; conductivity, 49 mS/cm; water retention capacity, 70% vol.).
5. Tobacco (*Nicotiana tabacum* L.) in vitro plant line PDB6.
6. Tobacco culture medium: to prepare 1 L of medium, mix 3.24 g of Gamborg B5 [11] ready to use powder containing basal media with Morel and Wetmore vitamins [12], 30 g sucrose, and 2.8 g Phytagel. The pH was adjusted with 1 M KOH to 5.7 before autoclaving (20 min, 121 °C).
7. Growth chamber at 25 °C under a under a 16 h photoperiod, 150 μmol/m^2/s of photosynthetically active radiation.

2.3 Vectors and Bacteria

1. TRV-based vectors pTRV1 (pYL192; GenBank accession AF406990) and pTRV2 (pYL156; GenBank accession AF406991) [13] (*see* **Note 1**).
2. pTRV2-*LuTT2*: the last 300 bp in 3′ of TT2 (Phytozome accession *Lus10033438*) was cloned into pTRV2 using *Xba*I and *Bam*HI restriction sites (*see* **Note 1**).
3. *Agrobacterium tumefaciens* strain GV3101 (C58C1) containing the disarmed plasmid pGV2260 [14] (*see* **Note 2**).

3 Methods

Unless specified all procedures are carried out at room temperature. A simplified complete workflow with the timing of the whole protocol is presented in Fig. 1.

3.1 Preparation of Tobacco Sap Extract

1. Start pre-cultures of the TRV1, TRV2 (*see* **Note 3**), and TRV2-*LuTT2 A. tumefaciens* strains from glycerol stocks or isolated single colonies in culture tubes containing 3 mL of selective YEB medium containing 50 μg/mL kanamycin, 30 μg/mL

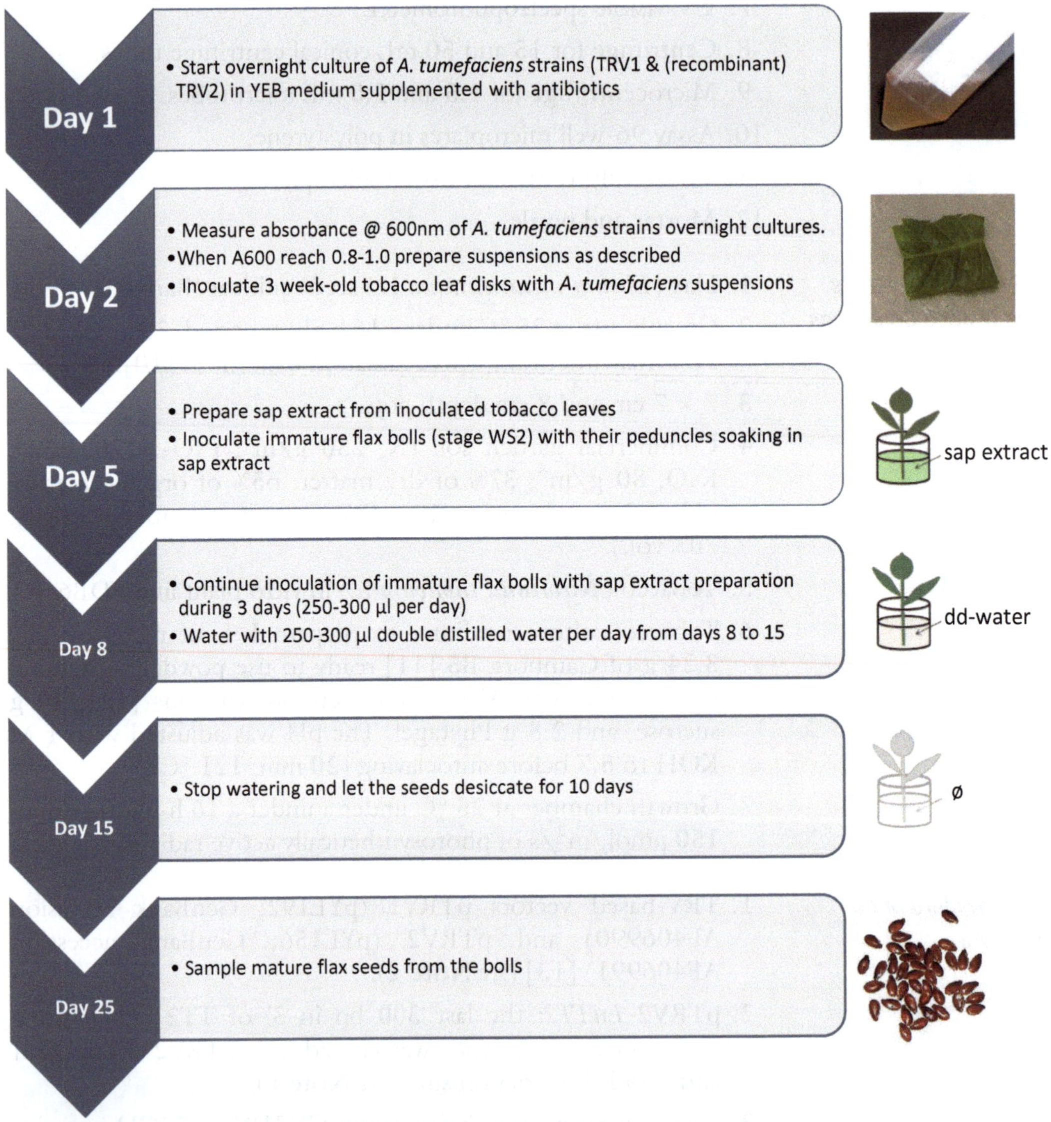

Fig. 1 Workflow of the optimized procedure (with timing) used for VIGS in flax seed coat

rifampicin, and 50 μg/mL gentamicin for 16 h at 28 °C in agitated (180 rpm) water bath (*see* **Note 3**).

2. Dilute these overnight pre-cultures ten times by mixing 1 mL of pre-culture with 9 mL of fresh YEB medium containing 20 μM acetosyringone (*see* **Note 4**), and let these bacterial suspensions grow to an OD_{600} of 0.8–1.0 at 28 °C in agitated (180 rpm) water bath.
3. Collect the bacterial cells by centrifugation at 4 °C at 4000 × *g*, and resuspend the *A. tumefaciens* inoculum in agroinduction media to an OD_{600} of 2.0.
4. Just prior to inoculation, prepare the *A. tumefaciens* inoculums by mixing the TRV1 cultures with TRV2 (or TRV2-*LuTT2*) cultures in a 1:1 ratio. Note that different controls can also be prepared (*see* **Note 5**).
5. Tobacco (*Nicotiana tabacum* L.) PDB6 line was used to produce sap containing recombinant viral particles from TRV1 and TRV2-TT2 vectors following *Agrobacterium tumefaciens* transformation. In vitro plants were maintained in tobacco culture medium and grown in a chamber at 25 °C.
6. 1 cm^2 leaf disks of tobacco cutted with a scalpel under sterile laminar flow hood are immersed in 2 mL of the freshly prepared *A. tumefaciens* inoculum in 6-well plate and incubated in the dark at 25 °C during 3 h (*see* **Note 6**).
7. Slightly dry each leaf disk on sterile absorbent paper and plate it on non-selective solid B5 medium. Incubate for 2 days at 25 °C in a growth chamber.
8. Grind one leaf disk in 3 mL of ice-chilled 50 mM phosphate buffer, pH 7.2, using mortar and pestle cooled on ice. Centrifuge at 4 °C at 4000 × *g* to collect cell debris. Use the supernatant as tobacco sap extract. Store it at 4 °C in the dark for the 3 next days.

3.2 Procedure for VIGS in Flax Seed Coat

1. The plantlets resulting from seedlings were grown in pots (one plantlet resulting from one seed per pot), filled with a commercial garden soil in a growth chamber at 25 °C under a 16 h photoperiod. Plants were irrigated using overhead mist irrigation once a day and one complete watering per week.
2. Plants were cultivated until seed bolls appeared (ca 2.5 months after seedling under greenhouse conditions). Seed bolls (containing seeds) at developmental stage WS2 were used for VIGS by recombinant tobacco sap inoculation. The seed development stages are illustrated in Fig. 2a, b [15].
3. Inoculate with 150 μl of freshly prepared tobacco sap extract, in one well of 96-well microplates, capsules containing immature developing flax seeds (development stage WS2) with a

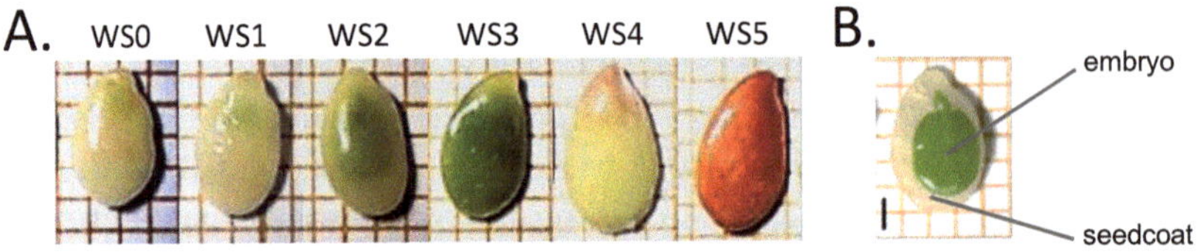

Fig. 2 Representative morphology of the flax seeds at different developmental stages [11] (**a**) and morphology of dissected seed coat and embryo at developmental stage 2 (WS2) used for VIGS experiments (**b**). Square represents 1 mm^2

Fig. 3 Picture (**a**) and schematic representation (**b**) of the experimental procedure used for VIGS in flax seed coat

remaining peduncle (around 1 cm) soaking directly in the tobacco sap extract as shown in Fig. 3 (*see* **Note 7**).

4. Continue inoculation following this procedure for 3 days in growth chamber (*see* **Note 8**). Tobacco sap extract is regularly added in the microplate to maintain peduncle in solution.
5. Continue the same procedure by replacing the tobacco sap extract with double distilled water during 8 days (*see* **Note 8**).
6. Stop watering and let the capsule and the seeds desiccate during 10 days. Dissect the capsules from each condition and collect the seeds for analysis.

3.3 Visualization of the VIGS in Flax Seed Coat

Here the advantage in using *transparent testa 2* gene as a selectable marker is to directly visualize the effective gene silencing as a result of testa depigmentation due to the disruption of anthocyanin production in this tissue (Fig. 4). With this protocol around 35% of silencing efficiency was observed (*see* **Note 9**).

4 Notes

1. An optimal length for VIGS is reported between 300 and 1500 bp for the fragment. To limit off-target gene, analyzing gene structure and excluding conserved regions (except if silencing of a whole gene family is expected) is a critical preliminary step. Identification of potential off-target genes can be identified using publicly available software such as http://

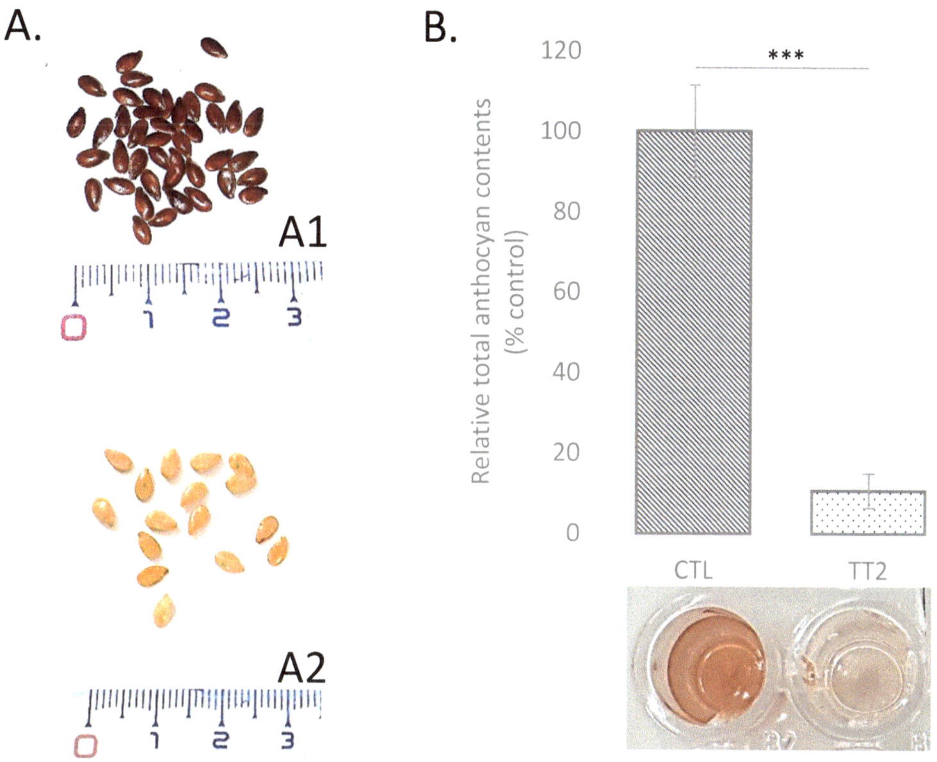

Fig. 4 Impact of *TT2*-VIGS in flax seed coat. (**a**) Representative morphology of the flax seed coats of control (**a1**) and *TT2* silenced (**a2**) using the optimized VIGS procedure. (**b**) Relative total anthocyanin contents in the control and *TT2* silenced flax seed coat extracts quantified spectrometrically [15], $***p < 0.001$

bioinfo2.noble.org/RNAiScan.htm and https://vigs.noble.org/blast/blast.php.

Here, for visualization of gene silencing in flax seed coat, the last 300 bp in 3′ of TT2 was used in order to exclude possible silencing of other MYB TFs (the R2R3 domain encoding region is located in the 5′ part of the gene). The TRV2-LuTT2 construct was done according to the strategy described in [16] using primers including the *Xba*I and *Bam*HI restriction sites (underlined) in their 5′ ends (forward primer: 5′-TGA<u>TCTAGA</u>ACCTCATCGTCTAACGAG-3′; reverse primer: 5′- TGA<u>GGATCC</u>CGCCTGATCAATGAC-3′. Note that the addition of three nucleotides upstream of the restriction sites is recommend in order to facilitate enzymatic digestion of the PCR-generated fragments.

2. The vector pGV2260 [16] carries the virulence genes (*virA* to *virG*) involved in the T-DNA transfer into plant cells. It is derived from the octopine Ti plasmid pTiB6S3 which T-DNA has been completely deleted (disarmed plasmid) and replaced by sequences of the plasmid pBR322 providing a gene for resistance to ampicillin and carbenicillin.

3. Agitation is here crucial to ensure good oxygenation of the bacterial culture and therefore an optimal growth of the latter. Note that the transformation efficiency of tobacco leaf disks requires a sufficiently high concentration of bacteria. Here, optimal transformation rate was obtained for an OD_{600} of 2.0 of the *A. tumefaciens* suspension.
4. Addition of acetosyringone significantly increased the virulence of *A. tumefaciens* leading to a higher transformation efficiency. The use of cutted (i.e., wounded) leaf disks further increases the *A. tumefaciens* transformation efficiency. Note that a short sonication treatment of the plant explants could also result in wounding of the tissue and increase the transformation rate. Nevertheless, if sonication treatment is chosen, a particular attention has to be paid to avoid killing the *A. tumefaciens* cells. Optimization using design of experiment approach could be used to identify the optimal conditions for each plant species.
5. Different transformation controls could be used: e.g., mock inoculated (i.e., agroinduction media without *A. tumefaciens*), empty vector pTRV2 with or without pTRV1, recombinant pTRV2 (e.g., here pTRV2-*TT2*) without pTRV1, etc.
6. Several parameters such as temperature or duration of the co-culture as well as the presence or absence of light during this co-culture have a great influence on the transformation efficiency. The conditions presented here are the optimal conditions for efficient VIGS in flax seed coat, but these conditions may vary from one plant to another; therefore we strongly recommend to evaluate these different parameters before starting a more complete analysis.
7. The secondary inflorescence stems of flax plants (development stage WS2) were harvested and immediately placed in deionized water. Capsules with a remaining peduncle of ca 1 cm were sampled from stems and immediately size sorted. Three fresh capsules were immediately placed in one well of 96-well microplates containing 150 μL of tobacco sap extract.

   The developmental stage (of flax capsule) was decisive for optimal results. In flax, WS2 was the best developmental stage to go up to mature seeds using the present protocol.
8. Around 250–300 μL per day are necessary for developing flax capsules. It is therefore critical to carefully follow the experiment in order to avoid drying the peduncles which would result in water stress causing the start of the desiccation phase for the seeds and cutting off the communication link between the seeds (through the capsule structure) and the mother plant (here the peduncle soaking in the inoculum or water). The incubation time with tobacco sap extract is crucial. Note that,

here, more than 3 days of incubation with the tobacco sap extract led to a significantly higher abortion level, whereas less than 3 days led to lower transformation efficiency. Watering phase was also critical to complete seed filling before desiccation and obtain high level of mature seeds. Lastly, a final period of 10 days without watering was required to obtain desiccated mature seeds.

9. Note that additional molecular or biochemical analyses to confirm the effectiveness of gene silencing and/or its consequence could be performed. As example, further characterization is possible by RT-qPCR to assess decrease in gene expression of the targeted genes. More global analysis like RNAseq could provide more in-depth information about the gene network controlled by TT2 in flax seed coat. It would be of interest to also analyze a possible impact of this transgenesis in embryo, despite VIGS is not supposed to have an impact in progeny.

Acknowledgments

This work was supported by Ligue contre le Cancer and Loire Valley Regional Council.

References

1. Francoz E, Lepiniec L, North HM (2018) Seed coats as an alternative molecular factory: thinking outside the box. Plant Reprod 31:327–342
2. Fliniaux O, Corbin C, Ramsay A et al (2014) Microwave-assisted extraction of herbacetin diglucoside from flax (*Linum usitatissimum* L.) seed cakes and its quantification using an RP-HPLC-UV system. Molecules 19:3025–3037
3. Corbin C, Fidel T, Leclerc EA et al (2015) Development and validation of an efficient ultrasound assisted extraction of phenolic compounds from flax (*Linum usitatissimum* L.) seeds. Ultrason Sonochem 26:176–185
4. Renouard S, Corbin C, Lopez T et al (2012) Abscisic acid regulates pinoresinol-lariciresinol reductase gene expression and secoisolariciresinol accumulation in developing flax (*Linum usitatissimum* L.) seeds. Planta 235:85–98
5. Garros L, Drouet S, Corbin C et al (2018) Insight into the influence of cultivar type, cultivation year, and site on the lignans and related phenolic profiles, and the health-promoting antioxidant potential of flax (*Linum usitatissimum* L.) seeds. Molecules 23:2636
6. Lamblin F, Aimé A, Hano C et al (2007) The use of the phosphomannose isomerase gene as alternative selectable marker for *Agrobacterium*-mediated transformation of flax (*Linum usitatissimum*). Plant Cell Rep 26:765–772
7. Lacomme C (2015) Strategies for alternative plant traits using virus-induced gene silencing technologies. In: Mysore KS, Senthil-Kumar M (eds) Plant gene silencing: methods and protocols, vol 1287. Springer Sciences+Business Media, New York
8. Chantreau M, Chabbert B, Billiard S et al (2015) Functional analyses of cellulose synthase genes in flax (*Linum usitatissimum*) by virus-induced gene silencing. Plant Biotechnol J 13:1312–1324
9. Lepiniec L, Debeaujon I, Routaboul J et al (2006) Genetics and biochemistry of seed flavonoids. Annu Rev Plant Biol 57:405–430
10. Corbin C, Drouet S, Mateljak I et al (2017) Functional characterization of the pinoresinol–lariciresinol reductase-2 gene reveals its roles in yatein biosynthesis and flax defense response. Planta 246:405–420
11. Gamborg OL, Miller RA, Ojima K (1968) Nutrient requirements of suspension cultures of soybean root cells. Exp Cell Res 50:151–158

12. Morel G, Wetmore RH (1951) Fern callus tissue culture. Am J Bot 38:141–143
13. Yoshida K, Ma D, Constabel CP (2015) The MYB182 protein down-regulates proanthocyanidin and anthocyanin biosynthesis in poplar by repressing both structural and regulatory flavonoid genes. Plant Physiol 167:693–710
14. von Lintig J, Kreusch D, Schroder J (1994) Opine-regulated promoters and LysR-type regulators in the nopaline (noc) and octopine (occ) catabolic regions of Ti plasmids of *Agrobacterium tumefaciens.* J Bacteriol 176:495–503
15. Hano C, Martin I, Fliniaux O et al (2006) Pinoresinol-lariciresinol reductase gene expression and secoisolariciresinol diglucoside accumulation in developing flax (*Linum usitatissimum*) seeds. Planta 224:1291–1301
16. Liu Y, Schiff M, Dinesh-Kumar SP et al (2002) Virus-induced gene silencing in tomato. Plant J 31:777–786

Chapter 18

Virus-Based microRNA Silencing in Plants

Jinping Zhao, Guanyu Wang, Haolang Jiang, Tingli Liu, Jingao Dong, Zonghua Wang, Baolong Zhang, and Junqi Song

Abstract

Virus-based microRNA silencing (VbMS) is an efficient, powerful, and high-throughput approach to screen and investigate the function of microRNAs (miRNAs) in plants. The VbMS system was originally developed in *Nicotiana benthamiana* and tomato (*Solanum lycopersicum*) and has been extended to various other plant species such as *Arabidopsis*, cotton, and wheat with different virus vectors. VbMS is generally designed to use virus vectors to direct the expression of miRNA target mimic (TM) molecules which can complementarily pair to target miRNAs and block their function. Here, we describe the TRV- and PVX-based VbMS approaches to silence endogenous miRNAs in *N. benthamiana* and tomato plants by *Agrobacterium* infiltration. This method can be further applied to other plant species using suitable virus vectors in combination with diverse TM strategies, which will facilitate functional studies of miRNAs in plants.

Key words Virus-based microRNA silencing (VbMS), Tobacco rattle virus (TRV), Potato virus X (PVX), microRNA (miRNA), Target mimic (TM)

1 Introduction

Plant microRNAs (miRNAs) are a class of regulatory small non-coding RNAs encoded by the nuclear genome with a typical length of 20–24 nucleotides [1]. miRNAs are well-programmed and involved in the regulation of a broad range of biological processes, including development [2, 3], nutrient and energy regulation [4, 5], environmental stresses [6, 7], hormone responses [8, 9], and plant immunity [10, 11] at posttranscriptional levels by either cleavage or translational repression of target mRNAs. In plants, miRNA-mediated cleavage of target mRNAs normally occurs between the opposite positions corresponding to nucleotides 10–11 of the miRNA within the base-paired binding site [1, 12]. Loss-of-function analysis of individual miRNA genes is challenging, because most miRNAs belong to multigene families and have considerable genetic redundancy [13].

Vincent Courdavault and Sébastien Besseau (eds.), *Virus-Induced Gene Silencing in Plants: Methods and Protocols*, Methods in Molecular Biology, vol. 2172, https://doi.org/10.1007/978-1-0716-0751-0_18,

To characterize the function of miRNAs, two reciprocal approaches are generally used to either enhance or suppress the expression of miRNAs [14, 15]. The artificial miRNA (amiRNA) technology can enhance the activity of miRNA by ectopic expression or overexpression of a mature amiRNA from an artificially designed primary transcription precursor termed pri-amiRNA in which the original miRNA/miRNA* pairing sequences are replaced with the amiRNA/amiRNA* sequences [14, 15]. Reciprocally, miRNA-resistant target blocks miRNA function through expression of an uncleavable mRNA target with mutations in the miRNA binding site [16–20] or the miRNA target mimic (TM) which has a mismatched bulge at the cleavage site [13, 21–24]. The bulge in the TM molecule is uncleavable and thus blocks the function of miRNA, which results in increased expression of the target mRNA [13, 21–23].

To date, multiple TM technologies have been developed to characterize the function of endogenous miRNAs. In Arabidopsis, the *IPS1* (*induced by phosphate starvation 1*) encodes a non-coding RNA which has been proved to be a functional endogenous miRNA target mimicry (eTM) with a 3 nt insertion in the center (between positions 10 and 11) of the miR399 complementary motif [23]. The eTM-based miRNA mimics (MIMs) can block the function of miRNA families by introducing the complementary motif of Ath-miR399 in *IPS1* into the miRNA of interest [21–23]. To inhibit a wide range of miRNA families in different plant species, an approach called short tandem target mimics (STTMs) has been established with the tandem repeated TMs separated by a spacer of 48–108 nt with an imperfect “stem-loop” structure [13, 25–27]. The miRNA sponge technology, which was developed to target all members of a miRNA family in animals, has been adapted to plants [28, 29]. In Arabidopsis, the miRNA sponges with central mismatches (cmSPs) consist of multiple TMs of a miRNA of interest, with each TM unit containing a central mismatch (CM) [30]. In addition, the TM cassette can be integrated either into the 3′ UTR of protein-coding sequences or into the backbone of non-coding RNA, both of which can sequester the activities of miRNAs [31]. Based on the TM technologies, several resources for functional interrogation of miRNAs in Arabidopsis and crop plants have been established. This provides a valuable platform for identifying and constructing miRNA-regulated gene networks in diverse biological processes [25]. However, these TM technologies are dependent on transformation and are labor-intensive and time-consuming, therefore, they are hardly applicable for transform-recalcitrant plant species.

To overcome the requirement of stable transformation, the virus-based miRNA silencing (VbMS) system has been established in plants. By integrating the TM strategies together with plant virus-based vectors, TM molecules can be expressed from a virus backbone to sequester endogenous miRNA. VbMS offers an efficient and powerful tool for high-throughput functional analysis of plant miRNAs [22, 32]. VbMS was first developed in *N. benthamiana* and tomato with tobacco rattle virus (TRV) [22, 32, 33] and has been extended to Arabidopsis, cotton, and wheat with potato virus X (PVX) [27], cotton leaf crumple virus (ClCrV) [34], cucumber mosaic virus (CMV) [35, 36], and barley stripe mosaic virus (BSMV) [37].

Here, we present a powerful and robust VbMS approach to evaluate the function of miRNAs in *N. benthamiana* and tomato with ligation-independent cloning (LIC)-adapted PVX [27] and TRV [22] vectors. We selected miR165/166 as an example to illustrate the VbMS protocol because they belong to a highly conserved miRNA family in plants and have been extensively studied for their roles in plant growth and development [13, 25, 38]. The miRNA165/166 family has a clearly defined set of target mRNAs, which encode class III homeodomain/leu zipper (HD-ZIP III) transcription factors. The *HD-ZIP* III genes regulate meristem development and organ polarity. Silencing of miR165/166 leads to upregulation of the expression of the *HD-ZIP* III genes, resulting in pleiotropic developmental defects manifested by loss of apical dominance and aberrant leaf morphology [13, 22, 27, 36]. We also included protocols to appraise the TM efficacy by real-time RT-PCR analysis of miRNAs and their corresponding target mRNAs. The STTM strategy has been proven as a simple and reliable approach to block the function of miRNAs [13, 22, 27, 39], therefore, we chose STTM as the TM design strategy. Since the TRV- and PVX-based vectors have been widely used in diverse plant species for virus-induced gene silencing (VIGS) and gene expression [40–43], this VbMS approach is also applicable to other plant species.

2 Materials

2.1 Plant Growth

1. *Nicotiana benthamiana*.
2. *Solanum lycopersicum* variety Moneymaker.
3. Soil (e.g., Sun Gro Horticulture, USA).
4. 3.5″ pots, trays, and transparent domes.
5. Greenhouse conditions: 24°C with a16 h light/8 h dark photoperiod and a light intensity of 125 μmol/m^2/s.

2.2 Plasmid Construction

1. DNA polymerase with dNTPs.
2. Restriction enzymes: *Pst*I and *Sma*I.
3. T4 DNA polymerase and buffer.
4. 10 mM dATP and 10 mM dTTP.

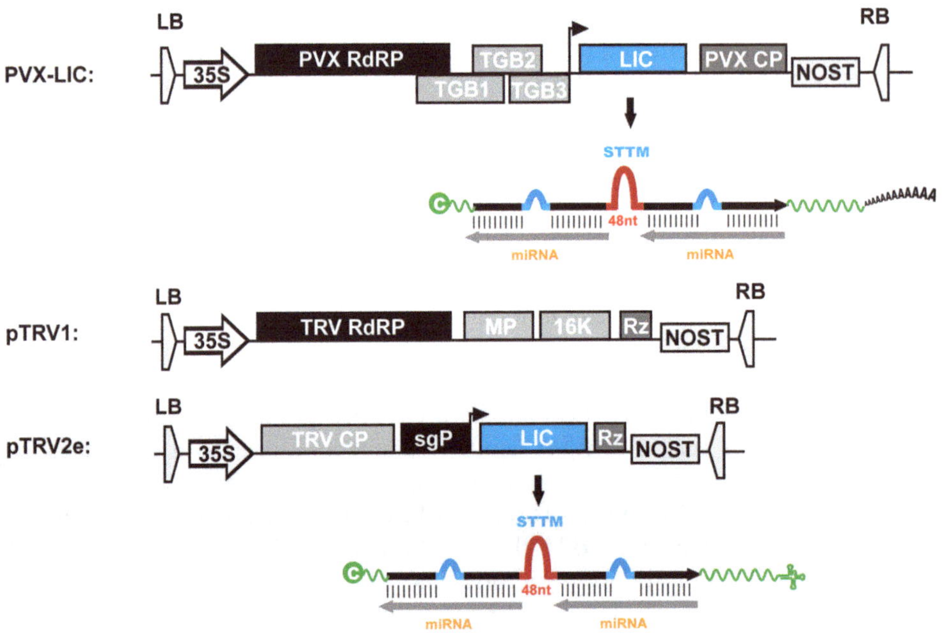

Fig. 1 Schematic diagram of PVX- and TRV-based VbMS vectors. *LB* T-DNA left border, *RB* T-DNA right border, *35S* cauliflower mosaic virus 35S promoter, *NOST* nopaline synthase terminator, *PVX RdRP* PVX RNA-dependent RNA polymerase, *TGB1* PVX triple gene block protein 1, *TGB2* PVX triple gene block protein 2, *TGB3* PVX triple gene block protein 3, *PVX CP* PVX coat protein, *LIC* LIC cassette, *TRV RdRP* TRV RNA-dependent RNA polymerase, *MP* TRV movement protein, *16K* 16-kDa Cys-rich protein, *TRV CP* TRV coat protein, *Rz* self-cleaving ribozyme, *sgP* PEBV CP subgenomic promoter, *48 nt* 48-nucleotide imperfect stem-loop linker

5. 1 M DTT.
6. *E. coli* DB3.1 for maintenance of PVX-LIC and pTRV2e containing the *ccdB* gene (*see* **Note 1**).
7. *E. coli* DH5α for the destination constructs generated by LIC cloning.
8. Ethanol.
9. 3 M sodium acetate, pH 4.
10. 25:24:1 (v/v/v) phenol/chloroform/isoamyl alcohol.
11. 24:1 (v/v) chloroform/isoamyl alcohol.
12. TRV VbMS vectors: pTRV1 (TRV RNA1) [44], pTRV2e (TRV RNA2) with a pea early brown virus (PEBV) subgenomic promoter [45] and a downstream LIC cassette for TM cloning (Fig. 1) [22, 32].
13. PVX VbMS vector: PVX-LIC with a LIC cassette for TM cloning [27].
14. An oligo corresponding to the 48-nt spacer of STTM 5′-GTTGTTGTTGTTATGGTCTAATTTAAATATGGTCTA AAGAAGAAGAAT-3′.

15. Specific miRNA-targeted STTM forward 5′-CgACgACAAgACCgTxxxxxxxxxxx**XXX**xxxxxxxxxxxGTTGTTGTTGTTATGGT-3′ and reverse 5′-gAggAgAagAgCCgTxxxxxxxxxx**XXX**xxxxxxxxxxxATTCTTCTTCTTTAGACCAT-3′ primers as design in Subheading 3.2.
16. LB liquid medium and LB plates containing antibiotics.
17. Antibiotics: 50 mg/mL chloramphenicol stock solution in 90% ethanol, 50 mg/mL gentamicin stock solution in water, 50 mg/mL kanamycin stock solution in water, and 50 mg/mL rifampicin stock solution in DMSO.
18. PCR machine.
19. Centrifuge.
20. Shaker and incubator.

2.3 Agrobacterium Infiltration

1. *Agrobacterium tumefaciens* strains: GV3101, GV2260, or EHA105.
2. Infiltration buffer: 10 mM $MgCl_2$, 10 mM MES pH 5.6, and 200 mM acetosyringone.
3. Dimethyl sulfoxide, DMSO.
4. Sterile 1 mL needleless syringe.
5. Sterile culture tubes.
6. Incubator, shaker, and centrifuge.
7. pH meter.
8. Spectrometer.

2.4 Expression Analysis

1. Total RNA extraction reagent (e.g., RNAzol RT, Sigma-Aldrich, USA).
2. Reverse transcription kit containing reverse transcriptase, oligo dT, and RNase inhibitor.
3. miRNA-specific reverse transcription primer 5′-GTCTCCTCTGGTGCAGGGTCCGAGGTATTCGCACCAGAGGAGACxxxxxx-3′ as designed in Subheading 3.6.
4. RNase-free water.
5. miRNA-specific forward primer for qPCR and universal qPCR miRNA reverse primer 5′-GTGCAGGGTCCGAGGT-3′.
6. Target gene primers for qPCR.
7. Reference gene primers (*see* **Note 2**).
8. Real-time PCR reagent: 2× SYBR Green Master Mix.
9. 10 mM dNTPs.
10. Real-time PCR plates and optical adhesive films.
11. Scale.

12. Pestles and mortars.
13. Liquid nitrogen.
14. Ethanol.
15. Isopropanol.
16. NanoDrop spectrophotometer.
17. Microcentrifuge.
18. PCR machine.
19. Real-time PCR machine.
20. Ultralow freezer (−80°C).

3 Methods

3.1 Plant Growth

1. Fill the 3.5″ pots with soil and soak the soil with water in a tray.
2. Sow seeds into soil and cover the tray with a dome to keep moisture.
3. Place the tray in a growth room for up to 10 days to allow seeds to germinate and seedlings to grow.
4. Transplant seedlings into 3.5″ pots filled with soil (*see* **Note 3**).
5. Maintain moisture for 1–2 days with a dome.
6. For *N. benthamiana*, allow seedlings to grow for an additional 2–3 weeks before VbMS. For tomato, 10-day-old seedlings with fully expanded cotyledons are subjected to VbMS.

3.2 STTM Primer Design

1. Plant miRNA sequences can be acquired from the miRBase database (http://www.mirbase.org/). The Nbe-miR166 sequence used in this study has been described previously [46].
2. Acquire the reverse complement sequence of the miRNA mature form, which represents the binding site to its corresponding target mRNA. The RNA sequence of the miR166 mature form is 5′-UCGGACCAGGCUUC AUUCCCC-3′ and the reverse complement DNA sequence of miR166 is 5′-GGGGAATGAAGCCTGGTCCGA-3′.
3. Design the TM module by inserting a mismatch bulge sequence into the binding sequence between the 10th and 11th corresponding nucleotides of the miRNA. The bulge sequence is a consensus trinucleotide 5′-**CTA**-3′ [23, 38]. The TM sequence for miR166 is 5′-GGGGAATG AAG-**CTA**-CCTGGTCCGA-3′.
4. The 48-nt spacer of STTM is 5′-GTTGTTGTTGTTATGG TCTAATTTAAATATGGTCTAAAGAAGAAGAAT-3′.

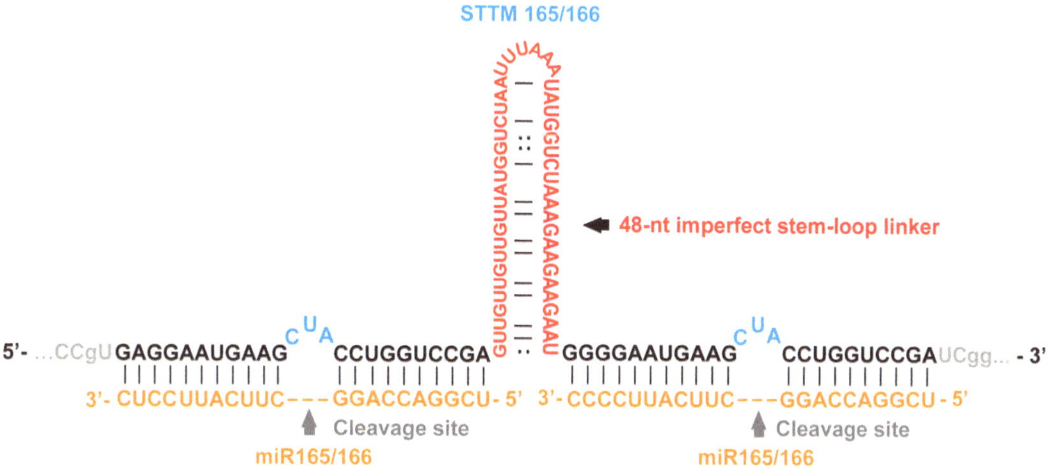

Fig. 2 Schematic diagram of the STTM165/166 structure. STTM165/166 consists of tandemly repeated target mimics of miR165/166 separated by a 48-nucleotide imperfect stem-loop linker (48 nt)

5. The forward primer consists of a LIC1 linker (5′-CgACgACAA gACCgT-3′), a TM sequence for target miRNA, and a partial 5′ sequence of the 48-nt spacer (5′-GTTGTTGTTGTTATGGT-3′). For STTM-miR166, the forward primer is 5′-CgACgA CAAgACCgT-GAGGAATGAAG-**CTA**-CCTGGTCCGA-GTTGTTGTTGTTATGGT-3′.
6. The reverse primer consists of a LIC2 linker (5′-gAggAgAagA gCCgT-3′), a reverse complement sequence of the TM, and a sequence that is reverse complementary to the 3′ region of the 48-nt spacer (5′-ATTCTTCTTCTTTAGACCAT-3′). For STTM-miR166, the reverse primer is 5′-gAggAgAagAgCC gT-TCGGACCAGG-**TAG**-CTTCATTCCCC-ATTCTTC TTCTTTAGACCAT-3′ (Fig. 2).

3.3 Construction of the VbMS Vectors

1. Digest 5 μg of PVX-LIC with *Sma*I or pTRV2e with *Pst*I.
2. To purify the digested PVX-LIC or pTRV2e vectors, add equal volume of phenol/chloroform/isoamyl alcohol (25:24:1) to the reaction, mix vigorously, and centrifuge at 14,000 × *g* for 10 min.
3. Transfer the supernatant to a new centrifuge tube, add equal volume of chloroform/isoamyl alcohol (24:1), vortex vigorously, and centrifuge at 14,000 × *g* for 10 min.
4. Transfer the supernatant to a new centrifuge tube, add 2.5 volumes of ethanol and 1/10 volume of 3 M sodium acetate (pH 4.0), mix vigorously, centrifuge at 14,000 × *g* for 10 min, and remove the supernatant.

Table 1
Treatment with T4 DNA polymerase

Digested plasmids	2.50 μL (~50 ng)	PCR products	2.50 μL (~50 ng)
10× NEBuffer 2.1	0.50 μL	10× NEBuffer 2.1	0.50 μL
100 mM dTTP	0.25 μL	100 mM dATP	0.25 μL
1 M DTT	0.05 μL	1 M DTT	0.05 μL
T4 DNA polymerase	0.10 μL	T4 DNA polymerase	0.10 μL
ddH_2O	1.60 μL	ddH_2O	1.60 μL
Total	5.00 μL	*Total*	5.00 μL

5. Add 1 mL of 70% ethanol, vortex vigorously, centrifuge at 14,000 × *g* for 10 min, and remove the supernatant (*see* **Note 4**).
6. Dissolve the digested plasmid with 100 μL of ddH_2O.
7. Amplify the specific STTM fragment in a total volume of 50 μL by PCR using DNA polymerase, the synthesized 48-nt DNA oligo template, and the specific forward and reverse STTM primers designed previously.
8. Add 2.5 volumes of ethanol and 1/10 volume of 3 M sodium acetate (pH 4.0) to the PCR products, mix vigorously, centrifuge at 14,000× *g* for 10 min, and remove the supernatant (*see* **Note 5**).
9. Add 1 mL of 70% ethanol, vortex vigorously, centrifuge at 14,000× *g* for 10 min, and remove the supernatant.
10. Dry the pellet and dissolve it in 20 μL of ddH_2O.
11. Set up the T4 DNA polymerase reaction in a final volume of 5 μL on ice by mixing 2.5 μL of DNA (~50 ng), 0.5 μL of 10× buffer, 0.05 μL of 1 M DTT, 0.25 μL of 100 mM dTTP or dATP, and 0.1 μL of T4 DNA polymerase (3 U/μL). Use dTTP for digested plasmids and dATP for PCR products (Table 1).
12. Incubate the mixture at 37°C for 15 min followed by heat treatment at 75°C for 20 min to inactivate T4 DNA polymerase.
13. For LIC cloning, mix 5 μL of the T4 DNA polymerase-treated PVX-LIC (or pTRV2e) plasmids with 5 μL of treated PCR products. Incubate at 70°C for 5 min, cool down to 22°C with a ramp at 0.1°C/s, and keep at 22°C for 30 min in a PCR machine.

14. Transform 5 μL of the LIC reaction products into the *E. coli* strain DH5α, plate cells on a LB plate containing 50 μg/mL kanamycin, and incubate overnight.
15. Pick up colonies and verify by PCR and sequencing (*see* **Note 6**).
16. Isolate plasmids from the positive clones for subsequent *Agrobacterium* transformation.

3.4 VbMS Assay with PVX on N. benthamiana Plants

1. Transform the PVX-STTM plasmids into *Agrobacterium* strain GV3101, GV2260, or EHA105 (*see* **Note 7**).
2. Spread the transformed cells on a LB plate containing 50 μg/mL kanamycin and 50 μg/mL rifampicin at 28°C for 2–3 days.
3. Verify colonies by PCR.
4. Inoculate 5 mL of liquid LB medium containing the same antibiotics as described above with a positive clone containing the PVX-STTM plasmid. Grow the culture in a shaker with 220 rpm at 28°C for ~16 h to reach an OD_{600} of 1.0.
5. Collect *Agrobacterium* cells by centrifuging at 3400× *g* for 10 min at room temperature.
6. Resuspend *Agrobacterium* pellet in 5 mL of infiltration buffer.
7. Adjust the culture with infiltration buffer to a final OD_{600} of 1.0 and incubate at room temperature for 2.5–4 h.
8. Select 2–3-week-old *N. benthamiana* plants and infiltrate *Agrobacterium* culture into the abaxial side of fully expanded leaves with 1 mL needleless syringe.
9. Grow infiltrated plants in a growth room at 24°C with a 16 h light/8 h dark photoperiod and a light intensity of 120 μmol/m^2/s.
10. Phenotypes associated with miRNA silencing typically appear in 2–4 weeks post-inoculation in the newly generated tissues.
11. Photograph plants with visible phenotypes.
12. Collect tissues and store them in an ultralow freezer at −80°C for expression analysis.

3.5 VbMS Assay with TRV on Tomato Plants

1. Transform pTRV1 and the pTRV2e-STTM plasmids separately into *Agrobacterium* strain GV3101, GV2260, or EHA105 (*see* **Note 7**).
2. Spread transformed cells on a LB plate containing 50 μg/mL kanamycin and 50 μg/mL rifampicin and incubate at 28°C for 2–3 days.
3. Verify colonies by PCR.
4. Inoculate 5 mL of liquid LB medium containing the same antibiotics as described above with a positive clone. Grow the

culture in a shaker with 220 rpm at 28°C for ~16 h to reach an OD_{600} of 1.0.

5. Collect the *Agrobacterium* cells carrying the pTRV1 or pTRV2e-STTM plasmids by centrifugation at 3400× *g* for 10 min at room temperature.
6. Resuspend *Agrobacterium* pellet with 5 mL of infiltration buffer and adjust the culture with infiltration buffer to an OD_{600} of 1.0.
7. Mix the *Agrobacterium* cultures carrying pTRV1 or pTRV2e-STTM in a 1:1 ratio and incubate at room temperature for 6 h.
8. Select tomato plants with fully expanded cotyledons for VbMS assay. Infiltrate the *Agrobacterium* culture into the abaxial side of cotyledons with a 1 mL needleless syringe (*see* **Note 8**).
9. Place infiltrated plants into a tray and cover them with a dome for 24 h to maintain moisture.
10. Grow the infiltrated plants in a growth room at 22°C with a 16 h light/8 h dark photoperiod and a light intensity of 120 μmol/m^2/s.
11. Phenotypes associated with miRNA silencing typically appear in 2–4 weeks post-inoculation in the newly generated tissues.
12. Photograph plants with visible phenotypes.
13. Collect tissues and store them in an ultralow freezer at −80°C for expression analysis.

3.6 Expression Analysis

1. Isolate total RNA from collected tissues using a RNA extraction reagent according to the manufacturer's instructions (*see* **Note 9**).
2. Quantify the RNA samples with a NanoDrop spectrophotometer and check RNA quality by electrophoresis.
3. Design the stem-loop reverse transcription primer for mature miRNA according to Chen et al. [47] and Varkonyi-Gasic et al. [48–50] (*see* **Note 10**). The stem-loop reverse transcription primers contain a universal 5′ backbone and a 3′ six-nucleotide extension of a miRNA of interest. The backbone 5′-GTCTCCTCTG<u>GTGCagggtccgaggt</u>attcGCACCAGAGG AGAC-3′ is designed to form a stem-loop structure (the upper case corresponds to inverted repeat sequences and the lower case the loop), part of which can serve as the reverse primer for PCR amplification (sequence underlined). The specificity of a stem-loop RT primer to an individual miRNA is provided by a six-nucleotide extension which is a reverse complement to the 3′ end of the miRNA (bold characters in the following sequence). For miR166 family, the stem-loop reverse

transcription primer is 5′-GTCTCCTCTGGTGCagggtcc-gaggtattcGCACCAGAGGAGAC**GGGG(A/G)A**-3′.

4. Set up the reverse transcription assays for miRNAs with the stem-loop reverse transcription primer and for mRNAs with a universal primer, respectively (*see* **Note 11**). For each reverse transcription reaction, mix 50–1000 ng of total RNA with 1 μL of reverse transcription primer (100 μM), incubate at 65°C for 10 min, immediately place the mixture onto ice, and keep on ice for 5 min. Add 2 μL of 10× buffer, 0.2 μL RNase inhibitor (40 U/μL), 0.25 μL dNTPs (10 mM Each), and 1 μL of reverse transcriptase (200 U/μL). Adjust the RNA-primer mixture with nuclease-free ddH_2O to a total volume of 20 μL.
5. Perform reverse transcription. For miRNA, use pulsed reverse transcription procedure. Incubate the reaction mixture at 16°C for 30 min, perform a temperature cycle of 30°C for 30 s, 42°C for 30 s, and 50°C for 1 s, for a total of 60 cycles, and heat to 85°C for 5 min to inactivate the reverse transcriptase. For mRNA, follow the standard reverse transcription procedure. Incubate the reaction mixture at 37°C for 60 min and heat to 85°C for 5 min to inactivate the reverse transcriptase.
6. For real-time RT-PCR analysis of miRNA expression, design the forward primer based on the sequence of miRNA, followed by adding a 5′ extension of 3–7 nucleotides to optimize the length, melting temperature, and GC content (*see* **Note 10**). The forward primer should not include sequences that overlap with the stem-loop reverse transcription primer. For miR166 family, the forward primer is 5′-CGGC**TCGGACCAGGC TT**-3′ and the universal reverse primer is 5′-GTGCAGGGTCC GAGGT-3′.
7. For mRNAs, predict target using the psRNATarget program (http://plantgrn.noble.org/psRNATarget/) [51] and design primers for target mRNAs with the Primer3Plus software (https://primer3plus.com/cgi-bin/dev/primer3plus.cgi) [52]. *NbActin1* and *SlTubulin* genes serve as internal controls for normalization in *N. benthamiana* and tomato, respectively (*see* **Note 2**).
8. Set up the real-time PCR reaction with the 2× SYBR Master Mix. The PCR program is as follows: 95°C for 3 min, 40 cycles of 95°C for 3 s, and 60°C for 30 s. The melting curve test is as follows: 95°C for 15 s, cool down to 60°C with 20°C/s, incubate at 60°C for 1 min, heat to 95°C with a ramp of 0.2°C/s, and keep at 95°C for 15 s (*see* **Note 12**). The Ct values are analyzed using the ΔΔCt methods.

4 Notes

1. Both the pTRV2e and PVX-LIC vectors harbor the LIC cassette for TM cloning. The LIC cassette of PVX-LIC contains a *ccdB* gene and a chloramphenicol-resistant gene, while the LIC cassette of pTRV2e contains only the *ccdB* gene. The two vectors need to be maintained and propagated in the *E. coli* strain DB3.1.
2. For *NbActin1*, the forward primer is 5′-CGCACCACCAGAG AGGAAAT-3′ and the reverse primer 5′-TGCTGAGGGAAG CCAAGATAG-3′. For *SlTubulin*, the forward primer is 5- ′-CACATTGGTCAGGCCGGTAT-3′ and the reverse primer 5′-ATCTGGCCATCAGGCTGAAT-3′.
3. For each VbMS construct, at least three to six plants need to be infiltrated for consistent and reliable results.
4. *Sma*I digestion of the PVX-LIC vector produces a 2.2-kb LIC insert and a 16.6-kb backbone. *Pst*I digestion of the pTRV2e vector produces a 0.8-kb LIC insert and a 12.3-kb backbone.
5. The average length of amplified STTM fragments is ~120–130 bp. Ethanol precipitation recovers STTM fragments more efficiently than most commercial PCR product purification kits.
6. To identify positive clones, perform colony PCR with the forward primer for PVX-LIC (5′-GTGTTGGCTTGCAAA CTAGAT-3′) or the forward primer for pTRV2e (5′-TGTTA CTCAAGGAAGCACGATGAGCT-3′) in combination with the specific reverse primer for STTM cloning. The sequences of STTM fragments need to be verified by DNA sequencing.
7. It is recommended to include TRV-GFP- or PVX-GFP-infiltrated plants as controls to monitor virus spread.

   The VbMS phenotype typically takes 10–20 days to appear after infiltration. However, the phenotype development depends on the properties of miRNA/target genes, plant growth conditions, and selection of VbMS vectors.
8. Tomato seedlings with fully expanded cotyledons but no true leaves are ideal for infiltration.
9. miRNAs and mRNAs isolated separately are suitable for expression analysis. Total RNA extracts containing both miRNAs and mRNAs can also be used for expression analysis with similar results.
10. The stem-loop reverse transcription primers and real-time PCR primers can also be designed using the miRNA Primer Design Tool (http://genomics.dote.hu:8080/mirnadesigntool/) [53].

11. The universal reverse transcription primer for mRNA is an anchored primer 5′-TTTTTTTTTTTTTTTTTTTTTTTTTTTTTVN-3′.
12. It is recommended to include no template and no reverse transcriptase controls for real-time PCR assays to monitor contamination and primer-dimer formation that may generate false-positive results.

Acknowledgments

We thank Dr. Yule Liu from Tsinghua University for providing the pTRV1 and pTRV2e vectors. This work was supported by a start-up fund from the Texas A&M AgriLife Research and a Hatch Project from USDA National Institute of Food and Agriculture to JS (TEX0-1-9675), a fellowship from China Scholarship Council to GW (201708130105), a fellowship from China Scholarship Council to HJ (201707877008), and an Outstanding Youth Fund of Jiangsu Province (BK20160016) and a fund from the National Natural Science Foundation of China to TL (31671980).

References

1. Axtell MJ, Meyers BC (2018) Revisiting criteria for plant microRNA annotation in the era of big data. Plant Cell 30(2):272–284
2. Chen X (2009) Small RNAs and their roles in plant development. Annu Rev Cell Dev Biol 25 (1):21–44
3. Rubio-Somoza I, Weigel D (2011) MicroRNA networks and developmental plasticity in plants. Trends Plant Sci 16(5):258–264
4. Chiou T-J, Lin S-I (2011) Signaling network in sensing phosphate availability in plants. Annu Rev Plant Biol 62(1):185–206
5. Sunkar R, Chinnusamy V, Zhu J et al (2007) Small RNAs as big players in plant abiotic stress responses and nutrient deprivation. Trends Plant Sci 12(7):301–309
6. Covarrubias AA, Reyes JL (2010) Post-transcriptional gene regulation of salinity and drought responses by plant microRNAs. Plant Cell Environ 33(4):481–489
7. Chinnusamy V, Zhu J, Zhu J-K (2007) Cold stress regulation of gene expression in plants. Trends Plant Sci 12(10):444–451
8. Damodharan S, Corem S, Gupta SK et al (2018) Tuning of SlARF10A dosage by sly-miR160a is critical for auxin-mediated compound leaf and flower development. Plant J 96(4):855–868
9. Nizampatnam NR, Schreier SJ, Damodaran S et al (2015) microRNA160 dictates stage-specific auxin and cytokinin sensitivities and directs soybean nodule development. Plant J 84(1):140–153
10. Wang S, Cui W, Wu X et al (2018) Suppression of nbe-miR166h-p5 attenuates leaf yellowing symptoms of potato virus X on Nicotiana benthamiana and reduces virus accumulation. Mol Plant Pathol 19(11):2384–2396
11. Canto-Pastor A, Santos BAMC, Valli AA et al (2019) Enhanced resistance to bacterial and oomycete pathogens by short tandem target mimic RNAs in tomato. Proc Natl Acad Sci U S A 116(7):2755
12. Li J, Reichel M, Li Y et al (2014) The functional scope of plant microRNA-mediated silencing. Trends Plant Sci 19(12):750–756
13. Yan J, Gu Y, Jia X et al (2012) Effective small RNA destruction by the expression of a short tandem target mimic in *Arabidopsis*. Plant Cell 24(2):415–427
14. Schwab R, Ossowski S, Riester M et al (2006) Highly specific gene silencing by artificial microRNAs in *Arabidopsis*. Plant Cell 18 (5):1121–1133
15. Tang Y, Wang F, Zhao J et al (2010) Virus-based microRNA expression for gene

functional analysis in plants. Plant Physiol 153 (2):632–641
16. Teotia S, Tang G (2015) To bloom or not to bloom: role of microRNAs in plant flowering. Mol Plant 8(3):359–377
17. Wu G, Poethig RS (2006) Temporal regulation of shoot development in *Arabidopsis thaliana* by *miR156* and its target *SPL3*. Development 133(18):3539–3547
18. McConnell JR, Emery J, Eshed Y et al (2001) Role of PHABULOSA and PHAVOLUTA in determining radial patterning in shoots. Nature 411:709
19. McHale NA, Koning RE (2004) MicroRNA-directed cleavage of *Nicotiana sylvestris* PHAVOLUTA mRNA regulates the vascular cambium and structure of apical meristems. Plant Cell 16(7):1730–1740
20. Baulcombe D (2004) RNA silencing in plants. Nature 431(7006):356–363
21. Todesco M, Rubio-Somoza I, Paz-Ares J et al (2010) A collection of target mimics for comprehensive analysis of microRNA function in *Arabidopsis thaliana*. PLoS Genet 6(7): e1001031
22. Sha A, Zhao J, Yin K et al (2014) Virus-based microRNA silencing in plants. Plant Physiol 164(1):36–47
23. Franco-Zorrilla JM, Valli A, Todesco M et al (2007) Target mimicry provides a new mechanism for regulation of microRNA activity. Nat Genet 39(8):1033–1037
24. Reichel M, Li J, Millar AA (2011) Silencing the silencer: strategies to inhibit microRNA activity. Biotechnol Lett 33(7):1285–1292
25. Peng T, Qiao M, Liu H et al (2018) A resource for inactivation of microRNAs using short tandem target mimic technology in model and crop plants. Mol Plant 11(11):1400–1417
26. Tang G, Yan J, Gu Y et al (2012) Construction of short tandem target mimic (STTM) to block the functions of plant and animal microRNAs. Methods 58(2):118–125
27. Zhao J, Liu Q, Hu P et al (2016) An efficient Potato virus X -based microRNA silencing in *Nicotiana benthamiana*. Sci Rep 6:20573
28. Wong G, Millar AA (2019) The use of microRNA decoy technologies to inhibit miRNA function in *Arabidopsis*. In: de Folter S (ed) Plant microRNAs: methods and protocols. Springer New York, New York, pp 227–238
29. Ebert MS, Sharp PA (2010) MicroRNA sponges: progress and possibilities. RNA 16 (11):2043–2050
30. Reichel M, Li Y, Li J et al (2015) Inhibiting plant microRNA activity: molecular SPONGEs, target MIMICs and STTMs all display variable efficacies against target microRNAs. Plant Biotechnol J 13(7):915–926
31. Ivashuta S, Banks IR, Wiggins BE et al (2011) Regulation of gene expression in plants through miRNA inactivation. PLoS One 6(6): e21330
32. Zhao J, Liu Y (2016) Virus-based microRNA silencing. Bio-protocol 6(2):e1714
33. Yan F, Guo W, Wu G et al (2014) A virus-based miRNA suppression (VbMS) system for miRNA loss-of-function analysis in plants. Biotechnol J 9(5):702–708
34. Gu Z, Huang C, Li F et al (2014) A versatile system for functional analysis of genes and microRNAs in cotton. Plant Biotechnol J 12 (5):638–649
35. Du Z, Chen A, Chen W et al (2014) Using a viral vector to reveal the role of microRNA159 in disease symptom induction by a severe strain of cucumber mosaic virus. Plant Physiol 164 (3):1378–1388
36. Liao Q, Tu Y, Carr JP et al (2015) An improved cucumber mosaic virus-based vector for efficient decoying of plant microRNAs. Sci Rep 5:13178
37. Jiao J, Wang Y, Selvaraj JN et al (2015) Barley Stripe Mosaic Virus (BSMV) induced microRNA silencing in common wheat (*Triticum aestivum* L.). PLoS One 10(5):e0126621
38. Teotia S, Zhang D, Tang G (2017) Knockdown of rice microRNA166 by Short Tandem Target Mimic (STTM). In: Kaufmann M, Klinger C, Savelsbergh A (eds) Functional genomics: methods and protocols. Springer New York, New York, pp 337–349
39. Teotia S, Singh D, Tang X et al (2016) Essential RNA-based technologies and their applications in plant functional genomics. Trends Biotechnol 34(2):106–123
40. Dommes AB, Herbert DB, Kivivirta KI et al (2018) Virus-induced gene silencing: empowering genetics in non-model organisms. J Exp Bot 70(3):757–770
41. Lacomme C, Chapman S (2008) Use of Potato Virus X (PVX)-based vectors for gene expression and Virus-Induced Gene Silencing (VIGS). Curr Protoc Microbiol 8 (1):16I.11.11–16I.11.13
42. Lim H-S, Vaira AM, Domier LL et al (2010) Efficiency of VIGS and gene expression in a novel bipartite potexvirus vector delivery system as a function of strength of TGB1 silencing suppression. Virology 402(1):149–163

43. Gleba Y, Klimyuk V, Marillonnet S (2007) Viral vectors for the expression of proteins in plants. Curr Opin Biotechnol 18(2):134–141
44. Liu Y, Schiff M, Marathe R et al (2002) Tobacco Rar1, EDS1 and NPR1/NIM1 like genes are required for N-mediated resistance to tobacco mosaic virus. Plant J 30 (4):415–429
45. Wang D, Macfarlane SA, Maule AJ (1997) Viral determinants of pea early browning virus seed transmission in pea. Virology 234 (1):112–117
46. Yin K, Tang Y, Zhao J (2015) Genome-wide characterization of miRNAs involved in *N* Gene-mediated Immunity in response to tobacco mosaic virus in *Nicotiana benthamiana*. Evol Bioinform Online 11(Suppl 1):1–11
47. Chen C, Ridzon DA, Broomer AJ et al (2005) Real-time quantification of microRNAs by stem-loop RT-PCR. Nucleic Acids Res 33 (20):e179
48. Varkonyi-Gasic E, Hellens RP (2011) Quantitative stem-loop RT-PCR for detection of microRNAs. In: Kodama H, Komamine A (eds) RNAi and plant gene function analysis: methods and protocols. Humana Press, Totowa, NJ, pp 145–157
49. Varkonyi-Gasic E, Wu R, Wood M et al (2007) Protocol: a highly sensitive RT-PCR method for detection and quantification of microRNAs. Plant Methods 3(1):12
50. Varkonyi-Gasic E (2017) Stem-loop qRT-PCR for the detection of plant microRNAs. In: Kovalchuk I (ed) Plant epigenetics: methods and protocols. Springer US, Boston, MA, pp 163–175
51. Dai X, Zhuang Z, Zhao PX (2018) psRNATarget: a plant small RNA target analysis server (2017 release). Nucleic Acids Res 46(W1): W49–W54
52. Untergasser A, Cutcutache I, Koressaar T et al (2012) Primer3—new capabilities and interfaces. Nucleic Acids Res 40(15):e115–e115
53. Czimmerer Z, Hulvely J, Simandi Z et al (2013) A versatile method to design stem-loop primer-based quantitative PCR assays for detecting small regulatory RNA molecules. PLoS One 8(1):e55168

INDEX

A

Acetosyringone ... 30, 32, 42, 44, 55, 68, 71, 79, 89, 114, 129, 132, 133, 143, 147, 169, 174, 203, 216, 219, 225, 227, 234, 235, 237, 240, 246
African cassava mosaic virus (ACMV) ... 55
Agrobacterium tumefaciens
  AGL1 ... 53
  EHA105 ... 127
  GV2260 ... 27
  GV3101 ... 27, 127, 172, 203, 235
  LBA4404 ... 127
Agro-infiltration ... 30
Alkaloids ... 75–91, 93–108
Apple latent spherical virus (ALSV) ... 124, 184, 188

B

Binary vectors ... 78, 157, 167
Bipartite virus genome ... 27, 40, 76, 112
Brachypodium distachyon ... 16, 18, 22

C

Capsicum spp. ... 27–37
Catharanthus roseus ... 93, 96, 97, 99, 100, 102–105, 107
Chenopodium quinoa ... 18, 20, 21, 184, 185, 190
Cloning ... 2, 3, 5, 7, 9, 16–20, 22, 27, 28, 32, 33, 41, 42, 47, 52, 53, 55, 65, 67, 70, 76, 78, 95–97, 107, 112, 113, 128, 130, 142, 145, 152, 167, 168, 170, 171, 177, 184, 188, 189, 191, 195, 200, 202, 205, 207, 225, 228, 245, 246
Coat protein (CP) ... 2, 16, 27, 97, 112, 113, 124, 156, 200, 205, 206, 218, 219
Cucumber mosaic virus (CMV) ... 2, 6, 7, 9, 12, 245

D

Diploid potato ... 40, 41, 44, 46, 48
DNA gel extraction ... 130
Double-stranded RNAs (dsRNAs) ... 15, 16

E

Electrophorese ... 5, 18–20
Electroporation ... 58, 68
Escherichia coli ... 3, 5, 18, 20, 28, 33, 42, 44, 47, 53, 67, 97, 101, 114, 117, 128, 130, 140, 145, 151, 158, 159, 161, 169, 186, 189, 190, 205, 209, 216, 219, 246

F

Flavonoids ... 195, 233, 234
Flowers ... 1, 2, 9, 12, 199–220, 223–231, 234
Foxtail mosaic virus (FoMV) ... 16–23
Fruits ... 28, 155–162, 173, 177, 223
Functional genomics ... 18, 39, 40, 65, 66, 123, 156, 166, 167, 183, 234

G

Gateway-based vector ... 31
GC-FID ... 143
Geminivirus agroclones ... 51–63
Gene of interest (GOI) ... 113, 115, 145, 151, 152, 166–168, 171, 175, 177, 179, 183, 186, 189, 199–201, 203, 206–209, 212, 214, 216–219
Gene-silencing ... 1–12, 15–24, 27–37, 39–48, 51–63, 65–73, 75–91, 93–108, 111–121, 123–135, 139–152, 161, 162, 165–179, 183–197, 199–220, 223–231, 233–241
Grafting ... 167, 225–229

H

Hordeum vulgare ... 16, 18, 19, 22

I

Inoculation methods
  agro-infiltration medium ... 40, 73, 84
  apical meristem inoculation ... 205, 210, 211
  biolistic ... 158, 160, 187, 193
  rub inoculation ... 185, 190, 191, 195, 196
  sap inoculation ... 40

Vincent Courdavault and Sébastien Besseau (eds.), *Virus-Induced Gene Silencing in Plants: Methods and Protocols*, Methods in Molecular Biology, vol. 2172, https://doi.org/10.1007/978-1-0716-0751-0,

Inoculation methods (*cont.*)
syringe infiltration ... 127
vacuum infiltration ... 224
In vitro plants ... 43, 235
In vitro transcription ... 6

L

LB medium ... 44, 46, 68, 70, 114, 116, 117, 169, 172, 174, 190
LC-MS/MS ... 81, 88–89
Leaf-sap ... 8, 9
LIC cloning ... 42, 43, 47, 246
Ligation ... 3, 5, 20, 37, 42, 70, 101, 112, 116, 118, 130, 145, 161, 189, 208, 209, 218, 228
Lilium leichtlinii ... 2, 10, 12
Linum usitatissimum ... 65–73
Linum usitatissimum L. ... 233–241

M

Malus x domestica L. ... 183–197
Manihot esculenta ... 51
microRNAs (miRNAs) ... 243–245
Movement protein (MP) ... 97, 112, 113, 156, 184, 188
MS medium ... 42, 43, 193

N

Nepeta cataria ... 112, 114–116, 119, 120
Nepeta mussini ... 112
Nicotiana benthamiana ... 2, 7, 8, 17, 43, 124, 200, 201, 217, 223, 245
Nicotiana tabacum L. ... 235, 237

O

Ocimum basilicum ... 112, 123–135, 167
Olea europaea ... 165, 169, 170, 173
Oryza sativa L. ... 18, 22

P

Papaver somniferum ... 76, 78, 89
Petunia x hybrida ... 202, 206
Photo-bleaching ... 12, 78, 124, 167
Plasmid purification ... 18, 20, 55, 67, 70, 142, 145, 158, 159, 187, 205
Polymerase chain reaction (PCR) ... 5–7, 9–11, 18–22, 28, 32, 33, 35, 36, 41, 42, 44, 45, 55, 67, 69, 70, 78, 80–82, 86, 88, 96, 97, 99–101, 104, 114, 116, 117, 128, 130, 132, 133, 142, 143, 145, 147, 158, 159, 162, 169–171, 175, 177, 186, 187, 189, 191, 195, 196, 203, 206, 208–210, 214–216, 218, 247
Posttranscriptional gene silencing (PTGS) ... 27, 111, 119, 156, 166, 167, 199–201, 217, 223
Potato virus X (PVX) ... 40, 42, 47, 75, 112, 124, 156, 245, 246
Prunus ... 155–162, 166
Prunus necrotic ringspot virus (PNRSV) ... 155–162

Q

Quantitative RT-PCR (qRT-PCR) ... 9, 73, 76, 78, 81, 86, 143, 148–149, 152

R

Rauwolfia tetraphylla ... 94, 96, 97, 99, 102–104, 107
Reference genes ... 9, 81, 86, 191, 206, 214, 215
Restriction enzymes ... 3, 5, 6, 10, 18, 30, 41, 55, 67, 97, 100, 128, 130, 145, 168, 186, 189, 205, 208, 218, 245
Reverse transcription ... 41, 43, 86, 128, 171, 175, 186, 187, 246
RNA-dependent RNA polymerase (RdRp) ... 97, 112, 113, 124, 125, 131
RNA extractions ... 4, 19, 21, 41, 43, 53, 68, 80, 85, 86, 90, 104, 115, 128, 142, 148, 152, 170, 175, 186, 188, 195, 196, 206, 214, 220, 246
RNA silencing ... 15, 124, 155, 156
Rosa chinensis ... 225

S

Secondary metabolism ... 94
Seedcoat ... 233–241
Senescence phenotypes ... 212
Setaria viridis ... 18
Silencing phenotypes ... 2, 9, 10, 37, 44, 46, 131, 134, 201, 211, 213, 220
Small interfering RNA (siRNA) ... 4, 9–11, 15, 75, 111, 125, 208
Solanum bulbocastanum ... 40–42, 48
Solanum lycopersicum ... 124, 201, 245
Solanum tuberosum ... 39, 40, 43, 201
Sorghum bicolor ... 18, 24
Sorghum sudanense ... 18
Stem-loop PCR ... 10, 252, 254
Stem-loop reverse transcription ... 252–254
STTM fragment ... 254

T

Targeted genes ... 21, 73, 96, 104, 108, 155, 156, 183, 195, 241
Target gene fragments ... 2, 3, 5, 9, 19, 82
Target mimic (TM) ... 244–246
Tetraploid potato ... 39–48

Tobacco rattle virus (TRV)..............................27, 28, 36, 37, 40, 42, 65, 66, 75, 76, 95, 97, 105, 112, 124, 140, 156, 167, 168, 171, 175, 177–179, 184, 200–202, 206, 207, 214, 219, 220, 224, 234, 245, 246
Trees..................................155–162, 165–179, 183–197
Tripartite virus genome 2
Triticum aestivum..16, 18, 22

V

VIGS reporter genes
- cellulose synthase (CESA) 66, 234
- *chalcone synthase* (CHS)...................... 200, 201, 204, 208, 211, 213, 216, 217, 219, 234
- 4-hydroxy-3-methylbut-2-enyl diphosphate reductase (NbIspH)..43
- phytoene desaturase (PDS) 200–203, 208, 211, 213, 217
- protoporphyrin IX magnesium chelatase subunit chlH (ChlH)... 107, 112, 113, 115, 116, 118, 152
- ribulose-1,5-bisphosphate carboxylase (rbcS)..184
- transparent testa 2 (tt2)................................ 233–241

VIGS vectors
- ACMV DNA-A and B vectors.................... 55, 59, 60
- pCaRNA1, 2 and 3 vectors 158–161
- pCFV vector...20, 23
- pCHL1, 2 and 3 vectors.. 3
- pEALSR1 and pEALSR2L5R5 vectors 184, 186, 190
- pTRV1 and 2 vectors.......................... 45, 77, 78, 97, 101, 124, 127, 128, 132, 140, 152, 168, 207, 225
- pTRV2e-STTM plasmids.......................................252
- PVX-LIC vector 254
- PVX-STTM plasmid...251
- TRV VbMS vectors 246

Viral RNAs 15, 65, 156
Viral symptoms.. 7, 24, 40, 196, 200, 201, 211, 216, 219
Virus-based microRNA silencing (VbMS) 244–246
Virus-induced gene silencing (VIGS).........................2, 9, 15–17, 19, 21, 24, 27–37, 40, 42–48, 51–63, 65, 66, 68, 72, 75, 76, 85, 89, 94–97, 99, 100, 102, 105, 107, 111–113, 123, 124, 126–129, 131, 133, 140, 144, 145, 147, 152, 156–159, 161, 162, 166–168, 173, 183–185, 199–202, 206, 212, 213, 217, 218, 220, 223–225, 229, 233–241, 245

W

Withania somnifera.. 139–153
Withanolides..................... 139, 140, 144, 149, 150, 152

Y

YEB medium 225, 227, 235–237
Yellowing phenotype..124

MIX
Papier aus verantwortungsvollen Quellen
Paper from responsible sources
FSC® C105338

If you have any concerns about our products,
you can contact us on
ProductSafety@springernature.com

In case Publisher is established outside the EU,
the EU authorized representative is:
Springer Nature Customer Service Center GmbH
Europaplatz 3, 69115 Heidelberg, Germany

Printed by Libri Plureos GmbH
in Hamburg, Germany